The Ray Society
INSTITUTED 1844

This volume is No. 156 of the series

LONDON

1984

To C. M. YONGE

BIOLOGY
OF
OPISTHOBRANCH
MOLLUSCS

VOLUME II

T. E. THOMPSON, D. Sc.
and
GREGORY H. BROWN, Ph.D.

Department of Zoology
University of Bristol

ISBN 0 903874 18 0

Sold by The Ray Society,
c/o British Museum (Natural History), Cromwell Road, London SW7 5BD

PRINTED BY JOHN WRIGHT AND SONS (PRINTING) LTD,
AT THE STONEBRIDGE PRESS, BRISTOL BS4 5NU

CONTENTS

The harvest truly is plenteous,
but the labourers are few.

Matthew 9:37

FOREWORD

This volume concludes our survey of the representatives of the gastropod subclass Opistho-branchia recorded from the shores and shallow seas around the British Isles. Volume I contained an account of the general biology of the opisthobranchs of the world's seas, as well as a systematic treatment of the British Bullomorpha, Aplysiomorpha, Pleurobranchomorpha, Acochlidiacea and Sacoglossa (Thompson, 1976). Volume II treats the largest opisthobranch order, the Nudibranchia, containing 108 species in 4 suborders.

The suborder Dendronotacea contains 6 families, 6 genera.
The suborder Doridacea contains 12 families, 22 genera.
The suborder Arminacea contains 3 families, 4 genera.
The suborder Aeolidacea contains 11 families, 16 genera.

The arrangement of the systematic material is simple. After a brief diagnosis of the order Nudibranchia a list is given of the 108 British species, grouped into 32 families and 48 genera. To use this volume to identify an unknown nudibranch the student may use one of two approaches. The first is the more scholarly: peruse the characters of the 4 suborders (p. 3), turn to the appropriate section and follow out the dichotomous key to species. The second method is quicker but less reliable, with the penalty that no insight is obtained into the rationale behind the scheme of classification: turn to the pictorial synopsis (p. 213), where line-drawings are assembled illustrating the British naked opisthobranchs. This synopsis can be scanned until the unknown is matched with a name. Original colour plates from life depict all but a very few of the most rare species.

In this compilation, as in volume I, every effort has been made to verify previous descriptions, to compare British representatives with their European and north American equivalents, and to present new illustrations wherever possible. In these tasks, the authors are grateful to the many zoologists who have generously supplied advice and specimens. Foremost among these has been the uniquely gifted biologist and collector, Mr Bernard Picton, formerly of the University of Bristol, now at the Ulster Museum, Belfast, who has explored nudibranch habitats at innumerable sublittoral sites from Spain to the Orkneys; his help and advice have been indispensable. We must also single out for mention the generous help of the officers of the Conchological Society of Great Britain and Ireland, who have allowed us access to the results of their Marine Census; Mrs Stella M. Turk and Mr D. R. Seaward have been enthusiastic mentors. Dr David Heppell of the Royal Scottish Museum has advised on several difficult nomenclatural problems.

Grateful acknowledgement must be made of the encouragement and laboratory facilities given by Professors James Brough, D. J. Crisp, F.R.S., F. W. R. Brambell, C.B.E., F.R.S., J. E. Harris, C.B.E., F.R.S., H. E. Hinton, F.R.S. and B. K. Follett, F.R.S. Financial support has been received at various stages of the work from the Science Research Council, the Royal Society, the Nuffield Foundation, the British Council, the University of Bristol, the Leverhulme Trust and the Natural Environment Research Council.

The research work comprised in volume II began at University College, Bangor in 1953 and has continued with brief interruptions until 1983 in Bristol. The preparation of much of the final draft was completed at the Luc-sur-Mer Laboratoire Maritime of the Université de Caen, and at the Zoological Laboratory of the University of Athens. We would like to record our gratitude to Miss Jane Reynolds and to Dr D. M. Crampton-Thompson, who prepared many of the line-drawings for publication. Dr Crampton-Thompson also carried out a major microscopical re-investigation of the nudibranch radula, following a generous subvention from the Natural Environment Research Council.

A considerable proportion of volume I was devoted to a review of the literature dealing with the biology of the world's opisthobranch molluscs. The period since 1976, when volume I was published, has seen a continued upsurge in interest in the Opisthobranchia. In the Epilogue to the present volume an attempt is made to summarize and evaluate some of the main lines of advance.

CHARACTERS OF THE ORDER NUDIBRANCHIA

These are opisthobranch gastropod molluscs which lack a shell and operculum in the adult phase, although calcareous spicules may be present in the skin. The body shape is variable; it may be smooth and limaciform, aeolidiform (with dorso-lateral processes called cerata), or flattened. If cerata are present, they may contain defensive apical cnidosacs. The head often bears both oral and rhinophoral tentacles, the latter often wrinkled, lamellate, or arborescent, and sometimes retractile into elaborate pallial pits or sheaths. The mantle cavity is absent and the respiratory organs may take the form of a set of folds along the sides of the body, or there may be a crescent or a ring of structurally complex contractile branchial appendages situated at the rear of the body. The foot is elongated, occasionally broad, closely united with the head and visceral mass.

Horny gastric plates are sometimes present, but calcareous plates are lacking. Potash-resistant jaws are often present. The radula is extremely variable, uniseriate to broad, with or without a rhachidian tooth in each row; abraded or broken teeth are usually discarded, not retained in a special sac. The pharynx sometimes has a muscular buccal suction pump. The digestive gland is compact, or it may be much divided, with tributaries from the head, foot and cerata (where present). The hermaphrodite reproductive system lacks an external autospermal (seminal) groove. The penis is sometimes armed with a stout, hollow, cylindrical stylet; rarely, a solid masculine stylet is present, functioning as a stimulatory organ. Small retaining hooks may be present on the penis or in the vagina. Impregnation is never hypodermic. The central nervous system is euthyneurous, forming a ganglionic ring around the foregut.

Nudibranchs are all carnivorous, subsisting chiefly upon macrobenthic invertebrates, rarely attacking nektonic or planktonic prey.

SYNOPSIS OF THE CHARACTERS OF THE SUBORDERS

Suborder I DENDRONOTACEA (6 families, 6 genera, 22 species in British waters) (Page 9). Body with a dorso-lateral ridge along either side, bearing simple or branched processes which sometimes contain tributaries of the digestive gland, but rarely (e.g. *Hancockia*) contain cnidosacs; rhinophoral tentacles retractile into well-developed basal sheaths; anus mid-lateral.

Suborder II DORIDACEA (12 families, 22 genera, 39 species in British waters) (Page 37). Body protected by a more or less ample spiculose dorsal mantle, sometimes having club-like dorso-lateral processes, but never containing cnidosacs or tributaries of the digestive gland; rhinophoral tentacles contractile but rarely retractile into elaborate basal sheaths; anus postero-dorsal, usually surrounded by a circlet of gill plumes.

Suborder III ARMINACEA (3 families, 4 genera, 5 species in British waters) (Page 94). Body sometimes protected by an ample non-spiculose dorsal mantle with complex branchial ridges under the rim, or bearing simple or branched dorso-lateral processes; these excrescences may contain tributaries of the digestive gland but never cnidosacs; rhinophoral tentacles contractile but never retractile into elaborate basal sheaths; anus antero-dorsal or antero-lateral, on the right side.

Suborder IV AEOLIDACEA (11 families, 16 genera, 42 species in British waters (Page 103). Body bearing dorso-lateral clusters or rows of elongated finger-like ceratal processes, each of which contains a cnidosac and a tributary of the digestive gland; rhinophoral tentacles contractile, but never retractile into elaborate basal sheaths; anus antero-lateral, on the right side.

LIST OF BRITISH SPECIES OF THE ORDER NUDIBRANCHIA

Suborder DENDRONOTACEA
Family **TRITONIIDAE**

Tritonia Cuvier, 1798
1. *Tritonia hombergi* Cuvier, 1803
2. *Tritonia lineata* Alder & Hancock, 1848a
3. *Tritonia manicata* Deshayes, 1853
4. *Tritonia nilsodhneri* (Marcus, 1983)
5. *Tritonia plebeia* Johnston, 1828

Family **LOMANOTIDAE**

Lomanotus Vérany, 1844
6. *Lomanotus genei* Vérany, 1846
7. *Lomanotus marmoratus* (Alder & Hancock, 1845a)

Family **HANCOCKIIDAE**

Hancockia Gosse, 1877
8. *Hancockia uncinata* (Hesse, 1872)

Family **DENDRONOTIDAE**

Dendronotus Alder & Hancock, 1845b
9. *Dendronotus frondosus* (Ascanius, 1774)

Family **SCYLLAEIDAE**

Scyllaea L., 1758
10. *Scyllaea pelagica* L., 1758

Family **DOTOIDAE**

Doto Oken, 1815
11. *Doto coronata* (Gmelin, 1791)
12. *Doto cuspidata* Alder & Hancock, 1862
13. *Doto dunnei* Lemche, 1976
14. *Doto eireana* Lemche, 1976
15. *Doto fragilis* (Forbes, 1838)
16. *Doto hystrix* Picton & Brown, 1981
17. *Doto koenneckeri* Lemche, 1976
18. *Doto lemchei* Ortea & Urgorri, 1978
19. *Doto maculata* (Montagu, 1804)
20. *Doto millbayana* Lemche, 1976
21. *Doto pinnatifida* (Montagu, 1804)
22. *Doto tuberculata* Lemche, 1976

Suborder DORIDACEA
Superfamily ANADORIDOIDEA (= PHANEROBRANCHIA)
Family **GONIODORIDIDAE**

Goniodoris Forbes & Goodsir, 1839
23. *Goniodoris castanea* Alder & Hancock, 1845a
24. *Goniodoris nodosa* (Montagu, 1808)
Okenia Menke, 1830

4

25. *Okenia aspersa* (Alder & Hancock, 1845b)
26. *Okenia elegans* (Leuckart, 1828)
27. *Okenia leachi* (Alder & Hancock, 1854a)
28. *Okenia pulchella* (Alder & Hancock, 1854a)
Ancula Lovén, 1846
29. *Ancula gibbosa* (Risso, 1818)
Trapania Pruvot-Fol, 1831
30. *Trapania maculata* Haefelfinger, 1960b
31. *Trapania pallida* Kress, 1968a

Family ONCHIDORIDIDAE

Adalaria Bergh, 1878b
32. *Adalaria loveni* (Alder & Hancock, 1862)
33. *Adalaria proxima* (Alder & Hancock, 1854a)
Onchidoris Blainville, 1816
34. *Onchidoris bilamellata* (L., 1767)
35. *Onchidoris depressa* (Alder & Hancock, 1842)
36. *Onchidoris inconspicua* (Alder & Hancock, 1851)
37. *Onchidoris luteocincta* (M. Sars, 1870)
38. *Onchidoris muricata* (Müller, 1776)
39. *Onchidoris oblonga* (Alder & Hancock, 1845a)
40. *Onchidoris pusilla* (Alder & Hancock, 1845a)
41. *Onchidoris sparsa* (Alder & Hancock, 1846)
Acanthodoris Gray, 1850
42. *Acanthodoris pilosa* (Müller, 1789)

Family TRIOPHIDAE

Crimora Alder & Hancock, 1862
43. *Crimora papillata* Alder & Hancock, 1862

Family NOTODORIDIDAE

Aegires Lovén, 1844
44. *Aegires punctilucens* (Orbigny, 1837)

Family POLYCERIDAE

Polycera Cuvier, 1817
45. *Polycera faeroensis* Lemche, 1929
46. *Polycera quadrilineata* (Müller, 1776)
Thecacera Fleming, 1828
47. *Thecacera pennigera* (Montagu, 1815)
Greilada Bergh, 1894
48. *Greilada elegans* Bergh, 1894
Palio Gray, 1857
49. *Palio dubia* (M. Sars, 1829)
50. *Palio nothus* (Johnston, 1838b)
Limacia Müller, 1781
51. *Limacia clavigera* (Müller, 1776)

Superfamily EUDORIDOIDEA (= CRYPTOBRANCHIA)
Family CADLINIDAE

Cadlina Bergh, 1878b
52. *Cadlina laevis* (L., 1767)

Family **ALDISIDAE**

Aldisa Bergh, 1878
53. *Aldisa zetlandica* (Alder & Hancock, 1854a)

Family **ROSTANGIDAE**

Rostanga Bergh, 1879c
54. *Rostanga rubra* (Risso, 1818)

Family **DORIDIDAE**

Doris L., 1758
55. *Doris sticta* (Iredale & O'Donoghue, 1923)
56. *Doris verrucosa* L., 1758

Family **ARCHIDORIDIDAE**

Archidoris Bergh, 1878b
57. *Archidoris pseudoargus* (Rapp, 1827)
Atagema Gray, 1850
58. *Atagema gibba* Pruvot-Fol, 1951

Family **DISCODORIDIDAE**

Discodoris Bergh, 1877
59. *Discodoris millegrana* (Alder & Hancock, 1854a)
60. *Discodoris planata* (Alder & Hancock, 1846)

Family **KENTRODORIDIDAE**

Jorunna Bergh, 1876
61. *Jorunna tomentosa* (Cuvier, 1804)

Suborder ARMINACEA
Superfamily EUARMINOIDEA
Family **ARMINIDAE**

Armina Rafinesque, 1814
62. *Armina loveni* (Bergh, 1860)

Superfamily METARMINOIDEA
Family **JANOLIDAE**

Janolus Bergh, 1884
63. *Janolus cristatus* (Chiaje, 1841)
64. *Janolus hyalinus* (Alder & Hancock, 1854a)
Proctonotus Alder, 1844
65. *Proctonotus mucroniferus* (Alder & Hancock, 1844)

Family **HEROIDAE**

Hero Alder & Hancock, 1855
66. *Hero formosa* (Lovén, 1841)

Suborder AEOLIDACEA
Family **FLABELLINIDAE**

Coryphella Gray, 1850
67. *Coryphella browni* Picton, 1980
68. *Coryphella gracilis* (Alder & Hancock, 1844)
69. *Coryphella lineata* (Lovén, 1846)

70. *Coryphella pedata* (Montagu, 1815)
71. *Coryphella pellucida* (Alder & Hancock, 1843)
72. *Coryphella verrucosa* (M. Sars, 1829)

Family TERGIPEDIDAE

Cuthona Alder & Hancock, 1855
73. *Cuthona nana* (Alder & Hancock, 1842)
74. *Cuthona amoena* (Alder & Hancock, 1845)
75. *Cuthona caerulea* (Montagu, 1804)
76. *Cuthona concinna* (Alder & Hancock, 1843)
77. *Cuthona foliata* (Forbes & Goodsir, 1839)
78. *Cuthona genovae* (O'Donoghue, 1926)
79. *Cuthona gymnota* (Couthouy, 1838)
80. *Cuthona pustulata* (Alder & Hancock 1854)
81. *Cuthona rubescens* Picton & Brown, 1978
82. *Cuthona viridis* (Forbes, 1840)
Tergipes Cuvier, 1805
83. *Tergipes tergipes* (Forskål, 1775)
Tenellia Costa, 1866
84. *Tenellia adspersa* (Nordmann, 1845)

Family EUBRANCHIDAE

Eubranchus Forbes, 1838
85. *Eubranchus cingulatus* (Alder & Hancock, 1847)
86. *Eubranchus doriae* (Trinchese, 1874)
87. *Eubranchus exiguus* (Alder & Hancock, 1848)
88. *Eubranchus farrani* (Alder & Hancock, 1844)
89. *Eubranchus pallidus* (Alder & Hancock, 1842)
90. *Eubranchus tricolor* Forbes, 1838
91. *Eubranchus vittatus* (Alder & Hancock, 1842)

Family CALMIDAE

Calma Alder & Hancock, 1855
92. *Calma glaucoides* (Alder & Hancock, 1855)

Family FIONIDAE

Fiona Alder & Hancock, 1851
93. *Fiona pinnata* (Eschscholtz, 1831)

Family PSEUDOVERMIDAE

Pseudovermis Périaslavzeff, 1891
94. *Pseudovermis boadeni* Salvini-Plawen & Sterrer, 1968

Family EMBLETONIIDAE

Embletonia Alder & Hancock, 1851
95. *Embletonia pulchra* Alder & Hancock, 1851

Family CUMANOTIDAE

Cumanotus Odhner, 1907
96. *Cumanotus beaumonti* (Eliot, 1906)

Family **FACELINIDAE**

Facelina Alder & Hancock, 1855
97. *Facelina annulicornis* (Chamisso & Eysenhardt, 1821)
98. *Facelina bostoniensis* (Couthouy, 1838)
99. *Facelina coronata* (Forbes & Goodsir, 1839)
100. *Facelina dubia* Pruvot-Fol, 1949
Caloria Trinchese, 1888
101. *Caloria elegans* (Alder & Hancock, 1845)

Family **FAVORINIDAE**

Favorinus Gray, 1850
102. *Favorinus blianus* Lemche & Thompson, 1974
103. *Favorinus branchialis* (Rathke, 1806)
Dicata Schmeke, 1967
104. *Dicata odhneri* Schmekel, 1967

Family **AEOLIDIIDAE**

Aeolidia Cuvier, 1798
105. *Aeolidia papillosa* (L., 1761)
Aeolidiella Bergh, 1867
106. *Aeolidiella alderi* (Cocks, 1852)
107. *Aeolidiella glauca* (Alder & Hancock, 1845)
108. *Aeolidiella sanguinea* (Norman, 1877)

Suborder I DENDRONOTACEA

This suborder contains 10 families of nudibranchs (6 of them British), all possessing a mid-lateral anal opening and distinct, often elaborate, sheaths for the rhinophoral tentacles. They feed upon medusae, hydroids and actinians, in the case of *Phyllirhoe*, *Lomanotus*, *Hancockia*, *Dendronotus* and *Doto*, or on alcyonarian and gorgonian corals in the world-wide species of *Tritonia* and *Marionia*.

Horny jaws may or may not be present and the radula shows a great deal of variation in the different families, from broad and multiseriate (Tritoniidae) to uniseriate (*Doto*), resembling the radula of some of the aeolidaceans. Mention must be made here of *Tethys* and *Melibe*, which occur on coasts all over the world, with the exception of the British Isles, and lack both jaws and radula; they feed by casting about in muddy eel-grass and other rich areas for small crustaceans which are captured and swallowed whole with the aid of a dilated fimbriated buccal hood.

The body has a dorso-lateral ridge on each side, frequently bearing arborescent ceratal processes which function as gills. Sometimes these processes contain tributaries of the alimentary canal and a dendronotacean such as *Doto* can be mistaken for a true aeolidacean by the unwary student (although the presence of dilated rhinophore sheaths in *Doto* is decisive evidence for its retention in the Dendronotacea). The larval shell may be of veliger types 1 or 2 (Thompson, 1961a).

For a discussion of the taxonomic subdivisions of the Dendronotacea, see Odhner (1936).

KEY TO THE BRITISH SPECIES OF DENDRONOTACEA

1. Rhinophores simple, smooth — *Doto* ... 6
Rhinophores divided or lamellate — **2**

2. Rhinophores branched elaborately — *Tritonia* ... 17
Rhinophores lamellate — **3**

3. Rhinophores bearing a few longitudinal lamellae — *Hancockia uncinata* (p. 21)
Rhinophores bearing numerous oblique or transverse lamellae — **4**

4. Frontal margin of the head bearing branched or unbranched processes — **5**
Frontal margin of the head lacking prominent excrescences — *Scyllaea pelagica* (p. 24)

5. Frontal margin bearing finger-like unbranched processes — *Lomanotus* ... 21
Frontal margin bearing arborescent processes — *Dendronotus frondosus* (p. 22)

6. Nearly all the ceratal tubercles contain a crimson or blackish subepidermal body — **7**
Ceratal tubercles usually lack such darkly pigmented bodies — **13**

7. Blackish bodies inside ceratal tubercles — **8**
Crimson or red-brown bodies inside ceratal tubercles — **9**

8. Small black-tipped tubercles form rows across the dorsum — *Doto tuberculata* (p. 35)
No such rows present — *Doto pinnatifida* (p. 35)

9. Patches of crimson pigment occur on the mesial faces of the ceratal bases — *Doto coronata* (p. 27)
No such patches present — **10**

10. Flanks and dorsum obscured by dense dark brown pigment — *Doto dunnei* (p. 30)

Sparse pigment present — **11**

11. Terminal papilla of each ceras lacking dark pigment spot — *Doto maculata* (p. 34)

Dark red pigment spot visible inside each ceratal papilla — **12**

12. Scattered superficial red pigment on cerata — *Doto millbayana* (p. 34)

Without such superficial pigment — *Doto eireana* (p. 31)

13. Ceratal tubercles slightly elevated, giving each ceras the likeness of a fir-cone — *Doto fragilis* (p. 31)

Ceratal tubercles produced to form finger-like projections — **14**

14. Each ceras is surmounted by a greatly elongated terminal digit — *Doto koenneckeri* (p. 33)

Terminal ceratal papillae not so produced — **15**

15. Rhinophore sheaths smooth-rimmed — **16**

Rhinophore sheaths with tuberculate rims — *Doto hystrix* (p. 32)

16. Small white papillae present on the central area of the dorsum — *Doto cuspidata* (p. 30)

Central dorsum smooth — *Doto lemchei* (p. 33)

17. Body length up to 20 cm; oral veil bilobed, each lobe bearing numerous short digitiform processes — *Tritonia hombergi* (p. 11)

Body length up to 34 mm; oral veil not markedly bilobed, but bearing a few (maximally 8) long digitiform processes — **18**

18. Body white with opaque white dorsal longitudinal markings — *Tritonia lineata* (p. 12)

Body yellow, or mottled brown, or pink, without such markings — **19**

19. Body bright pink — *Tritonia nilsodhneri* (p. 15)

Body yellow, or mottled brown — **20**

20. Dorsum pale yellow, with brown mottling — *Tritonia plebeia* (p. 16)

Dorsum pinkish white, with dark red, black, or olive green mottling — *Tritonia manicata* (p. 14)

21. Body length up to 94 mm; cerata vase-shaped and squat, often curved at the tip, bearing gill-like wrinkles — *Lomanotus genei* (p. 18)

Body-length up to 34 mm; cerata slender and elongated, without wrinkles — *Lomanotus marmoratus* (p. 20)

TRITONIIDAE

Limaciform, soft dendronotaceans, the body approximately quadrilateral in transverse section. The mantle is produced to form a latero-dorsal series of delicate branched pallial gills. An oral veil is present, often notched mid-anteriorly so as to form two antero-lateral lobes; each lobe bears finger-like anterior processes which are often subdivided. The marginal process on each side is inrolled. The rhinophores each have a terminal digitiform projection with numerous branched subterminal excrescences. Anal, renal and genital openings are on the right flank.

The digestive gland consists of 2 lobes, fused to varying degrees in the different genera. Triangular, horny gizzard-plates are sometimes present (*Marionia* Vayssière, 1877; *Paratritonia* Baba, 1949; *Marionopsis* Odhner, 1934). The radula is broad, with, in each row, a median tooth and numerous lateral teeth; the first lateral tooth in each half-row is often differentiated. A pair of strong jaws is present.

The ovotestis forms a layer over the surface of the digestive gland. The penis is unarmed. The larval shell is of type 1.

Tritonia Cuvier, 1798

(type *Tritonia hombergi* Cuvier, 1803; validated in I.C.Z.N. Opinion number 667)

The pallial gills are branched, often arborescent. The anus is situated approximately half way back along the right side of the body.

The two digestive gland lobes are fused so as to form a single mass in the adult. Gizzard-plates are lacking.

The penis is simple and conical.

In a recent revision of the Tritoniidae, Odhner (1963) repeats the division of the British species into *Tritonia* s.s. and *Duvaucelia* Risso, 1826 (type *Tritonia gracilis* Risso, 1826), based upon the following characteristics:

Tritonia	*Duvaucelia*
Body clumsy	Body slender
Gills very numerous	Gills at most 8 on each side
Anus in the middle of the body	Anus in the anterior half of the body
Genital opening behind the foremost gill-tuft	Genital opening below or in front of the foremost gill-tuft

We consider that these features are so much affected by the age of any tritoniid that they do not permit or support the subdivision. Accordingly, we regard *Duvaucelia* as an unnecessary genus and in the arrangement in the present volume all the British tritoniids are referred to the single genus *Tritonia*.

1. **Tritonia hombergi** Cuvier, 1803
   "Limace de mer palmifere" Dicquemare, 1785
   Tritonia atrofusca Macgillivray, 1843
   Sphaerostoma jamesoni Macgillivray, 1843
   Tritonia pustulosa Deshayes, 1853
   ? **Tritonia conifer** Dalyell, 1853
   ? **Tritonia divaricata** Dalyell, 1853
   Tritonia alba Alder & Hancock, 1854a
   Candellista alba Iredale & O'Donoghue, 1923

APPEARANCE IN LIFE (Plate 1; volume I, Plate 1). This is the largest British nudibranch and may reach 200 mm in length. The body varies in colour from white to dark purplish brown (generally darker with age), lighter on the ventral surface. The mantle bears numerous soft tubercles of various sizes. The pallial branchiae are much subdivided and appear arborescent. The 5 or 6 largest gills on each side are pedunculate and reflected towards the mid-line; other, generally smaller gills project laterally. The number of gills increases markedly with age. In juveniles, which were at one time considered to be a separate and distinct species (*Tritonia alba* Alder & Hancock), white pigment forms patches mesial to each pallial gill.

The oral veil is markedly bilobed, even in the juveniles, and each lobe may bear up to 40 finger-like processes.

ANATOMY (Figs 1a & b, 2; Plate 4a). Many features have been described in volume I, especially the circulatory, digestive and reproductive systems. The radula of *T. hombergi* is interesting for taxonomists because it exhibits a dental metamorphosis at a time when the juveniles reach about 25 mm in body-length. Smaller individuals have radulae of the type described for *alba* by Alder & Hancock (1854a), with subterminal spines on the slender, curved lateral teeth (Fig. 1b). Larger individuals have radulae which lack such subterminal spines. Intermediate radulae may be found in 25 mm specimens. It is noteworthy that the juveniles (below 30 mm) have fine comb-like denticles on either side of the median cusp of each central tooth; these, too, disappear from the radulae of older specimens. Radular formulae vary from $12 \times 5.1.5$ (body-length 0·6 mm) and $25 \times 27.1.27$ (6 mm) to $90 \times 158.1.158$ (80 mm) and $73 \times 230.1.230$ (140 mm) (all specimens from the waters around the Isle of Man). An intermediate *alba*/*hombergi* specimen 25 mm long had the formula $39 \times 78.1.78$.

The jaws are massive (Fig. 2), with a strong median hinge, sturdy (often with a saw-toothed cutting edge) and have flared wing-like processes for muscle attachment. They are durable and are often found in the stomachs of fish.

HABITS. This large nudibranch has an annual life cycle in British waters. It is invariably found in association with *Alcyonium digitatum*, down to 80 m. Large pieces of *Alcyonium* are sliced away using the jaws and the resultant chunks of food are transferred whole into the oesophagus by the broad radula. The food is not "torn to pieces on the radula" in the way claimed by Eliot (1910). *Tritonia hombergi* is the only British opisthobranch known to be harmful to man. Byne (1893) found its epidermal secretions to blister his hands; we have tried to confirm this but without success. It is a matter of simple observation, however, that the skin-secretion contains rhabdoid-like bodies which may be presumed to be defensive. Human subjects probably differ in their reaction to the toxin.

Spawning occurs in the spring and early summer months, each pinkish spiral spawn-band (illustrated by Kress, 1971) containing up to 52,000 ova. The ova each measure 190–210 µm in diameter. At 10 °C, hatching occurs 36–38 days after oviposition, yielding lecithotrophic larvae (development-type 2) with shells of veliger type 1. Many details of the ontogeny were given in volume I.

This is one of the many nudibranch species known to act as host for the ectocommensal copepod *Lichomolgus agilis* (Leigh-Sharpe, 1935).

DISTRIBUTION (Map 1). This large and often locally abundant species is recorded at depths to 180 m from as far north as the Faeroes (Lemche, 1938) and Norway, where its range extends into the Arctic circle to Grøtøy (Odhner, 1939). It is reported from all around the British Isles, although there are very few records from western Ireland. Bouchet & Tardy (1976) knew of no French Atlantic records south of Brittany, but there are reliable reports from Portugal (Nobre, 1896) and the Mediterranean coasts of Spain (Ros, 1975) and France (Haefelfinger, 1960a). *Tritonia hombergi* does not seem to be generally distributed through the western Mediterranean. In this context, it may be important to note that the prey, *Alcyonium digitatum*, is said to extend only as far south as Portugal (Manuel, 1980) and it has not been recorded from the Mediterranean Sea.

2. *Tritonia lineata* Alder & Hancock, 1848a

APPEARANCE IN LIFE (Plate 2). Among the British tritoniids, this is the most delicate in appearance, but it is not the smallest and may attain 34 mm in length. The body is white, sometimes with a pinkish blush, limaciform and slender. A conspicuous longitudinal opaque white line runs from the head back along each side of the dorsum. This line is said to be absent from some Roscoff specimens (Hecht, 1895), but doubt has been cast upon this old observation by Tardy (1963), who believes that Hecht may have made his observations not on *lineata* but on the distinct *nilsodhneri*. The white line communicates with the base of each gill and is united mid-anteriorly and mid-posteriorly with its partner of the other side. Up to 6 pairs of pallial gills may be present, erect and arborescent. The oral veil is produced to form 2 pairs of elongated finger-like processes.

ANATOMY (Fig. 1c; Plate 4b). The radula of a 10 mm specimen was examined; the formula was $31 \times 24.1.24$. Each median tooth is broad and tricuspid, whereas the laterals are smooth hooks,

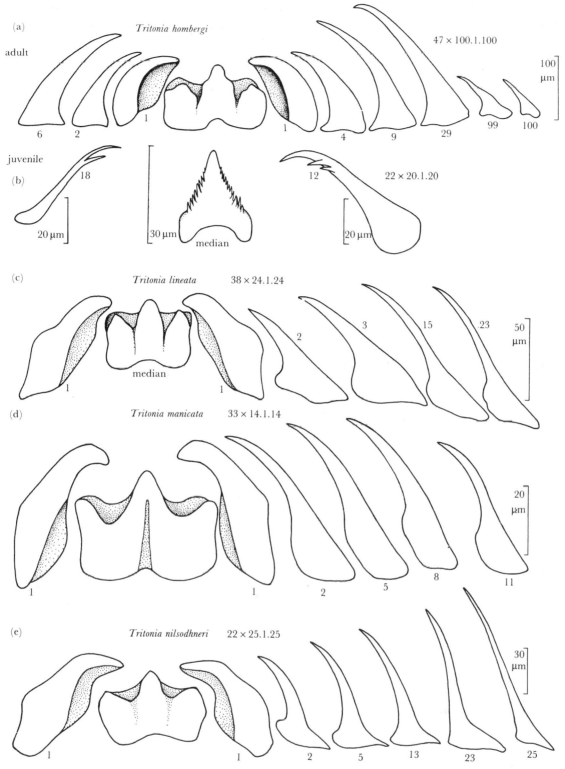

(a) *Tritonia hombergi*
adult

$47 \times 100.1.100$

$100\ \mu m$

6   2   1   1   4   9   29   99   100

juvenile

(b)

18   12 $22 \times 20.1.20$

$20\ \mu m$ $30\ \mu m$ median $20\ \mu m$

(c) *Tritonia lineata* $38 \times 24.1.24$

1   median   1   2   3   15   23 $50\ \mu m$

(d) *Tritonia manicata* $33 \times 14.1.14$

1   1   2   5   8   11 $20\ \mu m$

(e) *Tritonia nilsodhneri* $22 \times 25.1.25$

1   1   2   5   13   23   25 $30\ \mu m$

Fig. 1. Camera lucida drawings of dendronotacean radulae: *Tritonia* species. Teeth are numbered from the mid-line.

(a) *Tritonia hombergi*, preserved length 20 mm, Dale, S. Wales.
(b) *Tritonia hombergi*, length 4.5 mm, Isle of Man.
(c) *Tritonia lineata*, preserved length 10 mm, Dale, S. Wales.
(d) *Tritonia manicata*, preserved length 11 mm, Lundy Island.
(e) *Tritonia nilsodhneri*, preserved length 12 mm, Lundy Island.

13

becoming more elongated and slender towards the margins of the radula. The jaws appear in squash-preparation to consist of two elongated oval structures, united at the hinge and bearing along their mesial faces numerous minute denticles.

HABITS. There is uncertainty about the diet of *T. lineata*. Although the animal has a strong resemblance to *Alcyonium digitatum* it is rarely found on or near this cnidarian. The more likely prey is the inconspicuous and small octocoral *Sarcodictyon catenata* Forbes, which has often been collected together with *Tritonia lineata* around the British Isles; more information on feeding is still needed. Spawn is known for the month of August in Pembrokeshire; September in Yorkshire; the egg mass is white, closely resembling that of *T. plebeia*.

DISTRIBUTION (Map 1). This tritoniid is common in clean, shallow sublittoral localities all around British coasts. It is a northern Atlantic species, well known from southern Norway to the Mor-Braz region of Brittany (Bouchet & Tardy, 1976), down to 40 m. Most British records come from south-west England and Wales, or from the North Sea. Two specimens from the Bay of Naples (Schmekel, 1968) are the only records from the Mediterranean Sea.

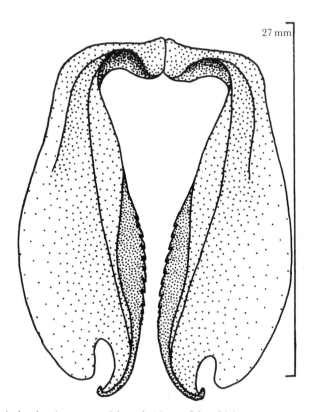

27 mm

FIG. 2. Jaws of *Tritonia hombergi*, preserved length 10 cm, Isle of Man.

3. **Tritonia manicata** Deshayes, 1853
   Nemocephala marmorata Costa, 1867
   Candiella moesta Bergh, 1884
   Candiella villafranca Vayssière, 1901
   Duvaucelia gracilis Pruvot-Fol, 1936 (*non* Risso, 1826)

APPEARANCE IN LIFE (Plate 2). This small species (not exceeding 13 mm in length), with its apparently vivid pattern, is nonetheless well camouflaged when in its natural habitat among benthic coelenterates. The ground colour of the body is opaque white with a pinkish blush due to the colour of the digestive gland within. A number of dark red, black, or olive green spots scattered over the dorsum have an optically disruptive effect, breaking up the symmetrical limaciform shape. The pallial gills are erect and arborescent; in an 11 mm specimen there were 4 pairs. The oral veil is produced to form symmetrical tentacles, 3–4 on either side of the mid-line. The genital opening is just in front of the level of the first pair of gills. The anal and renal pores lie near the right pallial margin between the first and second gill pairs. The flanks are dotted with dark pigment, but the pedal sole is pale.

ANATOMY (Fig. 1d). Haefelfinger's (1963) Mediterranean material included a 5 mm specimen with the radular formula 21 × 12.1.12. The radula of the sole British record, 11 mm in life when captured, has the formula 33 × 14.1.14. The median tooth is tricuspid, while the lateral teeth are smooth hooks, lacking subsidiary denticles. Stout jaws are present with 4–5 rows of fine serrations at the cutting edges.

HABITS. The diet in this case is unknown, but Schmekel (1968) reports that it feeds upon the alcyonacean *Cornularia* in the Bay of Naples.

DISTRIBUTION. This species, common in the Mediterranean Sea, has been found only once in British waters, at 12 m off Lundy Island in the entrance to the Bristol Channel. It has been found on the Turkish coast (Swennen, 1961a), in the northern Adriatic Sea at Trieste (Bergh, 1884), Bay of Naples (Schmekel, 1968), Mediterranean France (Haefelfinger, 1960a) and Spain (Ros, 1975), Portugal (Nobre, 1896), and along the open Atlantic coasts of Spain (Ortea, 1977), Morocco, France (Pruvot-Fol, 1953), and now from the British Isles (Brown, 1978). It had previously been recorded from as far north as Mor-Braz in Brittany (Bouchet & Tardy, 1976, as *Duvaucelia gracilis*). All records have been obtained at depths of less than 30 m.

DISCUSSION. Risso (1826) described a species (*Duvaucelia gracilis*) similar to the British animal illustrated in Plate 2, but Haefelfinger (1963) decided that this brief description probably referred to a juvenile *Mariona tethydea* (Chiaje, 1828). Pruvot-Fol (1936) asserted that, on the contrary, *D. gracilis* was identical with *Tritonia manicata* Deshayes, a species which was known from the author's drawing, published in 1853, his text having been lost. On balance, we favour Haefelfinger's decision.

4. ***Tritonia nilsodhneri*** Marcus, 1983
Duvaucelia odhneri Tardy, 1963
not ***Tritonia odhneri*** Marcus, 1959

APPEARANCE IN LIFE (Plate 1). This is an exquisitely camouflaged nudibranch, rose-pink like its alcyonacean prey. It is elongated and slender, up to 34 mm long, often coiled in a serpentine way around the cnidarian stems. There are up to 8 pairs of erect, arborescent pallial gills. The areas of the back mesial to the petioles are pale in colour. The oral veil bears 3 pairs of tentacular processes, the most lateral of which are always the longest.

ANATOMY (Fig. 1e). The radula of Tardy's 16 mm specimen had the formula 31.1.31. The median tooth was tricuspid; the first lateral exhibited a short extra denticle, while the remainder of the teeth were smooth, slender hooks. We have examined the radula and jaws of a specimen from Lundy Island, 12 mm in preserved length (Fig. 1e). The formula of this specimen is 22 × 25.1.25, with smooth, slender laterals without a trace of the extra denticle evident in Tardy's French material. The jaws have 4–6 rows of pointed denticles along the cutting edges.

HABITS. The diet consists of the gorgonian *Eunicella verrucosa* and its colouration and attitude enable it to elude recognition in this habitat. The colour-matching may be very exact. White varieties of *Eunicella verrucosa* bear white nudibranchs, and pink varieties of the cnidarian invariably give shelter to pink *T. nilsodhneri*. Spawning occurs in the spring and early summer (April to July), the pink egg mass being wound around a branch of *Eunicella*.

DISTRIBUTION (Map 1). This interesting species was only recently discovered, in the waste from a scallop-dredger operating near the Isle de Ré (Tardy, 1963). Since that time, it has been reported also from Roscoff, Brest, Jersey, northern Spain and from shallow waters (to 40 m) off Cornwall and western Ireland.

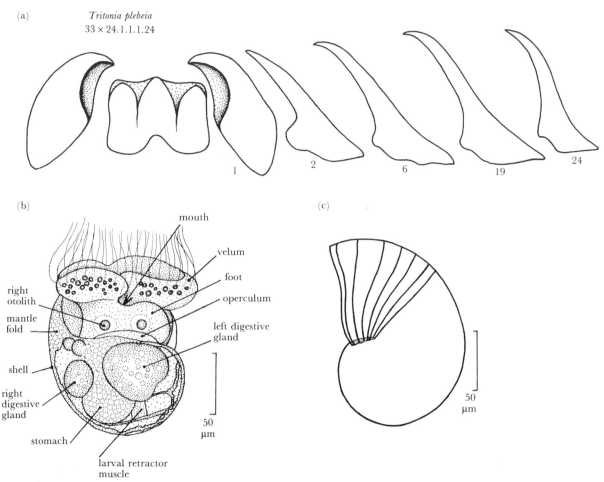

(a)

Tritonia plebeia
33 × 24.1.1.1.24

1 2 6 19 24

(b)

mouth
velum
foot
operculum
right otolith
mantle fold
left digestive gland
shell
right digestive gland
stomach
larval retractor muscle

50 μm

(c)

50 μm

Fig. 3. Camera lucida drawings of radular teeth and veliger larvae of *Tritonia plebeia*. Teeth are numbered from the mid-line.
  (a) Radula, length 14 mm, Lundy Island.
  (b) Newly hatched veliger larva, right latero-ventral aspect, Isle of Man.
  (c) Left lateral view of the larval shell several days after hatching, Isle of Man.

5. *Tritonia plebeia* Johnston, 1828
Tritonia pulchra Johnston, 1828
Candiella plebeia Gray, 1850

APPEARANCE IN LIFE (Plate 2). The body reaches 30 mm in length (usually 20 mm or less), and is pale yellow with brown mottling on the dorsum and flanks. Paler specimens prey upon white *Alcyonium* (Walton, 1908); the darker ones are usually found on the yellow variety. Larger individuals are generally darker than juveniles. White stippling occurs on the dorsal mantle, especially in paired areas mesial to the bases of the pallial gills. Two patches of darker pigment are usually present posterior to the rhinophore sheaths. The gills number 3 pairs in a 4 mm specimen, 4 in a 9·5 mm specimen, and 6 in a 16 mm specimen (Miller, 1958). Below 3 mm in body length, gills are frequently lacking; such individuals may be difficult to identify. The oral veil is produced to form smooth, finger-like processes, 2 pairs in a 4 mm specimen, up to 3 (rarely 4) pairs in adults 16 mm or more in length (Miller, 1958). Garstang (1889) described one adult individual with only 2 pairs of frontal processes, and another without any; he considered this to be a link with *Tritonia lineata*, but the two species are certainly separate and distinct.

ANATOMY (Fig. 3a). The radula, like that of *Tritonia hombergi*, undergoes a dental metamorphosis during early life. Juveniles less than 7 mm long possess lateral teeth which have 1–3 subsidiary denticles; these vanish later (Miller, 1958). Radular formulae are as follows: body-length 3 mm, formula 15 × 21.1.21; 7 mm, 20 × 33.1.33 (Miller, 1958); 6 mm, 26 × 50.1.50; 20 mm, 38 × 70.1.70 (Colgan, 1914). We have examined the radulae of two individuals for corroboration. One of these was from Lundy Island, measured 14 mm, and had the formula 33 × 24.1.24. The other had a preserved length of 18 mm, came from Kerrera, and had the formula 29 × 20.1.20. The masticatory borders of the stout jaws were serrated and denticulate.

HABITS. This tritoniid feeds upon *Alcyonium digitatum*, and is usually found around the bases of the colonies, in clear offshore waters all around the British Isles, down to 129 m. According to Swennen (1961), it may also take *Paramuricacea* in deeper waters. The spawn mass is white, ribbon-like and often coiled. The ova are 80–90 µm in diameter; each egg mass may contain up to 5000 embryos (Vestergaard & Thorson, 1938), and the embryonic period at 10 °C is 10 days (Thompson, 1967). The larval shell is of type 1 (Fig. 3c). Spawning has been recorded during many months of the year in the British Isles: January (personal), June (Miller, 1958), May and October (Alder & Hancock, 1845–55), August, October and November (Marine Biological Association, 1957).

DISTRIBUTION (Map 1). *Tritonia plebeia* reaches its northern limit at Trondheimsfjord in Norway (Odhner, 1939), and in the Faeroes (Lemche, 1938). It occurs in the North Sea and all around the British Isles, off the Atlantic coasts of France (Bouchet & Tardy, 1976), Spain and Portugal and throughout the western Mediterranean Sea (Ros, 1975; Pruvot-Fol, 1954) to the Aegean Sea (Swennen, 1961a).

DISCUSSION. Vestergaard & Thorson (1938) and Thorson (1946) based a classification of opisthobranch larvae on the shape of the hyperstrophic veliger shell at hatching. In this scheme they misinterpreted the shape of the shell of *T. plebeia* (Thompson, 1961a). This shell is illustrated in Fig. 3c. As was described in volume I, the veliger of *T. plebeia* has shell-type 1 (a spiral shell, forming $\frac{3}{4}$ to 1 whorl only).

LOMANOTIDAE

Elongated, soft dendronotaceans. The mantle has a conspicuous sinuous rim from the level of the rhinophore sheaths to the metapodial tip, where the two rims unite. The semilunar sinuosities are reflected alternately mesially and laterally. They bear a series of cerata but no gills, although gill-like wrinkles may be detectable on the largest mesial cerata. The bilobed oral veil bears 2 pairs of finger-like tentacles. The rhinophores are swollen and bear fine transverse lamellae, tipped by a small papilla. Each rhinophore issues from a large dilated sheath having finger-like processes on the rim. The genital apertures lie on the side, behind the right rhinophore. The anal and renal apertures are mid-lateral, on the right side. Strong lateral propodial tentacles are present.

The digestive gland ramifies, sending branches into the rhinophore sheaths, but does not always penetrate the adult cerata. The stomach lining is unarmed. The radula has the formula *n*.0.*n*. Stout jaws are present.

The ovotestis is attached to the rear of the digestive gland. The penis is unarmed. The larval shell is of type 2.

Lomanotus Vérany, 1844
(type *Lomanotus genei* Vérany, 1846)
With the characters of the family.
The world's species of this genus have been reviewed by Clark & Goetzfried (1976).

6. **Lomanotus genei** Vérany, 1846
   Lomanotus portlandicus Thompson, 1860
   L. hancocki Norman, 1877
   L. eisigii Trinchese, 1883
   L. varians Garstang, 1889 (part)

APPEARANCE IN LIFE (Plate 3). The adults may reach 94 mm in length. The pallial·cerata take the form in juveniles (up to 3 mm) of sparse discrete papillae along the pallial rim; in the adult, however, the rim carrying the cerata is enlarged to form a sinuous frill on either side of the body. The mesial cerata are the longest. The cerata are vase-shaped, more squat than those of *L. marmoratus*; the largest are often curved at the tip and may bear gill-like wrinkles.

The colouration varies considerably; three common patterns are shown in Plate 3. In the first (Plate 3a), the body lacks pigment, so that the yellowish digestive gland can be seen through the skin. In the second (Plate 3b), the bodily excrescences are conspicuously tipped by yellow or orange pigment situated subepidermally. In the third (Plate 3e), the body is cloaked with dark velvety red (occasionally purplish) superficial pigment, but the major excrescences contain terminal yellow bodies visible through the translucent epidermis. Earlier workers described a bewildering array of colour-varieties, for instance "transparent white suffused with pale orange red . . . the tips of the papillae being opaque white" (Farran, 1909, as *L. portlandicus*); "yellowish white suffused with brown" (Eliot, 1906); "rosy orange-coloured animal" with bright yellow rhinophores (Colgan, 1908); "deep purple or crimson . . . flecked with white spots" (Farran, 1909).

ANATOMY (Plate 4c; Fig. 4c–e). A serially sectioned specimen showed that the digestive gland does not penetrate the ceratal papillae in the adult. The digestive gland tributaries are withdrawn from the cerata during early life. A drawing of the reproductive system has been published by Schmekel (1970).

Strong, complex jaws are present (Fig. 4d), with small (10 µm long) polygonal elements near the masticatory edge (Fig. 4e). According to Odhner (1936), the radular formula may reach $42 \times 58.0.58$; all the teeth are denticulate with a formidable saw-edge on either side of the substantial cusp (Fig. 4c). We have examined the radula of a specimen from the Plymouth area, 13 mm long in preservative; it had a formula of $16 \times 18.0.18$. Up to 35 pointed denticles were present on either side of each of the teeth.

HABITS. This species is nearly always found feeding upon the calyptoblastic hydroid *Nemertesia ramosa* (Lamouroux) in coastal waters down to 90 m. It is normally slow-moving, but if abruptly disturbed may swim clumsily by means of lateral contractions of the whole body, the most posterior cerata helping to form a propulsive keel. Spawn has only rarely been seen; the only British record is for the month of September (Marine Biological Association, 1957). At Roscoff, spawning was noted in the Aquarium in the month of May (Cornet & Marche-Marchad, 1951); these authors also noted that *L. genei* is subject to infection with the internal parasitic copepod *Splanchnotrophus insolens* Scott.

DISTRIBUTION (Map 1). This flamboyant nudibranch is known from the Bay of Naples (Trinchese, 1883; Mazzarelli, 1903; Schmekel, 1968), from the type locality in the Gulf of Genoa, and from French·localities between Nice and Banyuls-sur-Mer (Pruvot-Fol, 1954). Many records from the European Atlantic coast require confirmation because of the taxonomic confusion with *Lomanotus marmoratus*. However, it is certain that *L. genei* occurs around the Brittany and Normandy coasts (Bouchet & Tardy, 1976) and all around the British Isles.

DISCUSSION. Several varieties of this species may prove to warrant the status of separate species. These varieties were described originally in the following terms:

(a) *portlandicus* Thompson, 1860: pellucid white, tinged with brownish yellow on the back and pale orange-red in front. Weymouth.

(b) *hancocki* Norman, 1877: transparent, light pinkish orange. Torbay.

(c) *eisigii* Trinchese, 1883; transparent white, marked with irregular opaque white blotches and red dots. Tips of all papillae orange-yellow. Naples.

It may be noted that Vérany's original description of *L. genei* from Genoa was: colour intense wine-red, dotted with white, variable by reason of its transparency, which allows the internal parts of a darker red to show through (translation by Colgan, 1908).

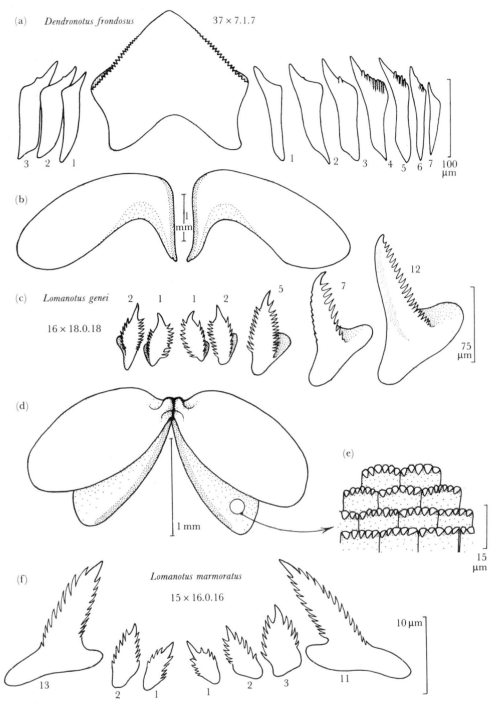

FIG. 4. Camera lucida drawings of dendronotacean jaws and radulae: *Dendronotus* and *Lomanotus*. Teeth are numbered from the mid-line.

(a) *Dendronotus frondosus* radula, preserved length 20 mm, Menai Straits.
(b) jaws of the same.
(c) *Lomanotus genei* radula, preserved length 13 mm, Plymouth, Devon.
(d) jaws of the same.
(e) detail of jaw.
(f) *Lomanotus marmoratus* radula, preserved length 6 mm, Sound of Mull.

7. **Lomanotus marmoratus** (Alder & Hancock, 1845a)
   Eumenis marmoratus Alder & Hancock, 1845a
   Lomanotus varians Garstang, 1889 (part)
   ?**Eumenis flavida** Alder & Hancock, 1846 (juvenile)

APPEARANCE IN LIFE (Plate 3). Adults may reach 34 mm in length; in the adult the simple discrete ceratal papillae of the juvenile (Plate 3g) have multiplied and are positioned along the edge of the enlarged, sinuous pallial rim (Plate 3f). The mesial cerata are the longest. The cerata are more slender and elongated than those of *L. genei* and, moreover, exhibit a typical sub-terminal swelling (Plate 3h). They never possess gill-like wrinkles, nor do they curve, both characteristics noted for *L. genei*.

The colour pattern is drab and consists of superficial speckles of chalk-white and dark brown, giving rise to a marbled or mottled effect. Scattered subepidermal brown pigment is usually present but tends to be masked by the brown digestive gland visible through the translucent skin.

ANATOMY (Fig. 4f). Serial sections through a 34 mm adult showed the continuing presence of digitiform tributaries of the digestive gland penetrating the cerata. This contrasts with the situation in *L. genei*, where this is simply a transient juvenile condition. Stout jaws are present. All the radular teeth are denticulate and, according to Odhner (1936), the formula may reach $20 \times 10.0.10$. An 8 mm specimen from Mull (measured after preservation) showed $24 \times 24.0.24$. A 6 mm specimen from the same collection, treated in the same way, exhibited the formula $15 \times 16.0.16$ (Fig. 4f); up to 12 pointed denticles were present on either side of each of the teeth. It was noteworthy that the masticatory borders of the jaw plates were composed of tiny polygonal plates, similar to those of *L. genei*.

HABITS. This uncommon species is usually found feeding on the calyptoblastic hydroid *Nemertesia antennina* (L.) in shallow subtidal localities. More information about its mode of life is needed.

DISTRIBUTION (Map 1). The only record of this species from outside the British Isles is from the Galicia region of northern Spain (Ortea, 1978). The type locality was Torbay in south Devon, and most subsequent finds were also in south-western parts of England and Ireland. There are a few records from the North Sea (Walton, 1908; Hamond, 1972; McKay & Smith, 1979).

HANCOCKIIDAE

Limaciform, soft dendronotaceans, the body approximately quadrilateral in transverse section. The mantle is produced to form a latero-dorsal series of delicate cerata of unique form; each ceras consists of a structure resembling a half-clenched human fist. An oral veil is present, forming two distinct lobes, each of which bears 3–8 digitiform projections. The rhinophores have clubbed tips and swollen stems bearing sparse longitudinal lamellae. Each rhinophoral tentacle issues from a tall, dilated pallial sheath. The genital apertures lie on the right side, between the rhinophore and the first ceras. The anal and renal openings lie close together on the right side, between the first and second cerata.

The digestive gland is much-divided, with diverticula within the cerata and the rhinophore sheaths. These connect with the stomach via a pair of anterior hepatic ducts (supplying the rhinophores and the first pair of cerata) and a median ventral posterior hepatic duct (supplying the remaining cerata). These three hepatic ducts unite and open into the floor of the stomach close to the entry of the oesophagus. Heavily cuticularized ridges are present in the stomach lining. The radula is triseriate. Strong jaws are present.

The ovotestis consists of numerous separate lobules lying above the posterior hepatic duct. The penis is usually unarmed (but has chitinous ridges in the Australian *Hancockia burni*).

Cnidosacs are present, opening to the surface on the flanks and in the cerata and rhinophore sheaths. The larval shell is of type 2.

Hancockia Gosse, 1877
> (type *H. eudactylota* Gosse, 1877 = *H. uncinata* (Hesse, 1872)
> With the characters of the family.

The world's species of this genus have been reviewed by Thompson (1972).

8. ***Hancockia uncinata*** (Hesse, 1872)
   Doto uncinata Hesse, 1872
   Hancockia eudactylota Gosse, 1877
   Govia rubra Trinchese, 1885
   Govia viridis Trinchese, 1885

APPEARANCE IN LIFE (Plate 5). This is an exceedingly rare species, reaching only 14 mm in length, while Hesse's original Brittany material did not exceed 5 mm. The colour is pale green or pink, dotted with white, and the body appears fragile and slender. There are up to 4 or 5 pairs of ceratal outgrowths, each ceras consisting of a structure resembling a half-clenched human fist. Tall rhinophore sheaths with crenulate or tuberculate rims give rise to swollen, knobbed rhinophoral tentacles bearing sparse longitudinal lamellae. The oral veil is markedly bilobed, each lobe extending to form 3 or 4 finger-like projections.

Cnidosacs are present in the tissues of the cerata, flanks, and rhinophoral sheaths; they often form elevated pustules visible externally. According to Eliot (1910), each cnidosac communicates with the lumen of the digestive gland and also has a pore to the exterior. It is not known whether the nematocysts within these cnidosacs are derived, as is well established in the aeolidacean nudibranchs, from cnidarian prey.

ANATOMY (Fig. 5a). The reproductive organs were described by Schmekel (1970). Stout jaws are present, bearing irregular but distinct denticles along the masticatory border. The radula has the formula 1.1.1. There are 31 rows of teeth in a 7 mm individual, but counts of up to 48 rows have been reported for older individuals at Arcachon (Cuénot, 1927). The median teeth each bear 4–5 denticles on either side of the cusp. We have examined a specimen from Sidmouth, 4·2 mm long in preservative, which has the formula 55 × 1.1.1 (Fig. 5a); details of the older descriptions (e.g. Eliot, 1906, pl. xi, fig. 10) were verified.

HABITS. Hesse (1872) found this species in Brittany, associated with the gymnoblastic hydroid *Tubularia* and the calyptoblast *Nemertesia*. Our observations do not support the idea that *Hancockia* feeds on the former, because a specimen collected near Sidmouth in south Devon in 1977 refused *Tubularia indivisa* L. when this was offered as food. This inference is supported by Schmekel's (1968) observation that *Hancockia uncinata* occurs in the Bay of Naples amongst the calyptoblast *Campanularia* attached to the sea-grass *Posidonia*. Spawn is unknown in British waters, but in Arcachon has been noted in the months of August and September (Cuénot, 1927a). Development to hatching occupies 9 days at 16–18 °C (Schmekel & Portmann, 1982).

DISTRIBUTION (Map 2). This species has been most frequently encountered along the shores of the Bay of Biscay (Bouchet & Tardy, 1976), but records exist also for Naples (Schmekel, 1968), Spain (Ortea & Urgorri, 1979) and southern France (Pruvot-Fol, 1954). The northern limit of distribution is in the north Celtic Sea.

DENDRONOTIDAE

Body limaciform, bearing dorso-lateral arborescent pallial cerata (4–8 pairs). These cerata lack cnidosacs but are penetrated by lobes of the digestive gland in the juveniles. An oral veil is present, bearing 2–5 pairs of finger-like processes which are usually subdivided. The rhinophores are elongated, swollen and bear 8–30 transverse lamellae. The free margins of the rhinophore sheaths are produced to form numerous branched processes. Another branched pedunculate process arises from the lateral face of each rhinophore sheath. Anal, renal and genital openings are on the right flank.

The digestive gland consists of 3 lobes, one large posterior part and two smaller anterior subdivisions, with lobular diverticula extending into the rhinophore sheaths and cerata (in juveniles only). Gizzard-plates are lacking. The radula is moderately narrow (6.1.6 to 21.1.21), with in each row a strongly cusped median tooth and elongated narrow pointed laterals. A pair of strong jaws is present.

The ovotestis is a large discrete structure ventral to the digestive gland; the penis is unarmed. The larval shell is of type 2.

Dendronotus Alder & Hancock, 1845
  (type *Doris arborescens* Müller, 1776)
  With the characters of the family.
Most of the world's species of *Dendronotus* have been reviewed by Robilliard (1970, 1972).

9. **Dendronotus frondosus** (Ascanius, 1774)
   Amphitrite frondosa Ascanius, 1774
   Doris arborescens Müller, 1776
   Tritonia arborescens; Cuvier 1805
   Tritonia reynoldsi Couthouy, 1838
   Tritonia lactea Thompson, 1840
   Tritonia felina Fabricius; Alder & Hancock, 1842
   Dendronotus arborescens; Alder & Hancock, 1853
   Campaspe pusilla Bergh, 1863
   Dendronotus luteolus Lafont, 1871
   Dendronotus arborescens aurantiaca Friele, 1879
   Campaspe major Bergh, 1886
   Dendronotus lacteus; Nordsieck, 1972
   ? **Tritonia felina** Fabricius, 1780
   ? **Doris cervina** Gmelin, 1791
   ? **Tritonia pulchella** Alder & Hancock, 1842
   ? **Dendronotus elegans** Verrill, 1880

APPEARANCE IN LIFE (Plate 5). This is a handsome nudibranch, reaching 100 mm in length (115 mm in Alaska, according to MacGinitie, 1959), translucent grey or white, marbled, speckled and streaked with white, orange, red and brown in varying proportions. Specimens from deep coastal water (below 25 m) are usually paler overall than those from shallower habitats (Odhner, 1926; Robilliard, 1970). There are also differences associated with age. Juveniles up to a length of 4 mm are pale, then from 4 to 30 mm the developing patterns may become vivid, forming superficial stripes along the flanks and dorsum and contrasting bands of colour on the cerata. In adults, and juveniles longer than 30 mm, the pattern once again appears drab. Elmhirst (1922) suggested that the body colour of *D. frondosus* was associated with the diet, so that specimens feeding on calyptoblastic hydroids were pale straw-coloured, and specimens found on the gymnoblast *Tubularia* were brick-red. There is undoubtedly some truth in this suggested association, but it is unlikely that there is a simple causal relationship, because the pigments are situated in the skin, not the alimentary canal.

The dorso-lateral pallial rim bears on each side of the body a row of up to 9 arborescent ceratal processes. Similar processes protrude from the mouths of the rhinophore sheaths, from the lower part of each sheath and from the frontal margin of the oral veil. The number and degree of branching of these excrescences increases with age, as Miller (1958) has ably documented. A specimen of 1 mm length had 4 (unbranched) oral veil processes and 3 pairs of cerata; at 6 mm the figures were 4 (branched) and 5; at 28 mm there were 4 and 6; finally, at 95 mm there were 4 pairs and 9. The rhinophores are swollen and spindle-shaped, each bearing up to 13 lamellae (25, according to Bergh, 1894). The foot is narrow and the whole body laterally compressed. The body can be lashed from side to side in a feeble swimming escape reaction. The genital openings are situated on the right side between the rhinophore and the first ceras; the anal papilla lies behind this, between the 1st and 2nd cerata.

ANATOMY (Fig. 4a & b; Plate 4). Various features have been illustrated by Alder & Hancock (1845–55), Herdman & Clubb (1889, 1892), Tchang-Si (1934), Odhner (1936) and Robilliard (1970). A controversy has surrounded the question of whether or not branches of the digestive gland penetrate the cerata and other processes in *D. frondosus*. Alder & Hancock stated that, from the main mass of the digestive gland "branches pass off into the branchiae (i.e. cerata) and tentacular sheaths". In 1889, Herdman & Clubb denied this, and published evidence based upon serial sections showing the absence of digestive gland lobules in the cerata. This point of disagreement was resolved when it was demonstrated that, during the life cycle, the arrangement of the digestive gland branches changes radically. In very young stages, lobules of the gland penetrate the cerata (just as they do in all the aeolidaceans), but they are withdrawn before adulthood (Thompson, 1960a).

The radular formula has been established in a range of Manx specimens by Miller (1958). At a body length of 1 mm, the formula was $34 \times 3.1.3$; at 11 mm it was $41 \times 7.1.7$; at 58 mm it was $34 \times 10.1.10$, and at 95 mm it was $35 \times 11.1.11$. According to Robilliard (1970), the largest radula on record for this species is $49 \times 14.1.14$. The stout median tooth bears up to 40 sharp denticles on either side of the cusp. Nearly all the lateral teeth, with the exception of 1 or 2 rudimentary marginals, bear regularly spaced denticles on the lateral margin. The jaws are large and strong (Fig. 4b).

Robilliard (1970) gave a description of the reproductive organs, confirming Odhner's (1936) account of the distinctive nature of the prostate gland which is a circular disc, composed of 5–12 ovoid alveolar compartments.

HABITS. A remarkable change in diet occurs during early benthic life. Juvenile stages feed upon calyptoblastic hydroids such as *Sertularia cupressina* (L.), *Dynamena pumila* (L.) and *Hydrallmania falcata* (L.), while the adults (50 mm or more in length) attack the gymnoblastic *Tubularia indivisa* (L.) and *T. larynx* Ellis & Solander (for a review, see Thompson, 1964). Spawning may occur after the attainment of a body-length of 13 mm (Miller, 1961). The white or pale pink ova are arranged in the form of a ribbon containing up to 16,000 embryos (Marcus, 1961) or 33,000 (Roginsky, 1962). This occurs in the spring and summer months around the British Isles, from March to August, but records exist of egg-laying in October (Dalyell, 1853), January (Swennen, 1961b) and February (Lemche, 1941) in certain years. On the east coast of the U.S.A., spawning has been noted in August (Lemche, 1929) and May (Clark, quoted by Robilliard, 1970). The uncleaved ova each measure 200–240 μm in diameter, and hatch after 32 days at 10 °C (Thompson, 1967). The veliger larvae (with shell-type 2) are usually lecithotrophic; Williams (1971) gives a detailed account of larval development.

Miller (1962) considers the normal life span to vary between $1\frac{1}{2}$ and 2 years. There is some evidence that the mode of development and the number of generations passed through each year may vary in different populations of *D. frondosus*: temperature undoubtedly has a significant effect. Runnström (1927) has shown that normal embryonic development will occur between 3·3 and 10·5 °C (but not at −1·4 or above 12 °C). The state of the veligers at hatching varies between planktotrophic and lecithotrophic extremes (i.e. between development-types 1 and 2). The number of generations passed through in a year is similarly variable; both Elmhirst (1922) and Miller (1962) have noted that in favourable years there may be 2 distinct generations, whereas in normal years there is only 1. It is possible, of course, that some of the variation reported in *D. frondosus* may be genotypic; this suggestion would harmonize with Robilliard's (1975) careful analysis of the colour varieties in *Dendronotus frondosus*, as a result of which he proposed that there are "three and probably four ecologically distinct populations".

Dendronotus was always refused by fishes in feeding tests in the Port Erin Aquarium (Thompson, 1960b), although it should be noted that Herdman & Clubb (1892) found it to be acceptable to intertidal shannies (*Blennius pholis* L.) in the same area.

DISTRIBUTION (Map 1). This boreo-arctic species is known from the east Siberian Sea to the Barents Sea (Novaya Zemlya, Spitzbergen and Bear Island), and southwards from Jan Mayen

Island to the East Greenland coast; Iceland, the Faeroes and Scandinavia (Lemche, 1929, 1938, 1941; Odhner, 1939; Roginsky, 1962; Sneli & Steinnes, 1975). It is common all around the British Isles but reaches a southern limit on the French Atlantic coast near Arcachon (Bouchet & Tardy, 1976). It is also amphi-Atlantic, having been recorded on the West Greenland coast around northern Canada and southwards along the eastern American seaboard as far as New Jersey (Franz, 1970a). Pacific coast records are numerous as far south as Los Angeles, California (Lance, 1961; Roller & Long, 1969).

The related *Dendronotus robustus* Verrill is reported from sea area Faeroes (Seaward, 1982); this is indicated in Map 1.

SCYLLAEIDAE

Elongated, laterally compressed, soft dendronotaceans. The mantle is produced dorso-laterally to form 1 or 2 pairs of strong, complex cerata. Each ceras bears more or less complex flimsy transparent gills on its mesial surface. A bilobed oral veil may be present (rudimentary in *Scyllaea*). The rhinophores are swollen and bear oblique lamellae, each tipped by a finger-like process. Each rhinophore issues from a large dilated sheath which bears, in *Scyllaea* and *Notobryon*, a posterior flap. The genital apertures lie on the right side, between the rhinophore and the first ceras. The anal and renal openings lie close together on the right side, near the level of the cerata. A median dorsal metapodial keel is present in *Scyllaea* and *Notobryon*.

The digestive gland is divided into 2–4 compact masses and does not penetrate the cerata or rhinophore sheaths. Cuticularised ridges or plates are usually present in the lining of the stomach. The radula has the formula *n*.0.*n*. or *n*.1.*n*. Stout jaws are present; they consist of a pair of curved, concave chitinous flaps with, on either side, a wing-like expansion which exhibits countless microscopic subunits.

The ovotestis consists of a number of separate lobules lying on either side of the digestive gland. The penis is unarmed. Larval shell-type is unknown.

Scyllaea L., 1758
(type *S. pelagica* L., 1758)

Two pairs of cerata are present. The bilobed oral veil is rudimentary. The posterior margin of each rhinophore sheath bears a large flap. A median dorsal metapodial keel is present. The radula has the formula *n*.1.*n*. Feeds upon calyptoblastic hydroids of the floating *Sargassum* community.

10. **Scyllaea pelagica** L., 1758
    Lepus pelagicus L., 1758
    Scyllaea ghomfodensis Forskål, 1775
    S. quoyi Gray, 1850
    S. hookeri Gray, 1850
    S. grayae Adams & Reeve, 1850
    S. edwardsi Verrill, 1878
    S. lamyi Vayssière, 1917

APPEARANCE IN LIFE (Plate 5). This species is common in the central parts of the ocean basins, but it is on rare occasions cast up on European shores. This has, unfortunately, happened so infrequently that it is necessary to base the following description mainly on Caribbean and Gulf of Mexico captures. The body length may reach 35 mm, pale speckled brown or green, with a darker edging to the mantle rim, and scattered white markings often aggregated to form white streaks on the flanks. There are conspicuous iridescent blue-green patches on the sides and dorsum. Sparse retractile finger-like epidermal papillae may be detectable. The whole body is laterally flattened and the foot is very narrow, an adaptation for clinging to *Sargassum* and the hydroids thereon. If abruptly disturbed, a swimming escape reaction is elicited, involving repeated side-to-side lashings of the whole body.

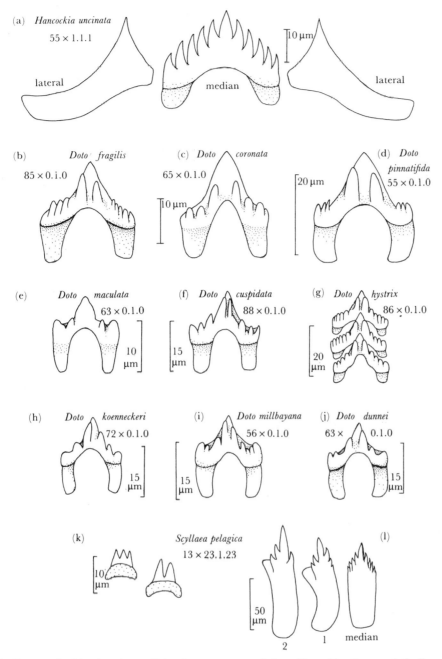

(a) *Hancockia uncinata*
55 × 1.1.1

lateral   median   lateral

10 μm

(b) *Doto fragilis*
85 × 0.i.0

(c) *Doto coronata*
65 × 0.1.0

10 μm

(d) *Doto pinnatifida*
55 × 0.1.0

20 μm

(e) *Doto maculata*
63 × 0.1.0
10 μm

(f) *Doto cuspidata*
88 × 0.1.0
15 μm

(g) *Doto hystrix*
86 × 0.1.0
20 μm

(h) *Doto koenneckeri*
72 × 0.1.0
15 μm

(i) *Doto millbayana*
56 × 0.1.0
15 μm

(j) *Doto dunnei*
63 × 0.1.0
15 μm

(k) *Scyllaea pelagica*
13 × 23.1.23
10 μm

(l)
50 μm
2   1   median

FIG. 5. Camera lucida drawings of dendronotacean radulae: *Hancockia*, *Doto* and *Scyllaea*. Teeth are numbered from the mid-line.

(a) *Hancockia uncinata*, preserved length 4·2 mm, Sidmouth, Devon.
(b) *Doto fragilis*, preserved length 8 mm, Lundy Island.
(c) *Doto coronata*, preserved length 8 mm, Strangford Lough, Ireland.
(d) *Doto pinnatifida*, preserved length 10 mm, Lundy Island.
(e) *Doto maculata*, preserved length 4 mm, Kilray Harbour, Ireland.
(f) *Doto cuspidata*, preserved length 12 mm, Strangford Lough, Ireland.
(g) *Doto hystrix*, preserved length 7 mm, C. Cork, Ireland.
(h) *Doto koenneckeri*, preserved length 4 mm, Donegal, Ireland.
(i) *Doto millbayana*, preserved length 2·1 mm, Lundy Island.
(j) *Doto dunnei*, preserved length 2 mm, Lundy Island.
(k) *Scyllaea pelagica*, length 13 mm, Corpus Christi, Texas, U.S.A., elements of the jaws.
(l) *Scyllaea pelagica*, as (k), representative radula teeth.

25

There are 2 pairs of dorso-lateral ceratal outgrowths, taking the form of flattened flaps held above the body or to the sides (the attitude is variable), bearing numerous, delicate, transparent, arborescent gills on the mesial faces. The tall rhinophore sheaths give rise to the lamellate rhinophoral tentacles, and bear on each posterior face a strong flap-like ridge. The cerata, rhinophoral sheaths and metapodial ridge all have a characteristic jagged appearance, enhancing the overall cryptic effect of the body-shape and coloration.

ANATOMY (Fig. 5k & l). The stomach lining bears numerous strong chitinous plates (up to 26 in number); in serial sections of Atlantic Ocean specimens, 2–3 digestive gland lobes and 6 ovotestis lobes were counted. The radular formula reached $24 \times 54.1.54$, all the teeth bearing conspicuous denticles. Parts of the jaws exhibit small toothed elements, each having a strong bifid or trifid cusp. A recent investigation of a 13 mm specimen from the Gulf of Mexico gave a radular formula of $12 \times 23.1.23$ (Thompson & Brown, 1981).

HABITS. *Scyllaea* is usually associated with drifting brown algae of the *Sargassum* type. The only British records are of rare specimens found at Falmouth on *Sargassum bacciferum* and *Saccorhiza polyschides*, and from incrustations on a sailing vessel in Plymouth docks (Marine Biological Association, 1957). It is clear that the prey consists of calyptoblastic hydroids living on the algae.

DISTRIBUTION (Map 2). Because of its drifting habitat, it is not surprising that this nudibranch has a cosmopolitan distribution. Most confirmed records, however, have come from warmer parts of the Atlantic Ocean although Vayssière (1917) collected it in the Antarctic, and Verrill (1878) found specimens on the coast of Massachusetts.

DISCUSSION. There has been some confusion about this species, most recently engendered by Barnard's (1927) misidentification of the common False Bay (South Africa) scyllaeid which is really a *Notobryon*. *Scyllaea* has a median tooth in each radular row, lacking in *Notobryon*. Unfortunately, Barnard did not examine the radula of his specimens and so was led into error. The False Bay dendronotaceans were recently re-investigated; they live under inter-tidal boulders, possess a longer metapodium than *Scyllaea*, and grow to a substantially greater size (up to 60 mm long). Both Odhner (1936) and Thompson & Brown (1976) at first accepted Barnard's claim, but it is no longer tenable, and one of the consequences of this is the need for a revision of the recorded size maxima for *Scyllaea pelagica*. It is now clear that no record in excess of 35 mm can be validated.

There is little danger of confusion between *Scyllaea* and *Notobryon*, on the one hand, and the third genus of the family, *Crosslandia*, on the other. *Crosslandia* contains bright green animals having only 1 pair of cerata and living in shallow tropical sea-grass beds.

DOTOIDAE

Body limaciform, bearing dorso-lateral pallial cerata, usually embossed with tubercles. These cerata are often stout and form a single row on each side, with the tubercles arranged in circlets, one above another. The cerata lack cnidosacs; in some species small gill-like excrescences (pseudobranchs) occur on the mesial faces of the largest cerata. The head bears lateral flap-like expansions. The rhinophores are smooth and finger-like, each arising from a tall flared pallial sheath. The genital openings lie on the right flank, beneath the 1st ceras, and the anal papilla (with the nephroproct close inside it) lies latero-dorsally, between the 1st and 2nd.

The digestive gland ramifies so that nearly all the lobules lie within the cerata. The stomach lining is unarmed. The radula has the formula $n \times 0.1.0$; frail jaws are usually present.

The ovotestis lies dorsal to the digestive gland ducts; the penis is unarmed. The larval shell is of type 1.

Recent observations by Mr Bernard Picton of the Ulster Museum suggest that dotoids do not attack and ingest the calyptoblastic hydroid polyps on which they are nearly always found, but feed instead by puncturing the perisarc using the exceedingly small and slender radular teeth and sucking out the caenosarc fluid.

Doto Oken, 1815
(type *Doris coronata* Gmelin, according to I.C.Z.N. opinion number 697)
With the characters of the family.

Several dubious genera of Dotoidae have been proposed, but none has been accompanied by adequate corroborative captures. They are *Gellina* Gray, 1850 (type *Tergipes affinis* Orbigny, 1837, from La Rochelle on the French Atlantic coast) with non-tuberculate cerata; *Iduliella* Thiele, 1931 (= *Dotilla* Bergh, 1878) (type *Doto pygmaea* Bergh, 1871a) with triseriate radula; *Caecinella* Bergh, 1870 (type *C. luctuosa* Bergh, 1870) with cerata which are for the most part small and simple, except for the most posterior pair, which are longer and tuberculated; and *Heromorpha* Bergh, 1873 (type *H. antillensis* Bergh, 1873) with a tentaculate oral veil and, so far as is known, smooth cerata. Thiele (1931) relegated these names to subgeneric status; Odhner (1936) adopted the same course.

A pioneering attempt by Odhner (1936) to revise the world's species of *Doto* has been overtaken by proposals for many new species. His revision was valuable, however, and he was the first to attempt to subdivide the dotoids into morphological subgroups.

Until Lemche (1976) published his revision of the northern European species of *Doto*, there had been 5 generally accepted species (Thompson & Brown, 1976). Thanks to Lemche, it is now certain that there are many more *Doto* species in British waters. These will be described in the following pages; Table 1 summarizes the salient features by which they may be distinguished.

11. **Doto coronata** (Gmelin, 1791)
(This name was validated in I.C.Z.N. Opinion number 697)
Doris coronata Gmelin, 1791
Melibaea coronata; Alder & Hancock, 1842

APPEARANCE IN LIFE (Plate 6). This species may attain a length of 15 mm, but usually does not exceed 12 mm. The body is pale yellow to white. The cerata (5–8 pairs) bear 3 or 4 concentric circlets of rounded tubercles, each containing a small but conspicuous crimson subepidermal body. The terminal tubercle of each ceras also exhibits this feature. Pseudobranchs are present in some individuals but are always rudimentary and often totally lacking. The dorsal surface of the body, the flanks and the head are streaked and blotched with superficial epidermal crimson pigment. Hepatic lobes penetrate the cerata, and their red or brown colour is conspicuous within the ceratal bases.

The dilated frontal margin of the head forms a pair of prominent quadrangular flaps (Plate 6c). This species differs from most of its congeners in lacking longitudinal ridges in front of the rhinophore sheaths. Each sheath is widely dilated, smooth, and without pigment on the rim.

The most conspicuous recognition feature is the crimson blotching, especially within the bases of the cerata.

ANATOMY (Fig. 5c). Much of our knowledge concerning the microanatomy and histology has come from the works of Kress (1968b) and Schmekel & Kress (1977). It is clear from their study of the reproductive organs that *Doto* forms a link between the Dendronotacea and the Aeolidacea. They also established that the penis is unarmed in *D. coronata*. Finally, they investigated the tiny radula and gave some formulae; at a body-length of 2·5 mm, the formula was $55 \times 0.1.0$; 4·0 mm, $60 \times 0.1.0$; 7·0 mm, $68 \times 0.1.0$. Each radular tooth had a slender erect cusp with 3–5 subsidiary denticles on either side. We have examined the radulae of 2 specimens, preserved lengths 5 and 8 mm, from British localities. The formulae were $83 \times 0.1.0$ and $65 \times 0.1.0$, respectively; a representative tooth is shown in Fig. 5c.

HABITS. This species is known to feed upon calyptoblastic hydroids in shallow waters around the British Isles, especially *Obelia geniculata* (L.), *Dynamena pumila* (L.) and *Sertularia argentea* (L.). Lemche (1976) states that it is also found on *Hydrallmania falcata* (L.) and *Abietinaria*

TABLE 1. Distinguishing features of British *Doto* spp.

Species: authority:	*coronata* (Gmelin, 1791)	*cuspidata* Alder & Hancock, 1862	*dunnei* Lemche, 1976	*eireana* Lemche, 1976	*fragilis* (Forbes, 1838)
maximum size: (mm)	12	14	25	7	34
no. of ceratal pairs: (adult)	5–8	8–10	7–8	4–5	8–10
head crests:	absent	small	prominent	small	prominent
pseudobranchs:	indistinct	indistinct	prominent	indistinct	prominent
body pigment:	irregular crimson blotches concentrated over head	often dense brown blotches always overlain with fine crimson speckling	very dense dark brown spots	small, sparse crimson spots	sparse white spots on brown
pigment continuous from dorsum to flanks between cerata:	present	present	present	absent	no spots
no. of circlets of tubercles per ceras:	3–4	4–6	6–7	3–4	8–12
apical pigment of ceratal tubercles:	large, round crimson spots	no spots but chalk-white granules with- in tubercles	black/red spots	small crimson spots	no spots
preferred diet species:	*Hydrallmania* *falcata* *Abietinaria abietina* *Dynamena pumila* *Obelia geniculata*	*Nemertesia ramosa*	*Kirchenpaueria* *pinnata*	*S. operculata*	*Nemertesia* *antennina* *N. ramosa*
other distinctive features:	red stripe on mesial surface of ceratal bases	a fine speckling of crimson spots all over body and cerata	fine streaks of pigment around bases of ceratal tubercles	rhinophore sheaths extended anteriorly but tight around rhinophore	irregular white lines along flanks

abietina (L.). Other records exist, of predation upon *Sertularia operculata* (L.) (Cornet & Marche-Marchad, 1951) and *S. cupressina* (L.) (Thompson, 1964). The spawn takes the form of a convoluted band, containing the white ova (Plate 6d). Each spawn mass contains up to 35,500 eggs, each ovum measuring 68–73 μm (Miller, 1958). The embryonic period is 16 days at 10 °C, 12 days at 15 °C and (Schmekel & Kress, 1977) 12 days at 15 °C. According to Miller (1962), there are 2, 3 or 4 generations each year; the normal life span is probably about 9 months. Spawning has been reported in most months of the year.

DISTRIBUTION (Map 2). It is impossible to ascertain whether many records of this species were correctly attributed. There is no doubt of its occurrence in the Bay of Naples (Schmekel & Kress, 1977), however, and other indubitable Mediterranean records include the Aegean and Adriatic Seas (Odhner, 1914) and southern France (Vincente, 1967). Its presence is also

hystrix Picton & Brown, 1981	*koenneckeri* Lemche, 1976	*lemchei* Ortea & Urgorri, 1978	*maculata* (Montagu. 1804)	*millbayana* Lemche, 1976	*pinnatifida* Montagu, 1804	*tuberculata* Lemche, 1976
12	8	10	9	14	29	19
5–7	4–5	4–6	4–5	5–6	8–10	4–8
prominent	absent	small	absent	prominent	prominent	prominent
prominent	indistinct	indistinct	indistinct	very prominent	indistinct	prominent
pale brown/ scattered white glandular patches	dense brown/ black comma-shaped dots	mottled brown over dorsum	sparse crimson spots	black/brown mottling	dense black mottling	none except on tubercle tips
present	absent	absent	present	absent	present	absent
4–6	2–3	4–5	3–4	4–5	5–6	3–4
no spots but white granules within tubercles	sub-apical black streaks	no spots but white granules within tubercles	small crimson spots except terminal tubercle	usually black/ brown spots	black spots	black spots
Schizotricha frutescens	*Aglaophenia pluma*	*Aglaophenia tubulifera*	*Plumularia catharina*	*Plumularia setosa* *Nemertesia ramosa*	*Nemertesia antennina*	*Sertularella gayi* *Abietinaria abietina*
elongated tubercles on body and cerata producing distinctive spiky appearance	elongated ceratal tubercles	—	—	irregular spots scattered all over cerata	lateral rows of tubercles below cerata	tubercles in transverse rows across body. Spawn laid in elongated spiral, not convoluted.

confirmed on the French Atlantic coast (Bouchet & Tardy, 1976) as well as in northern Spain and Portugal (Nobre, 1905). It is common all around the British Isles and has been reliably reported from Scandinavia, the Faeroes and Iceland to Spitzbergen and Murmansk (Lemche, 1938; Odhner, 1939; Sars, 1878). There are several records from the eastern American seaboard as far south as New Jersey (Loveland *et al.*, 1969), but these animals are distinctive and may be another species. Similarly, a record from the Red Sea (O'Donoghue, 1929) is not acceptable without confirmation.

Although often common intertidally, some of the more northerly records are from depths to 180 m.

DISCUSSION. Many of the older observations on *Doto coronata* (as then understood) were made before it was understood that this was an aggregate species. It is unfortunately not now possible to identify with precision the following dotoids: *Scyllaea punctata* Bouchard-Chanteroux, 1836; *Melibaea ornata* Alder & Hancock, 1842; *Meliboea arbuscula* Agassiz, 1850; *Doto forbesi* Deshayes, 1853.

Lemche (1976) believed that further subdivision of *Doto coronata* might be necessary; consequently, no neotype has been selected.

12. **Doto cuspidata** Alder & Hancock, 1862
    Doto aurea Trinchese, 1881
    ?**Doto cornaliae** Trinchese, 1881
    ?**Doto costae** Trinchese, 1881

APPEARANCE IN LIFE (Plate 7). This drab species may reach 14 mm, although most records have been of specimens 12 mm or less in length. The pale fawn body bears variable blotchy brown or black pigment and crimson speckles. The cerata (up to 10 pairs) bear 4–6 concentric circlets of blunt tubercles, each containing a terminal subepidermal cluster of chalk-white granules. The apical tubercle of each ceras is similar. Rudimentary pseudobranchs are present in the larger individuals.

The dilated frontal margin of the head forms a pair of prominent lateral flaps (Plate 7). There are small ridges or crests attached to the front of each rhinophore sheath; each sheath is flared, sometimes crenulated.

This species may be distinguished from other British forms by its white-tipped cerata and fine crimson speckling. Furthermore, the dorsal surface of the body exhibits characteristic small white epidermal tubercles.

ANATOMY (Fig. 5f). Descriptions of two radulae are on record. One was from a 6 mm specimen; the formula was $60 \times 0.1.0$ (Odhner, 1926). The other was from an 8 mm specimen, which had the formula $90 \times 0.1.0$ (Miller, 1958). More recently, we have examined the radula of a specimen from Strangford Lough, 12 mm long in preservative; the formula was $88 \times 0.1.0$. The teeth each had a slender erect cusp, flanked on either side by about 4 tiny denticles (Fig. 5f).

HABITS. It has been repeatedly reported that *Doto cuspidata* feeds upon the calyptoblastic hydroid *Nemertesia ramosa* (Lamouroux). Spawn has not been described.

DISTRIBUTION (Map 2). Early records of this species have recently been corroborated in south west England, in Ireland and in west Scotland. Jaeckel (1952) obtained specimens from 135 to 148 m in the North Sea. There are several records from the Norwegian coast as far north as Senja, where it is referred to as a rare inhabitant of depths around 300 m (Odhner, 1939). A record from the Asturias coast, north east Spain, is a first indication that the range extends south of the British Isles (Ortea & Urgorri, 1978).

13. **Doto dunnei** Lemche, 1976

APPEARANCE IN LIFE (Plate 8). This highly distinctive species reaches 25 mm in length. The dark brown or black body bears 7–8 pairs of cerata, with 7 concentric circlets of rounded tubercles, most of which exhibit both terminal and basal superficial dark brown spots. The terminal tubercle of each ceras· differs little from the other tubercles. Pigment-free areas of the epidermis surround the bases of all the cerata. Pseudobranchs are exceptionally well developed in *D. dunnei* (Plate 8b).

The frontal margin of the head is expanded laterally to form small semicircular flaps (Plate 8j). Prominent longitudinal ridges or crests are visible in front of the rhinophore sheaths. Each sheath bears dark brown spots, and is slightly flared, with a smooth, pale rim.

This species differs from British congeners by its broad shape, prominent pseudobranchs and dark pigmentation of the body. A further point, of uncertain significance, is that the anal papilla is placed nearer to the dorsal mid-line than is the case in other species of *Doto*.

ANATOMY (Fig. 5j). Lemche did not himself examine the radula of this species; we have investigated that of a Lundy specimen, 2 mm in preserved length. The formula was $63 \times 0.1.0$.

Each tooth had a slender, erect cusp, flanked on either side by 2–3 somewhat blunt, often asymmetrical denticles (Fig. 5j).

HABITS. This species feeds upon the calyptoblastic hydroid *Kirchenpaueria pinnata* (L.). The spawn takes the form of a convoluted band of white eggs, produced in May (western Ireland) or July (Lundy Island).

DISTRIBUTION (Map 2). Although only recently described from Galway Bay, this species is now known also from west Scotland (near Oban), from Northern Ireland and from Lundy Island in the Bristol Channel. Ortea & Urgorri (1978) described specimens from the Galicia region of northern Spain.

14. *Doto eireana* Lemche, 1976

APPEARANCE IN LIFE (Plate 8). This is a rare and diminutive species, not exceeding 7 mm in length. The body is translucent grey, with scattered small dark crimson spots, confined to the central part of the dorsum and to the lower flanks. There are 4–5 pairs of cerata, with 3–4 concentric circlets of rounded lateral tubercles and a prominent terminal tubercle, most of which exhibit an apical crimson spot. Rudimentary pseudobranchs may be detectable.

The dilated frontal margin of the head forms a pair of prominent quadrangular lateral flaps (Plate 8c). This species, like *D. coronata*, differs from British congeners in lacking longitudinal ridges or crests in front of the rhinophore sheaths. Each of these sheaths is slightly flared, pale, with a white-speckled rim.

Doto eireana may be distinguished by the two pale longitudinal areas of the back, linking the ceratal bases.

ANATOMY. The radula is not known.

HABITS. The diet consists of the calyptoblastic hydroid *Sertularia operculata* (L.). The spawn bands may contain up to 1500 white ova, and are produced in July and August on the western coasts.

DISTRIBUTION (Map 2). Around the British Isles, this small species is known only from the type locality, Ballyvaghan Bay, south west Ireland, and from Lundy Island in the Bristol Channel. Elsewhere, it is distributed southwards to north east Spain (Ortea & Urgorri, 1978).

15. *Doto fragilis* (Forbes, 1838)
Melibaea fragilis Forbes, 1838
?*Doto crassicornis* M. Sars, 1870
Doto pinnigera Hesse, 1872

APPEARANCE IN LIFE (Plate 7). This variable species may reach 34 mm in length. There are two principal varieties, associated with diet. One variety (Plate 7a) feeds upon *Nemertesia antennina* (L.) and *N. ramosa* (Lamouroux); it has a dark brown body with paler, more crowded cerata. The other feeds upon *Halecium muricatum* (Ellis & Solander); in this variety the body is transparent white with brown pigment only on the head, while the widely spaced cerata are pale cream or white. The dorsum of the latter variety usually bears a few small irregular soft excrescences, and a pattern of superficial white pigment, resembling snowflakes. The cerata (up to 10 pairs) bear up to 12 concentric circlets of pale tubercles. Pseudobranchs are borne on the mesial faces of the larger cerata. An important recognition feature is the invariable presence of an interrupted white line along each flank from the rhinophore sheath to the tail-tip.

The dilated frontal margin of the head is smoothly rounded. Prominent longitudinal ridges or crests are visible in front of the rhinophore sheaths. Each sheath is widely flared and exhibits white pigment incrusted around the lip.

ANATOMY (Fig. 5b). Some features of the internal structure were illustrated by Alder & Hancock (1845–55, Family 3, Plate 4), and the histology of the reproductive and other organs has been described by Dreyer (1912), Lloyd (1952) and Kress (1968b). The last author furnished a fascinating account of the regeneration of cerata following autotomy or experimental excision. She described a special granular layer of cells at the base of each ceras, assisting in the process of autotomy.

Published radular formulae vary from 62 × 0.1.0 (length 2·5 mm, Miller, 1958) to a maximum of 100 × 0.1.0 (Colgan, 1914). We have investigated the radula of a Lundy specimen, 8 mm in preserved length. The formula was 85 × 0.1.0; the median cusp was slender and erect, flanked on either side by up to 4 tiny denticles (Fig. 5b).

HABITS. Although typically found in association with *Nemertesia* and *Halecium*, as noted above, this species has been found also on the gymnoblastic hydroid *Tubularia larynx* Ellis & Solander (Walton, 1908). The life cycle is accelerated, according to Miller (1962), passing through "2 or possibly more" generations each year, and the smallest individual found to spawn was 5 mm long (Miller, 1958). The same author found the ova to measure 70–87 μm, and the largest spawn mass contained approximately 74,000 eggs. More detailed observations on the spawn of *D. fragilis* have been published by Kress (1975), who noted that the ova, 83 μm in diameter, developed to hatching in 18 days at 10–12 °C. A representative spawn mass is illustrated in Plate 7.

Spawning occurs during the greater part of the year at Plymouth (Marine Biological Association, 1957), in June and July in the Isle of Man (Miller, 1958), and from April to August in the Clyde (Elmhirst, 1922).

DISTRIBUTION (Map 2). This conspicuous species is common all around the British Isles, especially in the calmer seas preferred by *Nemertesia*. It is generally distributed throughout the north east Atlantic Ocean from Portugal and Spain (Ortea & Urgorri, 1978) to northern Norway and Iceland, at depths to 200 m (Odhner, 1939; Lemche, 1938). There are Mediterranean records, from Banyuls (Pruvot-Fol, 1953), and from Rovigno (= Rovinj) (Odhner, 1914), but corroboration is needed.

There are two subspecific varieties recorded from further afield. Marcus & Marcus (1969) described *Doto fragilis umia* from Greenland, and Baba (1971) described *Doto fragilis nipponensis* from Sagami Bay, Japan. In our opinion, it is preferable to regard these as separate and distinct species.

16. *Doto hystrix* Picton & Brown, 1981

APPEARANCE IN LIFE (Plate 9). When alive, the animals measure up to 12 mm in length and the general colour of the body is pale brown. Groups of white glands occur all over the dorsum and in an irregular line along the flanks, usually associated with the bases of body papillae. These glands also occur in the rhinophore sheaths and in the ceratal tubercles. The cerata (5–7 pairs) each bear from 4 to 6 concentric circlets of very elongated lateral tubercles. On the lower mesial faces of the cerata there are pseudobranchs which bear similar elongated tubercles. Rows of small tubercles are detectable on either side of the body, below the insertions of the cerata, and many similar tubercles are scattered over the central dorsum.

The frontal margin of the head bears two conspicuous dorso-lateral crests. The rhinophoral sheaths are flared and in large specimens the mouth of each sheath is subdivided into 7–9 pointed lobes.

A reliable recognition feature is the presence of long tubercles on the body and cerata (the specific name *hystrix*, which was taken from the identical generic name of the porcupine, indicates the spiky appearance of this recently discovered species).

ANATOMY (Fig. 5g). A specimen taken off the coast of Co. Cork, Ireland was dissected. The preserved length was 7 mm and the radular formula proved to be 86 × 0.1.0. Each tooth had a pointed median cusp, flanked by up to 5 blunt denticles on either side.

HABITS. This species feeds upon the calyptoblastic hydroid *Schizotricha frutescens* (Ellis & Solander). The spawn bands are usually attached to the rhachis of the hydroid, pink when first laid, turning to pale brown as development ensues.

DISTRIBUTION. The type description related to material collected off the Scilly and Sherkin Islands, near western British shores. Favourable localities were all below 12 m; the deepest record was from 25 m, off the Scillies (Picton & Brown, 1981).

17. *Doto koenneckeri* Lemche, 1976

APPEARANCE IN LIFE (Plate 6). This species can reach 8 mm in length. The ground colour of the skin is tranlucent white, overlying viscera which have an orange tinge. The dorsal surface of the body, the flanks and the head are blotched with superficial epidermal brownish black pigment. On either side of the body is a conspicuous thick longitudinal line of pigment-free skin linking the ceratal bases and extending forward to the rhinophore sheath. The cerata (up to 5 pairs) each bear 2–3 concentric circlets of elongated lateral tubercles and a prominent finger-like terminal tubercle. These ceratal tubercles occasionally bear rounded blackish terminal spots; there are usually similar, rod-shaped spots on the upper surface of each ceratal tubercle. Pseudobranchs are rudimentary or absent.

The pale frontal margin of the head forms a pair of rounded, flattened lobes. This species is unusual in the genus in that it lacks longitudinal ridges or crests in front of the rhinophore sheaths. Each sheath is narrow, slightly dilated, smooth and without pigment on the rim.

The most reliable recognition feature is the dark dorsum, broken by a pale longitudinal line on each side.

ANATOMY (Fig. 5h). Lemche's original description of this species did not mention the radula; we have examined the radula of a specimen from Donegal, 4 mm long, in preservative. The formula was $72 \times 0.1.0$. On either side of each erect slender cusp was an asymmetrical series of up to 3 subsidiary denticles (Fig. 5h).

HABITS. This species feeds upon the calyptoblastic hydroid *Aglaophenia pluma* (L.). The spawn is white or pinkish and contains up to 2000 eggs (Lemche, 1976); it has been observed only in the month of May.

DISTRIBUTION (Map 2). The type description by Lemche (1976) was based on material from southern Norway and from localities around Carna, Eire. Subsequently, this species has been recorded around Northern Ireland and along the north east coast of Spain (Ortea & Urgorri, 1978).

18. *Doto lemchei* Ortea & Urgorri, 1978

APPEARANCE IN LIFE (Plate 6). This is a little known but distinctive species, reaching 10 mm in length. The skin is translucent white, with the pale orange digestive gland showing through, except on the central dorsum from head to tail where blotchy brown pigment supervenes. The cerata (4–6 pairs) bear 3–5 concentric circlets of blunt tubercles, each containing a terminal subepidermal cluster of chalk-white granules. The apical tubercle in each ceras is stout and conspicuous. Pseudobranchs are absent.

The dilated frontal margin of the head forms a pair of prominent lateral flaps. Faint ridges or crests are detectable in front of the rhinophore sheaths. Each sheath is widely flared and smooth-rimmed, not markedly crenulate.

This species can be readily distinguished by its possession of white-tipped ceratal tubercles, and brown pigment on the dorsum (without red speckling).

ANATOMY. Nothing is on record; the original authors did not examine the radula and no specimens have come into our possession.

HABITS. This species feeds upon the calyptoblastic hydroid *Aglaophenia tubulifera* (Hincks). Spawn has been found in the month of August, and has been illustrated by Ortea & Urgorri (1978).

DISTRIBUTION (Map 2). The only British specimens were collected by Mr Bernard Picton on the north coast of Cornwall in 1977. It seems to be more common near the type locality, Verdicio in northern Spain. (A number of recent new British records are listed by Picton & Brown (1981).)

19. **Doto maculata** (Montagu, 1804)
     Doris maculata Montagu, 1804
     not **Doris maculata** Garstang, 1895

APPEARANCE IN LIFE (Plate 6). This rare species is small and inconspicuous, never exceeding 9 mm in length. The pale cream body bears sparse superficial crimson pigment. The cerata (4–5 pairs) bear 3 or 4 concentric circlets of rounded tubercles, each containing a minute crimson subepidermal body. The terminal tubercle of each ceras lacks this feature. Rudimentary pseudobranchs are present in the larger individuals; the crimson blotching of the pseudobranch area, so conspicuous in *D. coronata*, is lacking.

The dilated frontal margin of the head forms a pair of prominent semicircular lateral flaps (Plate 6f). There are no ridges or crests attached to the front of each rhinophore sheath. Each sheath is flared, with a tongue-like anterior extension; there is a little white pigment but no crimson.

The British species with which *D. maculata* might be confused are *D. coronata* and *D. eireana*, but *maculata* lacks crimson pigment on the inner faces of the ceratal bases, on the terminal tubercles of the cerata, and on the rhinophore sheaths.

ANATOMY (Fig. 5e). We have examined the radula of a specimen from Kilray Harbour, 4 mm long in preservative. The formula was $63 \times 0.1.0$; each tooth has a slender, erect, median cusp, flanked on either side by 2 subsidiary denticles.

HABITS. This species feeds upon the calyptoblastic hydroid *Plumularia catharina* Johnston, on which it may also deposit spawn (Lemche, 1976). The spawn ribbon is very small, with sparse eggs, approximately 7 per transverse row. Spawn has been reported in Irish and Scottish waters in the months of January, May, June and July.

DISTRIBUTION (Map 3). For most of this century, this has been considered to be a subspecies of *D. coronata*. The only reliable attribution before Lemche's (1976) redescription is the type description of a specimen from south Devon. Recent corroborative reports have come from south west Ireland, Strangford Lough and western Scotland.

DISCUSSION. None of Montagu's material has survived, but a neotype is housed in the Zoological Museum of the University of Copenhagen (Lemche, 1976).

20. **Doto millbayana** Lemche, 1976

APPEARANCE IN LIFE (Plate 8). This is a distinctive species, reaching 14 mm in length. The greyish body bears sparse tiny superficial dark crimson pigment spots of irregular shape. The cerata (6 pairs) bear 4–5 concentric circlets of rounded tubercles, each usually containing a minute dark crimson or black subepidermal body. The terminal tubercle of each ceras also exhibits this feature. There is in addition a sparse speckling of dark crimson over the general surface of each ceras. Pseudobranchs are exceptionally well developed in *D. millbayana* (Plate 8p).

The dilated frontal margin of the slender head forms a pair of small semicircular lateral flaps (Plate 8o). Prominent longitudinal ridges or crests are visible in front of the rhinophore sheaths. Each sheath is slightly flared, with a scalloped rim. The sheaths bear conspicuous crimson speckling, replaced by white pigment on the rims.

This species may be distinguished from other British representatives of the genus by its slender shape, well developed pseudobranchs and scattered dark crimson pigment on the cerata.

ANATOMY (Fig. 5i). Lemche's original description did not mention the radula; we have examined a specimen from Lundy, 2·1 mm long, in preservative. The formula was $56 \times 0.1.0$;

each tooth exhibits a slender, erect cusp flanked on either side by up to 4 subsidiary denticles (not always disposed symmetrically).

HABITS. This species has been reported to feed on the calyptoblastic hydroids *Plumularia setacea* (Ellis & Solander) and *Nemertesia ramosa* (Lamouroux) (Lemche, 1976). Spawning has been observed in May (western Ireland) and October (south Devon). The spawn ribbon (Plate 8) is broad in comparison with other British species of *Doto* and, according to Lemche (1976), contains rows of up to 12 eggs.

DISTRIBUTION (Map 3). Recorded localities include Kilkieran Bay in south west Ireland, Strangford Lough, the Sound of Mull and the type locality near Plymouth. The only record from outside the British Isles is from north east Spain (Ortea & Urgorri, 1978).

21. *Doto pinnatifida* (Montagu, 1804)
Doris pinnatifida Montagu, 1804
Doto armoricana Hesse, 1872

APPEARANCE IN LIFE (Plate 7). This is one of the largest British species of *Doto*, reaching 29 mm in length. The body is pale fawn in colour, with dark brown or black mottling above, darkest anteriorly. The coloration of the body darkens with age. There are conspicuous dark brown spots around the rim of each flared rhinophore sheath and on the tips of small tubercles down each side of the body. The cerata (up to 10 pairs) bear about 6 concentric circlets of elongated tubercles, each containing a small but conspicuous dark brown or black subepidermal body. The apical tubercle may on rare occasions lack this dark terminal body. Pseudobranchs are rudimentary and difficult to discern.

The dilated frontal margin of the head is smoothly rounded. Prominent longitudinal ridges or crests are visible in front of the rhinophore sheaths.

A conspicuous recognition feature is the presence of a number (2–9) of simple, black-tipped tubercles along each flank. One or two similar tubercles may occur also on the dorsum.

ANATOMY (Fig. 5d). Kress (1968b) and Schmekel & Kress (1977) presented observations on the structure of the reproductive organs. Radulae of specimens from the Isle of Man (Miller, 1958) had the general formula $n \times 0.1.0$. In an 8 mm specimen, n was 92, while in a 14 mm specimen it was 91. A 14 mm specimen from the Plymouth area had 100 teeth (Schmekel & Kress, 1977). We have examined a Lundy specimen; this measured 10 mm in preservative and had a radular formula of $55 \times 0.1.0$. Another British specimen had a preserved length of 16 mm and a formula of $63 \times 0.1.0$. The teeth each had a slender erect cusp, flanked on either side by up to 5 irregular, tiny, subsidiary denticles (Fig. 5d).

HABITS. This species feeds exclusively upon the calyptoblastic hydroid *Nemertesia antennina* (L.), in quiet, shallow, sublittoral localities. Spawning occurs in most months of the year in the Plymouth area (Marine Biological Association, 1957), in July and October in the Isle of Man. Each spawn mass may contain up to 26,500 ova (Miller, 1958). Each ovum measures 91–108 μm; development to hatching occupies 16 days at 10–12 °C.

DISTRIBUTION (Map 3). This relatively large and distinctive *Doto* is common on the west coasts of the British Isles, but on the North Sea coasts is known only from the Moray Firth. There are rare records from southern Norway, while Ortea (1978) has confirmed its presence in northern Spain. Corroboration is required of a yellow variety (Kress, 1968b) from Banyuls, on the Mediterranean coast of France. A record from Naples (Schmekel, 1968) was later re-examined and allocated to a new species, *Doto acuta* Schmekel & Kress, 1977; this does not appear to live in British waters.

22. *Doto tuberculata* Lemche, 1976

APPEARANCE IN LIFE (Plate 8). This species may reach a length of 19 mm. The body is translucent white or pale yellowish green. The cerata (4–8 pairs) bear 3 or 4 imperfectly concentric circlets of elongated tubercles, each containing a small but conspicuous black subepidermal body which is absent from the apical tubercle. Larger cerata exhibit pseudo-branchs which take the form of a ridge on the mesial face, ending in what Lemche (1976) described expressively as "wings".

The dilated frontal margin of the head forms a pair of auriform projections (Plate 8). Prominent longitudinal ridges or crests are visible in front of the rhinophore sheaths. Each sheath is widely dilated and scalloped; the rim may bear 1–3 black tubercles.

The most obvious recognition feature is the presence of up to 5 transverse rows of small black-tipped epidermal papillae (Plate 8e), connecting the two longitudinal rows of flank papillae.

ANATOMY. Nothing is known concerning the radula of this species.

HABITS. This species occurs in association with its prey, the calyptoblastic hydroid *Sertularella gayi* (Lamouroux), in shallow waters on the western coasts of the British Isles. Lemche (1976) states that it is also found on *Abietinaria abietina* (L.). The spawn mass is distinctive, the shape resembling that of a dorid nudibranch. It takes the form of a flattened ribbon moulded during the act of oviposition into a loose spiral of $\frac{1}{2}$ to $1\frac{1}{2}$ turns (Plate 8g). Each mass may contain up to 400 white ova. Spawning has been observed in June and July around Lundy Island; in the Plymouth area, Lemche (1976) recorded spawn in the month of October.

DISTRIBUTION (Map 3). This recently described species has now been reported from southern Norway, Sweden and the west coasts of the British Isles, as well as from north east Spain (Ortea & Urgorri, 1978).

Suborder II DORIDACEA

This is the largest nudibranch suborder and in it are placed about 25 families. Most of these families are distributed throughout the seas of the world, but some of them (Platydorididae, Asteronotidae, Baptodorididae, Actinocyclidae, Bathydorididae and Doridoxidae) are absent or rare in cooler northern hemisphere waters, while others are decidedly tropical in their preferences (Chromodorididae, Hexabranchidae). On the other hand, there are certain families (Corambidae, Goniodorididae, Onchidorididae, Polyceridae, Cadlinidae and Archidorididae) which seem to be centred in the cool temperate regions. Sponges, bryozoans (= polyzoans), acorn barnacles, ascidians or tube-dwelling polychaetes (in the exotic *Okadaia*) form the diet of most doridacean nudibranchs.

The dorids often possess cavities or pits into which the rhinophoral tentacles may be retracted, but they only rarely have external sheaths surrounding the bases. The anal opening is nearly always situated mid-posteriorly, under the mantle rim in *Corambe* and *Doridella*, and incidentally in all dorids at an early developmental stage, but situated dorsally in all the typical forms, such as *Doris*, *Archidoris*, *Polycera*, *Adalaria*, *Hexabranchus* and many others. A variable number of retractile foliaceous gill plumes emerge from the mantle surface close to the anal papilla. In typical dorids these gills are arranged in a circlet, and on alarm they can be retracted in a co-ordinated way into a dorsal pocket. This pocket may, in some exotic forms, be lobed in various ways to give greater resistance to an inquisitive predator. In more primitive dorids, such as *Polycera*, *Crimora*, *Adalaria* and *Acanthodoris*, no such pocket is present, and on alarm each gill is contracted separately down to the mantle surface. Sometimes large spiculose mantle papillae, often containing elaborate defensive glands and luminescent or coloured lures, give extra protection to the gills and rhinophores, and draw the attention of a predator away from the more fragile vital parts. Such papillae never contain lobes of the ovotestis or of the digestive gland in the Doridacea, and the digestive gland usually forms a more or less compact single mass close to the stomach. The radula is usually broad, and never uniseriate. Jaws are occasionally present; the oral canal may be cuticularized and this cuticle may exhibit honeycombing or other substructure. One successful family of exotic doridaceans, the Dendrodorididae, contains species which lack both radula and jaws and ingest sponges by an ingenious muscular sucking modification of the buccal mass. The larval shell is of veliger type 1 (Thompson, 1961a).

KEY TO THE BRITISH SPECIES OF DORIDACEA

1. Gills retractile into a single deep branchial cavity   **2**
   Gills contractile separately   **11**
2. Mantle smooth or faintly papillate   *Cadlina laevis* (p. 77)
   Mantle bears conspicuous rough papillae   **3**
3. Mantle papillae small and uniform   **4**
   Mantle papillae large, or various sizes   **6**
4. Strong tubercles surround the rhinophoral pits; radular teeth greatly elongated, with a fringe of denticles   *Aldisa zetlandica* (p. 78)
   No such specialized tubercles; most of the lateral radular teeth lacking a denticulate fringe   **5**
5. Body colour reddish; marginal radular teeth bifid   *Rostanga rubra* (p. 79)
   Body colour brownish; all radular teeth smooth hooks   *Jorunna tomentosa* (p. 91)
6. Some of the mantle papillae massive (3–4 mm diameter) and pendunculate   **7**
   Papillae not massive   **8**

7. Mantle papillae up to 3 mm in diameter, inter-connected by anastomosing ridges *Doris sticta* (p. 82)

Mantle papillae up to 4 mm in diameter, without inter-connecting ridges *Doris verrucosa* (p. 82)

8. Oral veil bearing finger-like tentacles **9**

Oral veil without tentacles *Archidoris pseudoargus* (p. 84)

9. Mantle with a median longitudinal elevated zig-zig ridge *Atagema gibba* (p. 85)

Mantle without such a ridge **10**

10. Mantle with uniform small tubercles *Discodoris millegrana* (p. 86)

Mantle with coarser tubercles and conspicuous stellate acid-gland openings *Discodoris planata* (p. 88)

11. Mantle rim ample, forming a skirt all around the back **12**

Mantle rim reduced **22**

12. Low, crenulated rhinophore sheaths present; mantle papillae conical, tall and soft *Acanthodoris pilosa* (p. 62)

No rhinophore sheaths; mantle papillae rounded and stiff **13**

13. Dorsal mantle yellow-edged with red markings centrally *Onchidoris luteocincta* (p. 58)

Dorsal mantle not so marked **14**

14. Dorsal mantle blotched with brown **15**

Dorsal mantle white or yellow **20**

15. Gill pinnules up to 29 *Onchidoris bilamellata* (p. 53)

Gills 12 or less **16**

16. Mantle tubercles very small, uniform, conical, spiculose *Onchidoris pusilla* (p. 60)

Mantle tubercles larger **17**

17. Mantle tubercles rounded **18**

Mantle tubercles conical **19**

18. Area around each rhinophore conspicuously pigmented *Onchidoris sparsa* (p. 62)

No such areas evident *Onchidoris inconspicua* (p. 56)

19. Body extremely depressed *Onchidoris depressa* (p. 54)

Body not markedly depressed *Onchidoris oblonga* (p. 59)

20. Mantle bearing some massive tubercles *Adalaria loveni* (p. 49)

Mantle tubercles varying, but never massive **21**

21. Mantle tubercles rounded; radular formula up to 2.1.1.1.2 *Onchidoris muricata* (p. 58)

Mantle tubercles more elongated and tapering; radular formula up to 13.1.1.1.13 *Adalaria proxima* (p. 50)

22. Body covered with uniform small papillae *Aegires punctilucens* (p. 66)

Body not covered with uniform small papillae **23**

23. Body smooth, lacking prominent dorsal papillae **24**

Body bearing papillae of various shapes, sometimes finger-like, sometimes compound **25**

24. Body colour milky white *Goniodoris nodosa* (p. 41)

Body colour brownish *Goniodoris castanea* (p. 40)

25. Body bearing several finger-like frontal or dorsal papillae **26**

Body bearing numerous branched or compound papillae along the frontal margin; without elongated finger-like processes along the lateral pallial margins *Crimora papillata* (p. 64)

26. Finger-like frontal papillae present but no dorsal processes of this type *Greilada elegans* (p. 72)
Elongated dorsal papillae present **27**

27. A pair of conspicuous unbranched papillae present near the gill circlet **28**
Papillae more numerous or more complex **32**

28. Dilated sheaths present around the bases of the rhinophores *Thecacera pennigera* (p. 71)
No rhinophore sheaths **29**

29. A prominent finger-like process attached laterally to the base of each rhinophore **30**
Rhinophore bases smooth **31**

30. Body white and yellow *Trapania maculata* (p. 48)
Body white *Trapania pallida* (p. 48)

31. Up to 6 finger-like processes on the frontal margin *Polycera quadrilineata* (p. 68)
6–12 such processes, usually 8 or more *Polycera faeroensis* (p. 67)

32. Numerous finger-like papillae on either side of the body **33**
Papillae elongated and compound near the gill circlet **38**

33. A pair of finger-like papillae attached to the base of each rhinophore *Ancula gibbosa* (p. 46)
Rhinophore bases smooth **34**

34. Frontal processes smooth **35**
Frontal processes feathery *Limacia clavigera* (p. 75)

35. Elongated dorsal processes along mantle rim and in the central area of the back **36**
No such processes in the centre of the back **37**

36. Body white, red and golden yellow *Okenia elegans* (p. 43)
Body white, suffused with pale pink *Okenia leachi* (p. 44)

37. Dorsal processes all along the mantle rim *Okenia pulchella* (p. 44)
Dorsal processes present only alongside the gill circlet *Okenia aspersa* (p. 42)

38. Dorsal and lateral papillae delicate, elevated and pointed *Palio dubia* (p. 73)
Dorsal and lateral papillae more broad and flattened *Palio nothus* (p. 74)

Superfamily ANADORIDOIDEA (= PHANEROBRANCHIA)

The mantle rim is usually reduced, sometimes indetectable, ample only in the Corambidae and Onchidorididae. Strong jaws are often present; the radula is usually narrow. A suctorial pump may be attached dorsally to the buccal mass. The penis may bear an armature of minute hooks. The gills are separately contractile and there is no common branchial pit for their reception and concealment.

GONIODORIDIDAE

Phanerobranchiate, soft, limaciform doridaceans with the mantle skirt reduced to form a rim which often bears elongated papillae; similar papillae sometimes arise from the rhinophores and from the central dorsum. Rhinophores are lamellate, without sheaths. The head has an oral veil which may bear lateral tentacles. They feed upon encrusting polyzoans and ascidians.

The penis bears numerous hooked chitinous spines. A dorsal buccal pump is present, sometimes pedunculate. Jaws, if present, are feebly developed. Radula 1.0.1, 1.1.0.1.1 or 1.1.1.1.1; first lateral highly differentiated.

Goniodoris Forbes & Goodsir, 1839
(type *Doris nodosa* Montagu, 1808)
Goniodorids with the mantle rim much reduced, lacking conspicuous elongated papillae; oral veil flattened, drawn out laterally to form tentaculiform lobes; buccal pump pedunculate; radula 1.1.0.1.1.

23. **Goniodoris castanea** Alder & Hancock, 1845a
Doris paretii Vérany, 1846
Goniodoris barroisi Vayssière, 1901
G. castanea pallida Dautzenberg & Durouchoux, 1913

APPEARANCE IN LIFE (Plate 10). This is a distinctive species, impossible to confuse with other British nudibranchs, although in appearance it approaches the South African *G. brunnea* Macnae, 1957. The colour pattern, while it may appear conspicuous in the laboratory, renders the animal cryptic in the field, with the red-brown body covered dorsally with white specks. The dorsum and the flanks bear small ridges and tubercles. The pallial rim and a mid-dorsal pallial ridge are especially notable, the latter being continued posteriorly into a dorsal metapodial keel. The overall length may reach 38 mm. Rare specimens may be pinkish white, shaded with yellow (Walton, 1908). The lamellate rhinophores and the gills are brown; the latter are tripinnate, up to 9 in number, encircling the anal papilla. Each rhinophore may exhibit up to 17 lamellae. The sides of the head are produced to form flattened oral tentacles.

ANATOMY (Plate 13b, Fig. 7c). This species is rarely found in great numbers and hence has not been used for physiological or life-history studies. Forrest (1953) gave some details about the structure of the alimentary canal, while Alder & Hancock (1845–55) and Woodland (1907) stressed the distinctness of the pallial spicules, more angular and nodulous than those of *G. nodosa*. More recently, some observations on the reproductive organs were presented by Schmekel (1970).

The radula was found to vary surprisingly little with age, with formula 28 × 1.1.0.1.1. (body-length 12 mm in preservative); 29 × 1.1.0.1.1 (14 mm in life); 28 × 1.1.0.1.1 (28 mm in life); 30 × 1.1.0.1.1 (18 mm in preservative). The principal lateral teeth have a smooth ridge along the cusp, and the small marginal teeth each possess a delicate cusp. In both these particulars there are profound differences from *G. nodosa*. It is noteworthy, too, that the lateral teeth of *G. castanea* are much smaller, about half the length found in *G. nodosa* of the same body-size.

HABITS. This goniodorid feeds principally on compound ascidians such as *Botryllus schlosseri* (Pallas) and *Botrylloides leachi* (Savigny). According to Forrest (1953), *Dendrodoa* may also be taken. When lying in a shallow excavation among these ascidians, *Goniodoris castanea* is extremely well hidden; we extricated with some difficulty 6 specimens which had penetrated the base of an *Ascidia mentula* Müller and burrowed deeply into the test.

In British localities, spawn has been noted in early spring (February), in the summer (June and July) and into the early autumn (September and October) in different years (Alder & Hancock, 1845–55; Marine Biological Association, 1957). Development to hatching occupies 12–13 days at 16 °C. (Schmekel & Portmann, 1982).

DISTRIBUTION (Map 3). This species has an interesting worldwide distribution, although it is apparently never common in any individual locality. The most northerly record is from southern Norway, while around Britain there are scattered records from all regions except western Ireland. Around British coasts, this species lives in shallow offshore waters, usually in depths of 25 m or less. At one extreme, intertidal specimens have been noted, while, at the other, Walton (1908) found 1 specimen in approximately 90 m, dredging in the North Sea. Swennen (1961b) cited infrequent examples from the Dutch coast, but it appears to be more common along the French Atlantic coasts (Bouchet & Tardy, 1976). Other specimens, later recognized as synonymous with *Goniodoris castanea*, were described from Italy (Vérany, 1846) and the Gulf of Marseilles (Vayssière, 1901). Recent corroboration of these Mediterranean records has been provided by Schmekel (1968) in Naples, and Vicente (1967) in Marseilles.

Further afield, this species is known from Suez (Eliot, 1908), Japan (Baba, 1955) and New Zealand (Eliot, 1910). This last report, on the keel of a ship in Otago Harbour, led to the suggestion that ship-borne populations may have reached the Indo-Pacific via the Suez Canal (O'Donoghue, 1929).

24. *Goniodoris nodosa* (Montagu, 1808)
Doris nodosa Montagu, 1808
Doris barvicensis Johnston, 1838a
Doris elongata Thompson, 1840
Goniodoris emarginata Forbes, 1840

APPEARANCE IN LIFE (Plate 10, Fig. 6a; volume I, Plate 4). This is a robust species, despite its superficial appearance of fragility. It may reach 27 mm in length, translucent white in colour, with tinges on the dorsum of yellow and pink. There are patches of opaque white pigment and also areas of especially transparent skin (notably around the bases of the rhinophores and behind the gills, forming in the latter case a false "pore"). The mantle edge is fairly well developed but does not form an ample skirt. A low keel runs down the middle of the back and on either side of this are small conical tubercles; a median keel is also visible on the dorsal surface of the metapodium. The yellow-tinged rhinophores are lamellate (12 lamellae on each rhinophore in the case of a medium-sized animal, 11 mm in length). There are up to 13 pinnate gills around the anal papilla. The head bears a pair of lateral, flattened, oral tentacles.

ANATOMY (Plate 13a, Fig. 7f). This is a common and accessible species which has been much investigated. Feeding and digestion were studied by Alder & Hancock (1845–55), and by Forrest (1953), the nervous system by Pelseneer (1906), and the reproductive system by Lloyd (1952). Many radulae have been examined, by Miller (1958) and ourselves. The formulae were $11 \times 1.1.0.1.1$ (body-length 2 mm); $22 \times 1.1.0.1.1$ (7·5 mm); $21 \times 1.1.0.1.1$ (12 mm); $21 \times 1.1.0.1.1$ (13 mm); $22 \times 1.1.0.1.1$ (14 mm); $22 \times 1.1.0.1.1$ (22 mm); $24 \times 1.1.0.1.1$ (15 mm in preservative). The principal lateral teeth have a ridge along the cusp, bearing up to 28 subsidiary denticles; the small marginal teeth are flattened discs without an elevated cusp of any kind. The teeth are much larger than those of *G. castanea*, as already noted, and this can be confirmed by comparing Figs 7e and f.

HABITS. This species feeds upon encrusting polyzoans, such as *Alcyonidium polyoum* (Hassall), *Callopora dumerili* (Audouin) and *Flustrellidra hispida* (Fabricius) when young, but the adults transfer their attentions to ascidians, especially *Diplosoma listerianum* (Milne Edwards), *Botryllus schlosseri* (Pallas) and *Dendrodoa grossularia* (van Beneden) (McMillan, 1942; Miller, 1961; Swennen, 1961b).

As early as 1846, it was noted by Reid that these animals aggregated for mating and spawning, and an early life history study was carried out on *G. nodosa* by Garstang (1890) in the Plymouth area. He concluded that it was an annual organism, spawning in the early spring and dying off after laying eggs. In exceptionally favourable years, an extra period of oviposition occurred in the autumn months. Later authors have verified and amplified these conclusions (Miller, 1962; Thompson, 1964). Each egg mass, illustrated by Kress (1971), may contain 13,900–32,500 eggs; the white ova each measured 73–103 µm in diameter. The larval shell was of type 1 (Pelseneer, 1911; Thompson, 1961a).

DISTRIBUTION (Map 3). Bouchet & Tardy (1976) suggested that the southern distributional limit was around Bas Poitou on the French Atlantic coast, but we have recently seen specimens from north west Spain collected by Mr Bernard Picton of the Ulster Museum. It is a common species on shores and in the shallow sublittoral (to 20 m) all around the British Isles, although in other sea areas it has been reported down to 120 m. It reaches its northern limit around the Faeroes and on the southern Norwegian coast (Lemche, 1929) where it is found at depths to 120 m.

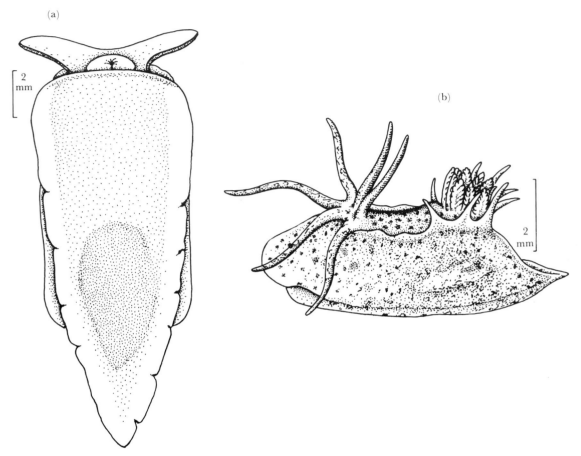

(a)

(b)

Fig. 6. Details of doridacean morphology.
(a) *Goniodoris nodosa*, ventral view.
(b) *Okenia aspersa*, lateral view (after Alder & Hancock, 1845–55).

Okenia Menke, 1830
(type *Idalia elegans* Leuckart, 1828)
Goniodorids having a relatively broad body; mantle ridge much reduced, but bearing a number of elongated papillae; similar papillae often arise from the central dorsum; oral veil flattened, lacking tentacles; buccal pump conspicuous but not pedunculate; radula 1.1.0.1.1.

25. *Okenia aspersa* (Alder & Hancock, 1845b)
Idalla caudata Oersted, 1844
Idalia aspersa Alder & Hancock, 1845b
Idalia inaequalis Forbes & Hanley, 1851

APPEARANCE IN LIFE (Fig. 6b). This delicate and beautiful species is uncommon and imperfectly understood. The body may reach 22 mm in length, red or yellow in colour, speckled with orange and brown. There are 2 pairs of long, tapering, anteriorly directed pallial papillae in front of the rhinophores, and up to 4 pairs of shorter processes alongside the gills. The rhinophores each bear numerous (up to approximately 35) fine lamellae. There are up to 12 simple pinnate gills, forming a circle around the anal papilla.

ANATOMY (Fig. 7g). The oral cuticle is thickened to form a pair of areas of strong hooked spines. The radula consists of 35 rows of 1.1.0.1.1 in a 22 mm specimen (Colgan, 1914). Each inner lateral tooth has a denticulate ridge along the cusp; the marginal teeth each bear an erect hooked spine. Other aspects of the gross anatomy are dealt with by Alder & Hancock (1845–55). These authors also described the calcareous spicules, which are stout, elongated, pointed at each end, and bent in the middle where there is a ring of nodules. We have examined a 10 mm preserved specimen with the radular formula 25 × 1.1.0.1.1. Each lateral tooth exhibited a row of up to 17 blunt denticles; the marginal teeth were smooth, elongated hooks. The buccal mass had a dorsal, sessile, buccal pump and the oral armature exhibited a complex denticulate substructure. The parasitic copepod *Splanchnotrophus gracilis* Hancock & Norman has been recorded from this nudibranch (Pruvot-Fol, 1954, quoting Hancock & Norman, 1863).

HABITS. The prey consists of ascidians such as *Molgula* (M'Intosh, 1874) and *Ascidiella* (Marine Biological Association, 1957, but perhaps this citation would be more correctly attributed to *O. elegans*). Spawning has so far been noted only in the month of October (Alder & Hancock, 1845–55; Marine Biological Association, 1957). The greatest depth from which *O. aspersa* has been brought is 60 m (off Holy Island in the North Sea by Walton, 1908).

DISTRIBUTION (Map 3). This rare species is known to occur as far north as Finmarken, Norway, and shows many resemblances to *Okenia modesta* (Verrill, 1875) from New England. British records come from the Shetlands, the North Sea and the Celtic Sea. Bouchet & Tardy (1976) included records of this species from Atlantic France as far south as Arcachon, but (mistakenly) under the name *Okenia quadricornis* (see below).

DISCUSSION. Pruvot-Fol (1954) considered this species to be embraced by Montagu's (1815) name *Doris quadricornis*. A scrutiny of Montagu's description and line-drawing (which, incidentally, do not match well) leads us to the conclusion that *quadricornis* Montagu, 1815 cannot be equated with any other taxon.

26. **Okenia elegans** (Leuckart, 1828)
   Okenia elegans "Leuckart" Bronn, 1826 (*nomen nudum*)
   Idalia elegans Leuckart, 1828
   Idalia laciniosa Philippi, 1841
   ? **Idalia neapolitana** Chiaje, 1841
   Idalia cirrigera Philippi, 1844
   ? **Idalia dautzenbergi** Vayssière, 1919

APPEARANCE IN LIFE (Plate 11; volume I, Plate 8). This, although it is among the most beautiful of the British nudibranchs, often lives much of its life concealed in burrows excavated into living ascidians. It may reach 80 mm in length. The body is white, freckled or suffused with pink. Conspicuous elongated pallial papillae project from the front (2, rarely 4), the edges of the mantle and the median dorsal area; there may be up to 35 papillae in all, orange with yellow or white tips. The rhinophores are rosy coloured, with golden yellow lamellated tips. The pinnate gills (up to 21 in number) are similarly marked. The foot is bordered with orange-yellow.

ANATOMY (Fig. 7i). The terminal region of the vas deferens can be seen in sections to be lined by a strong chitinous lining which bears short elevated cusps, 13·5 µm in height.

The oral cuticle forms a complete thickened ring, bearing tiny hooked spines. We examined an 11 mm preserved specimen from the Plymouth area; the radular formula was 27 × 1.1.0.1.1. Each lateral tooth bore a smooth ridge along the slender, curved cusp, but no subsidiary denticles. The marginal teeth each exhibited a small erect cusp. The muscular buccal pump opened widely into the dorsal buccal mass.

HABITS. The prey consists of ascidians such as *Molgula* and *Ciona* (Pruvot-Fol, 1954), into which it burrows. In southern Britain, it is found on sublittoral rocks, among *Molgula* and *Dendrodoa grossularia* (Beneden). Spawn was found in July and August in S. W. Britain, a pinkish wavy band attached to the adult prey.

DISTRIBUTION (Map 3). Despite its distinctive coloration, there are very few recent records of this large species. The earliest collections were from the western Mediterranean near Marseilles (Leuckart, 1828) and around Sicily (Philippi, 1841, 1844), more recently from Banyuls (Pruvot-Fol, 1954), Naples (Schmekel & Portmann, 1982) and the Bay of Biscay (Bouchet & Tardy, 1976). Most British records are from south west coasts of England, with rare reports from Anglesey and from Portrush in Northern Ireland. It is, however, also known from the southern North Sea (Swennen, 1961b; Hamond, 1972) and the Kattegat (Jaeckel, 1952). It occurs sublittorally to 28 m depth.

27. *Okenia leachi* (Alder & Hancock, 1854a)
Idalia leachi Alder & Hancock, 1854a

APPEARANCE IN LIFE (Plate 9). This rare species may reach 25 mm in length; the body is white, suffused with pink. Our knowledge still derives principally from the description of the type, a single dead specimen. In this individual, the pallial rim bore a number of long, stout, tapering papillae, 4 in front of the rhinophores, 7 on either side of the body. The two most posterior processes were bifid. Inside the pallial rim on each side were numerous shorter but still conspicuous tubercles. The rhinophores bore numerous fine lamellae. Eleven pinnate gills formed an incomplete circle around the anal papilla.

Recently, we received for examination a 16 mm (preserved) specimen from the Celtic Sea. This also had 4 elongated frontal papillae projecting from the mantle margin, but 8 on either side of the body (including the long, bifid posterior processes characteristic of the type). Inside the pallial rim, similar processes were arranged in 5 rows, as shown in Plate 9. Our specimen had 12 simple pinnate gills with approximately 18 lamellae on the largest. The rhinophores each bore 43 fine lamellae.

ANATOMY (Fig. 7h). In serial sections through the genitalia, it was found that the terminal region of the vas deferens was lined by a stout chitinous layer bearing broad-based, elongated, curved cusps, 24 µm in length.

The radula of the type had the formula 29 × 1.1.0.1.1. Each lateral tooth had a denticulate ridge and a long, curved cusp; the marginals possessed only rudimentary cusps. We examined the buccal mass of a 16 mm preserved specimen from the northern Celtic Sea; the radular formula was 30 × 1.1.0.1.1. Each lateral tooth exhibited a row of up to 42 sharp denticles along the cutting edge of the slender cusp; the marginal teeth each had a small, erect smooth cusp. The buccal pump was more slender and finger-like than is customary in the genus, but cannot be described as pedunculate. The oral cuticle bore small, blunt papillae.

HABITS. The diet is unknown. There appear to be no 20th century observations on its behaviour or mode of life, and the older records are all too vague (for example, Jeffreys (1863–69) says it occurs in "rather deepish water"). More information is urgently needed.

DISTRIBUTION (Map 3). This species appears to have a restricted boreal distribution, with records from Devon, the North Sea and the Hebrides (Alder & Hancock, 1854), Connemara, the Shetlands and southern Norway (Jaeckel, 1952). A recent collection came from 95 m in the Celtic Sea (Hartley, 1979).

28. *Okenia pulchella* (Alder & Hancock, 1854a)
Idalia pulchella Alder & Hancock, 1854a
Idalia pulchella fusca Odhner, 1907

APPEARANCE IN LIFE (Plate 11). This is the rarest of the British species of *Okenia*. Our knowledge until recently was based upon a single animal from north Cornwall preserved in 1839 and examined later by Alder & Hancock, with some later notes on Swedish material (Odhner, 1907). In these drab specimens, the body varied in colour from pale pink to dull brown; the

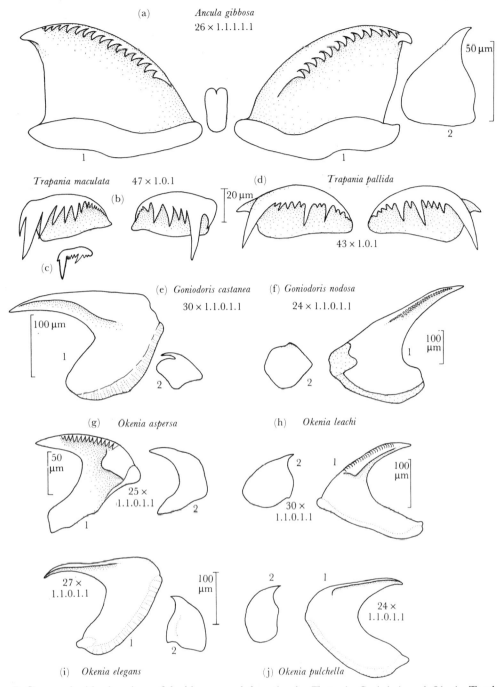

FIG. 7. Camera lucida drawings of doridacean radulae: *Ancula, Trapania, Goniodoris* and *Okenia*. Teeth are numbered from the mid-line.

(a) *Ancula gibbosa*, preserved length 8 mm, Orkney.
(b) *Trapania maculata*, length 15 mm, Portland, Dorset.
(c) *Trapania maculata*, radular tooth after Haefelfinger (1960).
(d) *Trapania pallida*, preserved length 5.5 mm, Lundy Island.
(e) *Goniodoris castanea*, preserved length 18 mm, Helford, Cornwall.
(f) *Goniodoris nodosa*, preserved length 15 mm, Kerrera, Scotland.
(g) *Okenia aspersa*, preserved length 10 mm, near Fair Isle.
(h) *Okenia leachi*, preserved length 15 mm, off the Cork coast.
(i) *Okenia elegans*, preserved length 11 mm, Plymouth, Devon.
(j) *Okenia pulchella*, preserved length 8·5 mm, Lundy Island.

length was maximally 19 mm. The head was said to be produced in front so as to form a shield-like projection. The pallial rim bore numerous elongated papillae, the most posterior of which was bifid in the Swedish material. Four long, tapering processes projected forward from the mantle in front of the lamellate rhinophores. There were 10 or 11 pinnate gills around the anal papilla.

Recent examination of specimens from Lundy, measuring up to 9 mm alive, has amplified many details. The body was darkly pigmented over the dorsum and the upper lateral surfaces, fading ventrally. The colour pattern is built up from discrete patches of brown, white and yellow pigment, in different proportions. The gills have a pale band about one-third of the way down from the free tip.

ANATOMY (Fig. 7j). The radula of the type had the formula 25–26 × 1.1.0.1.1. The teeth were said by Alder & Hancock to resemble those of *O. aspersa*, except that the marginals were shorter and the bases of the laterals were more slender and elongated. We examined the buccal mass of an 8·5 mm preserved specimen from Lundy Island; the radular formula was 24 × 1.1.0.1.1. The lateral teeth each bore a smooth ridge along the slender cusp, and the marginals each possessed a tiny erect cusp. The dorsal buccal pump was broadly sessile.

HABITS. The diet and mode of life are unknown. Alder & Hancock gave the habitat as "among corallines". Recent captures from Lundy in 1976 were obtained by searching carefully through mixed bottom material brought from 13 m. More information is urgently needed.

DISTRIBUTION (Map 4). The most southerly record is a specimen dredged off St Ives in Cornwall in 1853 (Alder & Hancock, 1854a). Recent specimens have come from Lundy in the Bristol Channel and from the Celtic Sea. It may be more common off the Norwegian coast, where it has been reported in various localities well into the Arctic Circle (Odhner, 1907). Lemche (1941) believed that a record from the east coast of north America related to this species, but corroboration is wanting.

Ancula Lovén, 1846
(type *Polycera cristata* Alder, 1841)

Goniodorids having a slender body; the mantle ridge indetectable except for a series of elongated pallial papillae on either side of the gill circlet; similar papillae arise from the bases of the rhinophores and project forwards; oral veil bearing a pair of tiny tentacles; buccal pump stout, sessile, not pendunculate; radula 1.1.1.1.1, the first lateral teeth denticulate, the marginals smooth.

29. *Ancula gibbosa* (Risso, 1818)
Tritonia gibbosa Risso, 1818
Polycera cristata Alder, 1841
Miranda cristata; Alder & Hancock, 1847
Ancula cristata; Alder & Hancock, 1847
Ancula sulphurea Stimpson, 1854
Ancula pacifica MacFarland, 1905

APPEARANCE IN LIFE (Plate 12). To the casual observer, on the shore or under shallow water, this species has a resemblance to the 2 British species of *Polycera*, but a closer inspection reveals important differences, especially in the number and positioning of the finger-like pallial tubercles.

The body of *A. gibbosa*, up to 33 mm in length, is white, elongated, and slender. At the level of the gills the mantle is produced on either side into strong, dorsally directed, pallial tubercles (up to 7 on each side). A pair of similar but more slender tubercles project forward from each rhinophore stem and the sides of the head are produced into short, finger-like, oral tentacles. All these processes are usually tipped with yellow or orange. Occasional all-white individuals have been reported, from Hilbre Island (Herdman & Clubb, 1889), Galway (Farran, 1903) and Den Helder in the Netherlands (Swennen, 1961b). Herdman & Clubb (1892) believed that paler individuals were usually the largest adults, but later authors have disagreed, citing 4 and 6 mm albino juveniles. Specimens received from Orkney in 1974 lacked yellow pigment only on the gills and rhinophores.

The rhinophores each bear up to 12 coarse lamellae (in a 20 mm specimen from the Isle of Man, described by Miller, 1958); specimens from Northumberland had fewer lamellae (6 in a 19 mm individual from Cullercoats, observed personally in 1962). The gills are tripinnate, 3 in number, protected and shielded by the defensive pallial papillae. A row of white spots is often visible along the rhachis of each gill.

ANATOMY (Fig. 7a). This is an uncommon species, sometimes found in pockets of abundance, but never predictably so. This explains why it has only rarely been used in anatomical studies. All that is known is that the spicules are very variable in shape, and may in Kiel specimens be absent (Meyer & Möbius, 1865), that "wandering cells" are present in the tissues (Millott, 1937), and that the penis is armed with small spines.

Radular formulae up to a maximum of $27 \times 1.1.0.1.1$ have been reported (Meyer & Möbius, 1865), but the overall length of that individual is not known. Records from the Isle of Man by Miller (1958) and the present authors are better documented: 18 rows (body-length 5 mm in preservative); 21 rows (9 mm alive); 14 rows (11 mm alive); and 17 rows (20 mm alive). We recently examined an 8 mm preserved specimen from the Orkneys; the radular formula was $26 \times 1.1.1.1.1$. The lateral teeth have a blunt, erect cusp with a subterminal row of slender, hooked, subsidiary denticles; the marginals exhibit rudimentary erect cusps. A feature missed by previous authors is the presence of a minute median tooth in each radular row. The cuticle of the oral canal is thickened and differentiated so that it exhibits minute projecting hooked elements, up to 33 µm in length.

HABITS. Contrary to Swennen's assertion (1961b, quoting Jaeckel), this species is not herbivorous, but feeds like other nudibranchs on animal material, in this case tunicates such as *Botrylloides leachi* (Savigny), *Botryllus schlosseri* (Pallas) and *Diplosoma listerianum* (Milne Edwards). Its occurrence is sporadic; Alder & Hancock (1845–55) reported that it was one of the most abundant nudibranchs of the Northumberland coast, and, near Den Helder, Swennen (1959) found *A. gibbosa* at densities of 15/m² over a 20 m² study area in March 1952. These fortunate experiences have not been shared by other naturalists.

Spawning has been noted in February to July (Den Helder, Swennen, 1961), March (Clyde area, Elmhirst, 1922), May (Greenisland, McMillan, 1944) and August (St Andrews, M'Intosh, 1863). In Northumberland, Alder & Hancock recorded spawn in the months of May, June, July, August, September and November. The larval shell is of type 1 (Pelseneer, 1911).

Adults are sometimes infected with ectoparasitic *Lichomolgus* (Leydig, quoted by Leigh-Sharpe, 1935) or endoparasitic *Splanchnotrophus* (Garstang, 1890).

DISTRIBUTION (Map 4). This species has one of the widest geographical ranges of any nudibranch of the north Atlantic. Recorded from the western Mediterranean near Nice (Pruvot-Fol, 1954), it also occurs along the length of the European Atlantic coast from the Bay of Biscay (Bouchet & Tardy, 1976) to the White Sea and Murmansk (Roginsky, 1962). Lemche (1929, 1938, 1941) recorded specimens from the Faeroes, Iceland and west Greenland, while Franz (1970a) published records from the New England American coast. It has been found at depths to 110 m.

DISCUSSION. Specimens found along the Californian coast, described under the name *Ancula pacifica* MacFarland, 1905, have extra pigment patches on the dorsum, but apparently lack other morphological differences.

Trapania Pruvot-Fol, 1931

(type *Drepania fusca* Lafont, 1874)

Goniodorids having a slender body; mantle ridge indetectable except for a pair of elongated pallial papillae at the level of the gill circlet; a similar papilla arises from the base of each rhinophore; strongly developed oral and propodial tentacles are present; buccal pump lacking; radula 1.0.1, all the teeth denticulate.

(A synopsis of the world's species of *Trapania* has been published by Kress, 1968a)

30. *Trapania maculata* Haefelfinger, 1960b

APPEARANCE IN LIFE (Plate 12). The limaciform body may reach 17 mm in length, white with numerous conspicuous yellow or orange markings. The numerous paired processes are, from front to rear, the slender, tapering, oral tentacles, the recurved propodial tentacles, the lamellate rhinophores, the anterior pallial tubercles (issuing from the rhinophoral bases) and the posterior pallial tubercles (on either side of the gills). All these processes, including the gills, either have yellow or orange tips, or are so coloured all over. This pigment also forms a characteristic pattern on the body, especially a median dorsal line, sometimes interrupted, and mid-lateral flank patches. The rhinophores bear up to 9 lamellae. There are 3 tripinnate gills around the anal papilla, the median gill being the largest. The pedal sole exhibits a characteristic median longitudinal groove when the animal is detached from the substratum.

ANATOMY (Fig. 7b, c). Little is known about the internal features, but Brown & Picton (1976) established that the penis was armed with tiny hooked spines. The oral cuticle is thickened to form 2 pads of reticulate elements. There is a curious discrepancy in published descriptions of the radulae of the type (from Villefranche-sur-Mer in the Mediterranean) and of the recent British finds. In the type material, a 15 mm specimen proved to have the formula $15 \times 1.0.1$, whereas in the British material the formula for 15 and 17 mm specimens was $47 \times 1.0.1$. There are also differences in the shape of the individual teeth. The specimen illustrated (Fig. 7b) is from the Dorset coast of England; a tooth of the Villefranche type material (Haefelfinger, 1960b) is shown in Fig. 7c.

HABITS. It is strange that such a conspicuous animal should have remained so rare. The sparse records are all from shallow water, 15 m or less, sometimes associated with arborescent polyzoans. Details of the diet are unknown. Haefelfinger (1960b) has described the spawn ribbon, about 17 mm in length, containing approximately 4000 white ova.

DISTRIBUTION (Map 4). This species approaches its northern distributional limit in the English Channel. It appears to be fairly common on the Brittany coast of France (Bouchet & Tardy, 1976), where it has been considered (mistakenly, in our opinion) to be a colour variant of *Trapania pallida* Kress.

Haefelfinger's type came from Villefranche, and other Mediterranean records have included Barcelona (Ros, 1975) and Naples (Schmekel, 1968).

31. *Trapania pallida* Kress, 1968a

APPEARANCE IN LIFE (Plate 12). This is another rare species, but one which is, unlike *T. maculata*, inconspicuous. The body is limaciform and recorded specimens have reached only 15 mm, uniformly cream in colour but with iridescent white patches on the tentacles, gills, metapodial tip and pallial tubercles. There are numerous distinct pairs of dorso-lateral processes; these are similar in position to those of *T. maculata*, but they are shorter and totally lacking in yellow or orange pigment. The rhinophores each bear up to 11 lamellae and there are 3 tripinnate gills around the anal papilla.

ANATOMY (Fig. 7d). Kress (1970) established that the penis bears numerous curved spines about 25 µm long, and gave other observations on the anterior genital mass. The oral cuticle is thickened on either side to form a pair of denticulate pads. The radular formula (Kress, 1968a, 1970) was $30 \times 1.0.1$ (9 mm specimen from Brittany) or $48 \times 1.0.1$ (15 mm specimen from Wembury, near Plymouth). We examined a 5·5 mm preserved specimen from Lundy Island; the radular formula was $43 \times 1.0.1$. Differences from *T. maculata* are evident (compare Figs 7b and d) but more information is needed about the range of radular variation in the genus.

HABITS. Very little is known about the mode of life of this species. It has been suggested that the diet may consist of small entoprocts such as *Loxocalyx*, which are often taken with the nudibranch sublittorally.

DISTRIBUTION (Map 4). First discovered in 1968, this species is now known from the west coast of Ireland, south west England, and the Atlantic coasts of France and Spain.

ONCHIDORIDIDAE

Phanerobranchiate, spiculose, flattened, ovoid doridaceans with an ample mantle skirt. Rhinophores lamellate, without well-developed sheaths. Head with flattened oral lobes forming a veil anterior to the mouth. Feeding upon encrusting ectoproct polyzoans.

The penis may be unarmed (*Onchidoris, Adalaria*) or may bear numerous hooked spines (*Acanthodoris*). The buccal mass has a muscular, often pedunculate pump. Oral and buccal cuticular thickenings are weak or absent. Radula $n.1.0.1.n$ or $n.1.1.1.n$ ($n =$ up to 13). The 1st lateral teeth are large and hooked, while the others are weak, flattened plates, usually lacking erect cusps. Median teeth, where present, are vestigial.

Adalaria Bergh, 1878b
  (type *Doris loveni* Alder & Hancock, 1862)
  Onchidorids in which the dorsal mantle bears abundant rounded or club-like tubercles; gills simple pinnate; radula $n.1.1.1.n$ ($n =$ up to 13); penis unarmed; buccal pump pedunculate.

32. **Adalaria loveni** (Alder & Hancock, 1862)
  Doris muricata variety Lovén, 1846 (not Müller, 1776)
  Doris loveni Alder & Hancock, 1862

APPEARANCE IN LIFE (Plate 14, Fig. 10c). This is a distinctive species yet it is one of the rarest of the British nudibranchs. It may reach 32 mm in length, sometimes pale yellow, but always white in the most northerly European specimens (Loyning, 1927), like its congener *A. proxima*. Unlike that species, however, its dorsal mantle bears widely spaced, large spiculose tubercles, with a few intervening smaller but similar processes. The centre of each tubercle is packed with bundles of calcareous spicules, but the peripheral area is clear. There are up to 12 simple pinnate gills, and each rhinophore may bear up to 25 lamellae (Bergh, 1880).

ANATOMY (Fig. 9a). The radular formula of a 16 mm preserved individual from Kerrera, Scotland was $45 \times 12.1.1.1.12$. Sars (1878) found the maximum number of marginals to be 12 on either side, in agreement with Friele & Hansen (1876). Bergh (1880) claimed that frequently the marginal plates were lost during early life and that only the most posterior 2–4 rows of the radula may be complete. Bergh's specimens from Bergen had the formulae $42–46 \times 12.1.1.1.12$. A detailed investigation was made of our 16 mm preserved specimen from Kerrera. The large 1st lateral teeth were smoothly hooked and had a distinctive root-shape. The most mesial marginal teeth were rudimentary slivers of chitin whereas marginals 6–12 were more substantial and bore tiny, rounded cusps (see Fig. 9a). The oral cuticle exhibited honeycomb-like patterning.

A short account of the nervous and genital systems was provided by Bergh (1880).

HABITS. Loyning's (1927) statement that this species is found on polyzoans was confirmed by our recent collections from *Securiflustra securifrons* (Pallas). A spawn mass deposited on this polyzoan by a 32 mm individual contained an estimated 5000–7000 eggs. The ova (measured in preserved material) had a diameter of 80 µm, much smaller than those of *Adalaria proxima*.

DISTRIBUTION (Map 4). There have been several collections from the Swedish coast, the Skagerrak (Jaeckel, 1952) and from Norway as far north as Trondheimsfjord (Odhner, 1939). Records from the Firth of Forth and off the Yorkshire coast (Jaeckel, 1952) require corroboration. Thirteen specimens were collected by aqualung divers between 5 and 20 m depth near Oban, W. Scotland during 1979. The depth range in the Norwegian fjords is from 0 to 200 m.

DISCUSSION. This species could be confused with a juvenile *Doris verrucosa* L., but the latter is cryptobranchiate (i.e. the gills can be retracted into a capacious gill cavity), whereas in *A. loveni* the gills are phanerobranchiate (separately contractile).

33. **Adalaria proxima** (Alder & Hancock, 1854a)
   Doris proxima Alder & Hancock, 1854a

APPEARANCE IN LIFE (Plate 14, Figs 8, 10b). This is a variable species, often confused with *Onchidoris muricata*, with which it is found on British shores. The body-size is, however, greater and reaches 17 mm (up to 25 mm in other European localities), whereas *O. muricata* rarely reaches 14 mm in the field. Other differences are listed below.

The body of *A. proxima* varies from white (in the most northerly and the most easterly localities) to yellow-orange. The ample mantle bears abundant rounded (pedunculate in places) spiculose tubercles which are taller and more slender towards the edge of the skirt. The rhinophores each bear up to 19 lamellae and are usually darker in colour than the rest of the body. Up to 12 pinnate gills are present, surrounding the anal papilla.

The ontogeny of body form is well known in this species (Thompson, 1958b) and can be used as an illustration of the rate of appearance and growth of external features in a doridacean (volume I, Fig. 45a–n). It can be seen that calcareous spicules make their appearance at a body-length of 0·30 mm, pallial tubercles at 0·5 mm, and, furthermore, that gills appear at 1·0 mm, rhinophore lamellae at 2·0 mm. Later development of gills and rhinophores is summarized in Fig. 8.

ANATOMY (Fig. 9b and c). The growth and wear of the radula were studied by Thompson (1958a). At a body-length of 0·5 mm, the formula was $15 \times 1.1.0.1.1$; 1·0 mm, $20 \times 2.1.0.1.2$; 2·0 mm, $22 \times 4.1.0.1.4$. After this stage, the median tooth makes its appearance and the formulae progress as follows: length 4·8 mm, formula $32 \times 4.1.1.1.4$; 8·0 mm, $37 \times 8.1.1.1.8$; 9·0 mm, $47 \times 9.1.1.1.9$; 10·0 mm, $39 \times 9.1.1.1.9$. The largest radula found was $50 \times 13.1.1.1.13$. A second abrupt change was noticeable at a body-length of 10 mm when new teeth became smooth, lacking the pectinate arrangement of denticles along the cusp of the large 1st lateral tooth on either side. The significance of these transitions during benthic life remains mysterious. There is no comparable halt or change in the growth of the radula as a whole; this proceeds smoothly (Fig. 8). The shape of the individual teeth differs significantly from *A. loveni* (compare Figs 9a and b). The oral cuticle lacks microscopic substructure.

HABITS. This species feeds preferentially upon the common intertidal polyzoan *Electra pilosa* (L.), which usually thrives on the alga *Fucus serratus*. Other polyzoans, such as *Flustrellidra hispida* (Fabricius), *Membranipora membranacea* (L.) and *Alcyonidium polyoum* (Hassall), will be attacked if *Electra* is not available. The life cycle is an annual one, with spawning in February to May, followed by rapid senescence and death; the maximal life span is about one year. The spawn is attached to fucoids and forms a cream-coloured spiral ribbon containing 180–2470 cream ova, each 165–195 µm in diameter. The embryonic period is 36–42 days at 9–10 °C, development is lecithotrophic (type 2 of Thompson, 1967), and the larval shell is of type 1 (Thompson, 1961a). The reproductive energetics of this species were studied by Todd (1979).

DISTRIBUTION (Map 4). This boreo-arctic species appears to be restricted to depths less than 60 m and is frequently encountered intertidally. It is known from East Greenland (Lemche, 1941) and southwards along the north American seaboard to Massachusetts (Franz, 1970a). It is also common around Iceland, the Faeroes (Lemche, 1929, 1938), and north along the Norwegian coast to Lofoten (Odhner, 1939). There are Baltic records and some from the Kattegat and Skagerrak (Jaeckel, 1952), indicating tolerance to lowered salinity. Northern and North Sea localities are numerous in Britain, while the most southerly confirmed reports come from the Bristol Channel. Mis-identifications may have led to records from the Plymouth area (Garstang, 1894) and the Channel Islands.

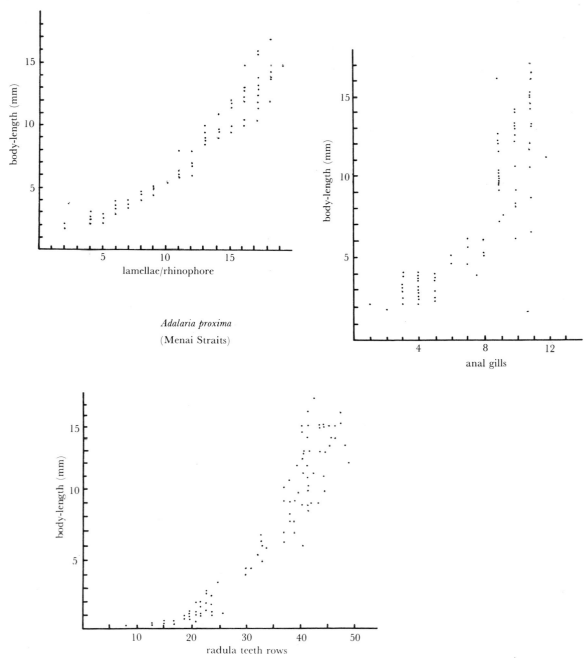

FIG. 8. *Adalaria proxima*: variation with body-size of three anatomical features, the number of lamellae on each rhinophoral tentacle, the number of gills around the anal papilla, and the number of rows of teeth in the radular ribbon. Based upon observations in the Menai Straits between Anglesey and the mainland of North Wales, 1954–57.

DISCUSSION. This species is easily confused with *Onchidoris muricata* (Müller). *Adalaria proxima* has a generally deeper colour (but not in extreme northern localities, like Orkney), the rhinophores are more bluntly tipped, and the digestive gland (visible through the translucent pedal sole) extends further forwards. But yellowish specimens of *O. muricata* can only be distinguished from pale *A. proxima* by examination of the radula.

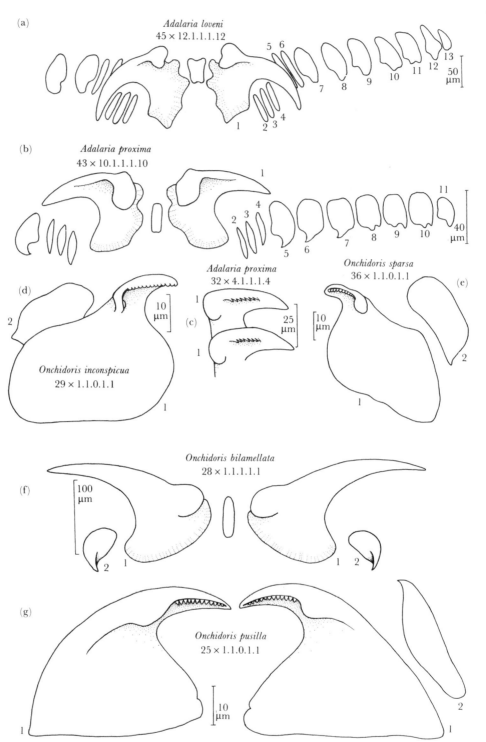

FIG. 9. Camera lucida drawings of doridacean radulae: *Adalaria* and *Onchidoris*. Teeth are numbered from the mid-line.

(a) *Adalaria loveni*, preserved length 16 mm, Kerrera, Scotland.
(b) *Adalaria proxima*, preserved length 8 mm, Kilve, Somerset.
(c) *Adalaria proxima*, length 4·8 mm, Menai Straits.
(d) *Onchidoris inconspicua*, preserved length 3 mm, Isle of Man.
(e) *Onchidoris sparsa*, length 6 mm, Orkney.
(f) *Onchidoris bilamellata*, length 29 mm, Orielton, S. Wales.
(g) *Onchidoris pusilla*, preserved length 4 mm, Lundy Island.

Onchidoris Blainville, 1816
 (type *Onchidorus leachi* Blainville, 1816)
= ***Lamellidoris*** Alder & Hancock, 1865
 (type *Doris bilamellata* L., 1767)
 Onchidorids in which the dorsal mantle bears abundant rounded or club-like tubercles (more elongated only in *O. luteocincta*); gills simple pinnate; radula 1.1.0.1.1 to 2.1.1.1.2; penis unarmed; buccal pump pedunculate.
 The name of the genus has been spelt in various ways, listed by Russell (1971).

34. ***Onchidoris bilamellata*** (L., 1767)
   Doris bilamellata L., 1767
   Doris fusca Müller, 1776
   Doris verrucosa Pennant, 1777 (not L., 1767)
   Doris elfortiana Blainville, 1816
   Onchidorus leachi Blainville, 1816
   Doris obvelata Bouchard-Chantereaux, 1836
   Doris affinis Thompson, 1840 (not Gmelin, 1791)
   Doris liturata Beck in Möller, 1842
   Doris vulgaris Leach, 1847
   Doris coronata Agassiz, 1850 (not Gmelin, 1791)
   Lamellidoris bilamellata praecedentis Mörch, 1868

APPEARANCE IN LIFE (Plate 14, Fig. 11a). This is one of the commonest British intertidal doridaceans, and also one of the largest, reaching 40 mm in length. The dorsum is dull white, mottled with brown, especially in a broad area down the middle. Juveniles all pass through a pure white stage; brown pigment appears on the stems of the rhinophores and around the bases of the gills at about 2 mm, and at 5 mm it has appeared also on the dorsum. Rare adult individuals may be white all over (Walton, 1908; Colgan, 1914; Pelseneer, 1922; Swennen, 1961b). The ample mantle bears abundant spiculose club-like tubercles of various sizes; brown pigment is usually absent from these tubercles. The pH of the skin secretion is acidic, varying from 1 to 3 (Edmunds, 1968). There may be as many as 29 separate simple pinnate gills with, characteristically, a few tubercles within the branchial circlet. Each rhinophore may bear up to 16 lamellae.

ANATOMY (Fig. 9f). Accounts of the general anatomy, with special reference to the alimentary canal, appear in the works of Hancock & Embleton (1852), Bergh (1880), Forrest (1953) and Marcus (1961). Observations on gamete translocation and the structure of the reproductive system were published by Thompson (1966), and an investigation of the functional anatomy of the buccal mass described by Crampton (1977). Potts (1981) has described the anatomy and histology of the gills and epidermal glands.
 The radula of a 2·25 mm specimen had the formula $20 \times 1.1.1.1.1$; 4 mm, $23 \times 1.1.1.1.1$; 12 mm, $26 \times 1.1.1.1.1$; 20 mm, $27 \times 1.1.1.1.1$; 29 mm, $29 \times 1.1.1.1.1$. The median teeth are small, frail and ovoid; the principal functional teeth are the first laterals, large and hooked, having smooth cusps; the marginal teeth are tiny, with rudimentary cusps. A scanning electron micrograph showing details of the teeth was published in Thompson & Bebbington (1973, plate 8). Figure 9f illustrates the radula of an adult specimen with the formula $28 \times 1.1.1.1.1$; the row shown consists of more or less perfect teeth, but nearer the functional end of the radula there is much breakage and abrasion evident. The lining of the oral canal exhibits microscopic honeycombing.

HABITS. The diet of this species was a mystery until Barnes & Powell's (1954) report that the prey consists of species of the barnacles *Balanus* and *Elminius*. Before that time it had been thought that *O. bilamellata*, like its congeners, attacked encrusting ectoproct polyzoans, or even that it ate carrion. It is now certain that the principal item of food is acorn barnacles, although it remains an open question whether or not polyzoans may be eaten by the juveniles.

Examination of dead barnacles showed that the cirri and other remains of the chitinous exoskeleton had been left behind; the dorids had taken mainly the soft parts (Barnes & Powell, 1954). The mode of action of the radula and buccal pump is discussed fully in the detailed paper of Crampton (1977).

Much of our knowledge of the ecology of *O. bilamellata* has come from an investigation on the south Kent coast by Potts (1970). In this area the chief prey is *Balanus balanoides*; even juveniles as small as 2 mm were seen to feed on these barnacles. This onchidorid population did not prove to be self-sustaining and required re-stocking through larval settlement emanating from elsewhere. Settlement occurs on the under-sides of intertidal boulders (we have recently verified this in the Oban area); from here, the growing dorids eventually move to the upper surfaces of the rocks where they grow to sexual maturity, spawn, then die.

Spawn has been reported from some British locality during every month of the year. The most thorough life-history study was that of Miller (1962) in the Isle of Man, who concluded that the normal life span was 12–16 months; spawning occurred twice during each lifetime. This confirmed the more sketchy observations made in Scotland by Evans & Evans (1917) and in Ireland by Renouf (1915).

The white spawn ribbons each contain up to 125,000 ova, 80–100 µm in diameter, taking 17–19 days to hatch at 10 °C (Thompson, 1967). The larval shell is of type 1 (Thompson, 1961a).

Some interesting data have recently appeared (Bleakney & Saunders, 1978) as the result of a study of *O. bilamellata* in Nova Scotia. These authors confirm that this species is unusual in that it must grow to near-maximum size before spawning. This is unusual because many opisthobranch molluscs first copulate at a body-length of only a few mm, and begin to spawn as they grow.

Perhaps the most startling phenomenon ever to be described in the literature of the Opisthobranchia was claimed for *O. bilamellata* by the distinguished Belgian malacologist Pelseneer (1922). At Wimereux in August 1922, he noted populations with a density up to $1000/m^2$. These countless hordes of dorids he watched as they migrated up and down the shore with the changing tide. No-one has confirmed this astonishing story, although SCUBA divers off Holy Island on the coast of Anglesey have sent us underwater photographs showing comparably dense aggregations at the start of the breeding season there.

We recently tried without success to confirm Pelseneer's claims, visiting Wimereux during July and August 1981. Copulating *O. bilamellata*, 2·5–20·0 mm in length, were found abundantly on the boulders of the lower shore, together with their spawn. Counts ranged from 18 to $92/dm^2$ $(1800–9200/m^2)$ on favourable boulders. Although watched at intervals over several days, no migratory behaviour was seen. We paid particular attention to the period on either side of the turn of the low tide. These observations cast further doubt on Pelseneer's attractive but implausible anecdote.

DISTRIBUTION (Map 4). This is a boreo-panarctic species reaching a southern limit in Europe on the French coast near Le Croisic (Bouchet & Tardy, 1976). It can be locally abundant in all sea areas around Britain, and northwards along the Norwegian coast to the White Sea and Spitzbergen (Lemche, 1929, 1938, 1941; Roginsky, 1962), as well as westwards to Iceland, Greenland and the eastern American seaboard as far south as New England (Franz, 1970a). Bergh (1880) examined specimens from Hagmeister Island in the Bering Sea, while O'Donoghue (1926) described material from Alaska and Puget Sound. Bergh (1894) considered specimens from the Pacific coast of central America to be sufficiently distinct to be described as a subspecies, *pacifica*.

35. *Onchidoris depressa* (Alder & Hancock, 1842)
Doris depressa Alder & Hancock, 1842

APPEARANCE IN LIFE (Plate 15). This flattened species reaches 9 mm in length and is uncommon. The mantle is so translucent that the eyes and the glistening calcareous spicules

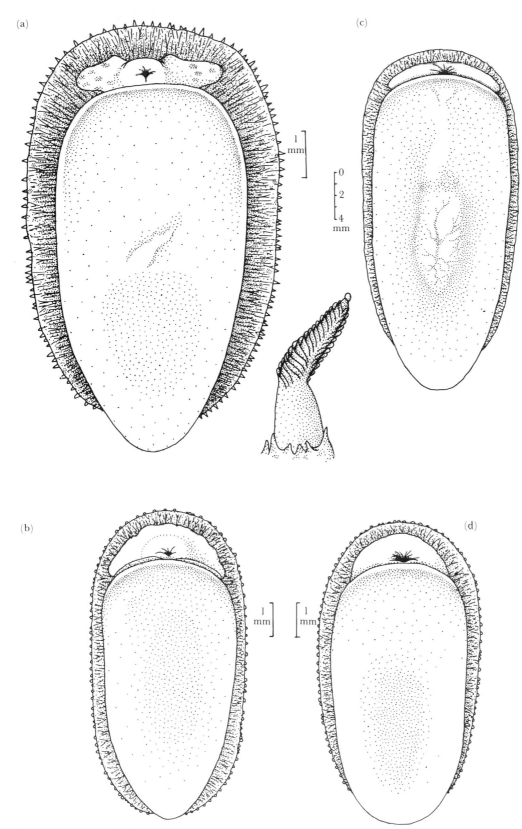

Fig. 10. Details of doridacean morphology; ventral views of (a) *Acanthodoris pilosa* (with rhinophore detail inset), (b) *Adalaria proxima*, (c) *Adalaria loveni*, (d) *Onchidoris muricata*.

may be discerned with a lens. The mantle is pale brown, with a pattern of scattered orange (sometimes purple-brown) spots tending, towards the mid-line, to possess a central black spot. A conspicuous and constant feature is the presence on the mantle of evenly spaced, uniformly long, soft tubercles. Up to 12 simple pinnate gills are present around the anal papilla. In some specimens there is a ring of low tubercles around the pallial openings of the rhinophores.

ANATOMY (Fig. 12b). The radula is distinctive, having only very small cusps on the principal teeth. The formula of a 6 mm specimen was $34 \times 1.1.0.1.1$. The small cusps of the lateral teeth each exhibit a stunted denticulate ridge. A careful examination was recently made of a 5 mm preserved specimen from Orkney, which had the radular formula $33 \times 1.1.0.1.1$ (Fig. 12b). Each lateral tooth is distinctively elongated, quite unlike that of any other onchidorid. The minute marginal teeth each possess a tiny erect cusp. The oral armature exhibits microscopic honeycombing.

HABITS. Feeding has been reported on the ectoproct polyzoan *Schizomavella linearis* (Hassall). The spawn is pinkish and has been found only during the month of September (Alder & Hancock, 1845–55).

DISTRIBUTION (Map 5). Schmekel (1968) obtained specimens which she attributed to *O. depressa* (see below) from the Bay of Naples between 5 and 45 m depth. The only other known Mediterranean locality is Banyuls (Pruvot-Fol, 1954), although Bouchet & Tardy (1976) listed several sites along the French Atlantic coast. British records include scattered localities on the west coast and in the Irish Sea, while several reports (mainly intertidal, but also down to 15 m) have come from the North Sea, especially along the Yorkshire coast and around the Orkneys.

DISCUSSION. This species is close to *O. neapolitana* (Chiaje, 1841) as redescribed by Schmekel & Portmann (1982); there is close correspondence too in the shape of the lateral teeth of the radulae. Schmekel's (1968) *Lamellidoris depressa* appeared to have a significantly different radula, so she later transferred this material to *Onchidoris bouvieri* (Vayssière, 1919) (Schmekel & Portmann, 1982).

36. *Onchidoris inconspicua* (Alder & Hancock, 1851)
Doris inconspicua Alder & Hancock, 1851

APPEARANCE IN LIFE (Fig. 11c). This flattened species may attain a length of 12 mm. The ample mantle is white or pale brown, with a tinge of purple, sprinkled with brown. The most distinctive external feature is the presence on the dorsal mantle of abundant tiny rounded tubercles. There may be up to 10 simple pinnate gills around the anal papilla. Despite its name, it is no more inconspicuous than other onchidorids.

ANATOMY (Fig. 9d). The radula of a 3 mm specimen (after preservation) was examined; the formula was $29 \times 1.1.0.1.1$. The principal teeth are the first laterals, and each bears a short cusp exhibiting a denticulate ridge. The marginal teeth are rudimentary.

HABITS. This species feeds on the encrusting ectoproct polyzoan *Cellaria sinuosa* (Hassall) (Miller, 1961), in shallow sublittoral localities. Breeding characteristics are unknown, but Alder & Hancock (1845–55) record spawn in the month of March.

DISTRIBUTION (Map 5). The species appears to have a restricted distribution between southern Scandinavia (Gullmarfjord in Sweden (Jaeckel, 1952)) and the coasts of Brittany and Normandy (Bouchet & Tardy, 1976). Most British records are from the Irish Sea, although the type locality was on the Northumberland coast. A Mediterranean record from Banyuls, suggested by Pruvot-Fol (1954) and included by Nordsieck (1972), is without foundation.

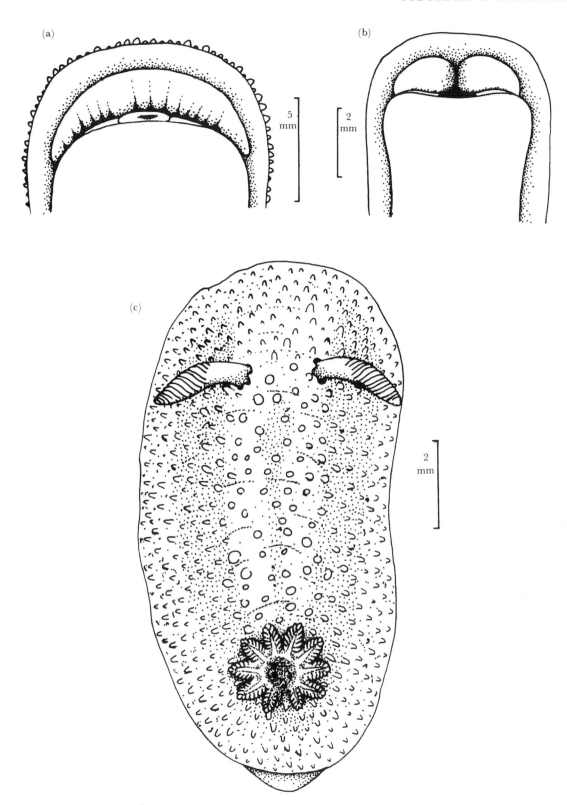

Fig. 11. Details of doridacean morphology.
  (a) *Onchidoris bilamellata*, ventral view of the head.
  (b) *Onchidoris luteocincta*, ventral view of the head.
  (c) *Onchidoris inconspicua*, dorsal view (after Alder & Hancock, 1845–55).

37. **Onchidoris luteocincta** (M. Sars, 1870)
   Doris luteocincta M. Sars, 1870
   Doris beaumonti Farran, 1903
   Diaphorodoris luteocincta; Iredale & O'Donoghue, 1923
   Diaphorodoris luteocincta reticulata Portmann & Sandmeier, 1960
   Diaphorodoris luteocincta alba Portmann & Sandmeier, 1960
   Diaphorodoris luteocincta; Schmekel & Portmann, 1982

APPEARANCE IN LIFE (Plate 15, Fig. 11b). This is one of the most colourful and attractive of the British dorid nudibranchs, rivalling the chromodorids of warmer seas. It reaches 11 mm in length, white in colour, with vivid crimson blotching on the tuberculate dorsal mantle. The red pigment is very superficial, and tends to rub off. The pallial tubercles are sparse, white and conical. Around the mantle skirt at a little distance from the edge is a narrow band of lemon-yellow. The metapodium projects some way behind the mantle and has a yellow edge and a median white keel. The mantle edge is sometimes uneven and may be crenulated. There may be up to 7 gills forming a circle around the anal papilla.

ANATOMY (Fig. 12d). Each radular row had the formula 1.1.0.1.1. In a 2·5 mm specimen (measured in preservative) there were 31 tooth-rows; length 6 mm, 23 tooth-rows. A specimen investigated by Eliot (1910) had the formula 34 × 1.1.0.1.1.

Two specimens were singled out for careful study. One was a 7 mm, preserved specimen from west Wales (radular formula 26 × 1.1.0.1.1), the other was from Donegal and measured alive 9 mm (formula 28 × 1.1.0.1.1). Each principal tooth had a denticulate ridge with up to 16 components. The rudimentary marginal teeth each had a tiny pointed cusp. The oral armature was flimsy and lacked substructure.

Portmann & Sandmeier (1960) described the arrangement of the ganglionic ring, the alimentary canal and the genital organs.

HABITS. The diet consists of the sublittoral ectoproct polyzoans, *Smittina reticulata* (Macgillivray), *Cellepora pumicosa* L., and more than one species of *Crisia*. Spawning occurs in March, according to Miller (1958), and each egg mass may contain up to 120 ova. The spawn is flattened, white and coiled; the ova are 70 μm in diameter.

DISTRIBUTION (Map 5). Although this is a common Mediterranean species, with records from Naples, Villefranche, Banyuls (Portmann & Sandmeier, 1960) and Spain (Ros, 1975), it has been reported from as far north as Norway (M. Sars, 1870). It is one of the commonest sublittoral onchidorids around western Britain (to 40 m depth) but it is rare along North Sea coasts. Bouchet & Tardy (1976) cited records from Brittany, while Ortea (1979) published the first record from northern Spain.

DISCUSSION. Portmann & Sandmeier (1960) described onchidorids from Villefranche and Banyuls on the French Mediterranean coast, which lacked the crimson blotching and had only few pallial papillae. These specimens they called the variety *alba*. More recently, we have found the extremes and all intergradations between crimson and white, in living material from south west Britain.

38. **Onchidoris muricata** (Müller, 1776)
   Doris muricata Müller, 1776
   Doris aspera Alder & Hancock, 1842
   Doris diaphana Alder & Hancock, 1845a
   Doris ulidiana Thompson, 1845
   Doris pallida Agassiz, 1850 (*nomen nudum*)
   Onchidoris pallida; Verrill, 1870

APPEARANCE IN LIFE (Plate 14, Fig. 10d). This species is widespread and common and in many northerly localities occurs with *Adalaria proxima*; when this happens the 2 species may be difficult to distinguish. *Onchidoris muricata* is generally smaller, reaching only 14 mm in the field (although pampered individuals raised in laboratory isolation can reach 20 mm (Thompson, 1961b)); *A. proxima* can attain a length of 17 mm in the field. Colour is an unreliable distinguishing feature; both species may be white or cream-yellow. In *O. muricata* the most northerly specimens are often yellow, whereas in the Menai Straits and the Bristol Channel the most frequent body-colour is white. Rare individuals can have brown dorsal freckling.

The ample mantle bears abundant rounded (pedunculate in places) spiculose tubercles which are taller and more slender towards the edge of the skirt. The rhinophores rarely bear more than 12 lamellae. There are up to 11 simple pinnate gills around the anal papilla. The digestive gland is visible through the ventral pedal epithelium; it does not extend anteriorly as far as in *Adalaria proxima*.

ANATOMY (Fig. 12a). The radula is distinctive and enables *O. muricata* to be separated infallibly from the superficially similar *Adalaria proxima*. The number of rows at 1 mm is 20; 2·5 mm, 27 rows; 4 mm, 29 rows; 6 mm, 36 rows; 11 mm, 36 rows. Each row has the formula 1.1.1.1.1, but in larger individuals this becomes 2.1.1.1.2. The median tooth is oblong and rudimentary. The marginal teeth are also rudimentary and the principal dentition consists of the large hooked first laterals, each of which has a denticulate ridge with up to 13 denticles. A noteworthy feature is the presence on each lateral tooth of a prominent, massive, spur-like subsidiary cusp (Fig. 12a). The first marginal bears a slender, curved, erect cusp, but the second marginal (detectable only in adults) is rudimentary and probably useless. The oral armature is robust and exhibits microscopic honeycombing. Histological observations on the reproductive organs are detailed by Todd (1978a).

HABITS. This species feeds upon a wide variety of encrusting ectoproct polyzoans, including *Alcyonidium polyoum* (Hassall), *Celleporella hyalina* (L.), *Cryptosula pallasiana* (Moll), *Electra pilosa* (L.), *Escharella immersa* (Fleming), *Membranipora membranacea* (L.), *Microporella ciliata* (Pallas), *Porella concinna* (Busk), *Schizomavella linearis* (Hassall), *Schizoporella unicornis* (Johnston), *Smittina reticulata* (Macgillivray), *Umbonula littoralis* Hastings (Miller, 1961; Swennen, 1961; Thompson, 1964). According to MacDonald & Nybakken (1978) a species from California identified as *O. muricata* attacks the polyzoan *Reginella mucronata*.

The life cycle is an annual one, with a maximal life span of 1 year (Behrentz, 1931; Miller, 1958; Thompson, 1961b). Egg masses are produced during a spring breeding season (January to May in the Menai Straits (Thompson, 1961b)), but continuing into the summer in more northerly areas. Each of the white ribbon-like spirals of spawn contains up to 53,000 ova, 73–100 μm in diameter. The embryonic period is 14 days at 9–13 °C (Behrentz, 1931; Thompson, 1967). The larva (Thompson, 1964) is planktotrophic and possesses a shell of type 1 (Thompson, 1961a). The reproductive energetics of this species have been studied by Todd (1978a & b, 1979).

DISTRIBUTION (Map 4). Lemche (1938) described *Onchidoris muricata* as a boreo-Arctic species; the distribution encompasses the White Sea to Murmansk, Spitzbergen, east and west coasts of Greenland, Iceland, the Faeroes and the west coasts of Europe. Tolerance of low salinity allows this species to penetrate the Baltic Sea as far as Lübeck in Germany. The southern limit around Europe is variously quoted as Cape Finistère (Lemche, 1929), Arcachon (Swennen, 1961b) or Wimereux (Bouchet & Tardy, 1976). In the western Atlantic Ocean it occurs around Nova Scotia and southwards to Connecticut (Clark, 1975), where breeding occurs during the late spring. Hurst (1967) was able to extend Pacific Ocean records from Alaska (Bergh, 1880) to include the San Juan Islands in Washington. The maximum depth was between 180 and 250 m, in a Norwegian fjord.

39. *Onchidoris oblonga* (Alder & Hancock, 1845a)
Doris oblonga Alder & Hancock, 1845a

APPEARANCE IN LIFE (Plate 15). This onchidorid may exceptionally grow to a length of 8 mm; most records have been of animals which did not exceed 5 mm. It was thought until recently to be an exclusively British species. It often adopts a discoid shape when at rest, resembling a small colony of an encrusting polyzoan. The mantle colour is usually pale brown with pink-red or purplish brown markings. The dorsum bears numerous small rounded spiculose tubercles which often have a basal ring of dark pigment and an apical spot. The pigmentation of the pallial epithelium around the rhinophores is usually different from the remainder, sometimes darker, more usually paler, and conspicuous pallial tubercles may arise from such areas; they may not be bilaterally symmetrical (Farran, 1903). The central orange-brown digestive gland is usually visible through the dorsum as well as ventrally through the foot. Up to 10 simple pinnate gills are present, arranged in a distinct horse-shoe configuration around the anus. The slender digitiform rhinophores each bear up to 9 lamellae.

ANATOMY (Fig. 12c). Radula preparations were made of a number of Manx specimens: length 2·5 mm, formula 12 × 1.1.0.1.1; 4·5 mm, 21 × 1.1.0.1.1; 7 mm, 28 × 1.1.0.1.1. Eliot mentions an 8 mm specimen from the English Channel which had 45 rows. The principal teeth, the 1st laterals, had long smooth cusps, each of which exhibited a denticulate ridge and a spur-like lateral prominence. The oral armature consisted of delicate microscopic honeycombing.

HABITS. This species is well camouflaged on its prey, the sublittoral ectoproct polyzoans *Cellaria fistulosa* Hincks and *C. sinuosa* (Hassall). Spawning takes place in the spring, from February to April (Marine Biological Association, 1957; Miller, 1958). The white ova are 64–73 µm in diameter, up to 520 per mass.

DISTRIBUTION (Map 5). The type locality of Torbay remains the most southerly record, with rare collections from Plymouth, Lundy, the Irish Sea and southern Norway. A suggestion by Pruvot-Fol (1954) that it may occur at Banyuls was included by Nordsieck (1972) but needs corroboration.

40. *Onchidoris pusilla* (Alder & Hancock, 1845a)
Doris pusilla Alder & Hancock, 1845a

APPEARANCE IN LIFE (Plate 15). This inconspicuous onchidorid reaches only 9 mm in adult length. During early life (between 2·5 and 4·5 mm) it undergoes a rapid change in body coloration. The smallest juveniles (Miller, 1958) are very pale and transparent, so that the eyes and the subepithelial spicules are visible with a hand-lens; little or no pigment is present on the central dorsum and only a very few tiny tubercles are present, with bristly bundles of projecting spicules. After 4·5 mm the adult coloration develops and mature *O. pusilla* are the darkest of the British onchidorids. The dorsum becomes covered with dark brown spots, most dense near the mid-line. An especially valuable recognition feature is the presence, all over the mantle, of small, almost microscopic, spiculose, conical tubercles. These are smaller and less elongated than the pallial tubercles of *O. depressa* or *O. oblonga*. Another characteristic feature is the white colour of both the rhinophores and the simple pinnate gills (up to 9 in number, forming a crescent anterior to the anal papilla).

ANATOMY (Fig. 9g). Each radular row always conforms to the pattern 1.1.0.1.1. In a 2 mm specimen there were 13 tooth-rows; 3 mm, 21 rows; 4 mm, 29 rows; 5 mm, 29 rows; 6 mm, 22 rows. The principal 1st lateral teeth have elongated cusps, each of which bears a denticulate ridge (up to 16 denticles), and a prominent lateral spur. The 4 mm preserved specimen from Lundy illustrated in Fig. 9g had the radular formula 25 × 1.1.0.1.1.

HABITS. This species was thought to be uncommon until Miller (1958) obtained over 100 specimens around the Isle of Man. As with other "rare" species, a precise knowledge of the ecology of this onchidorid had been lacking. Miller discovered that the prey consisted of the encrusting ectoproct polyzoans *Escharella immersa* (Fleming), *Microporella ciliata* (Pallas), *Escharoides coccineus* (Abildgaard) and *Porella concinna* (Busk), and was thus able to obtain abundant material. Spawning occurs in the spring, from February to April or May (Alder & Hancock, 1845–55; Miller, 1958). There is one generation per year and the normal life-span does not exceed 14 months. Each egg mass contains up to 480 ova.

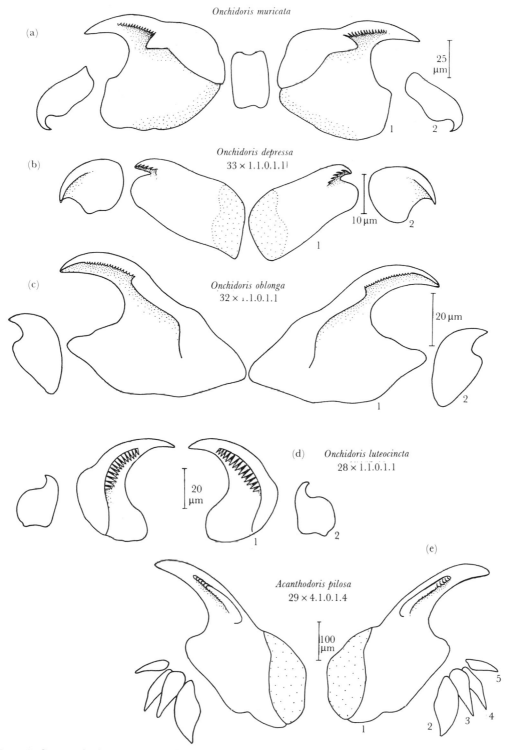

FIG. 12. Camera lucida drawings of doridacean radulae: *Onchidoris* and *Acanthodoris*. Teeth are numbered from the mid-line.

    (a) *Onchidoris muricata*, preserved length 8·5 mm, Kerrera, Scotland.
    (b) *Onchidoris depressa*, preserved length 5 mm, Orkney.
    (c) *Onchidoris oblonga*, preserved length 6 mm, Lundy Island.
    (d) *Onchidoris luteocincta*, length 9 mm, Donegal, Ireland.
    (e) *Acanthodoris pilosa*, preserved length 20 mm, Lundy Island.

DISTRIBUTION (Map 5). Ortea (1979) extended the known range to include the Spanish Biscay coast, and Bouchet & Tardy (1976) have cited records from the French Atlantic seaboard. There are sporadic reports from all around the British Isles and from Sweden, and Norway as far north as Kristiansund (Jaeckel, 1952).

41. *Onchidoris sparsa* (Alder & Hancock, 1846)
Doris sparsa Alder & Hancock, 1846

APPEARANCE IN LIFE (Plate 15). This flattened onchidorid exceptionally grows to 8 mm in length; most records have been of animals which did not exceed 5 mm. It was thought until recently to be an exclusively British species.

O. sparsa often adopts a rounded shape when at rest, resembling a small colony of an encrusting polyzoan. The mantle colour is usually pale brown with pink-red or purplish brown blotching. The dorsum bears numerous small, rounded, spiculose tubercles which often have a basal ring of dark pigment and an apical spot. The pigmentation of the pallial epithelium around the rhinophores is usually different from the remainder, sometimes darker, more usually paler, and conspicuous pallial tubercles may arise from such areas, which may not be bilaterally symmetrical (Farran, 1903). The central orange-brown digestive gland is usually visible through the dorsum as well as ventrally through the pedal sole. Up to 10 simple pinnate gills are present, arranged in a distinct horseshoe configuration around the anus. The slender, digitiform rhinophores each bear up to 9 lamellae.

ANATOMY (Fig. 9e). Two Orkney specimens were investigated, one 6 mm long (radular formula $36 \times 1.1.0.1.1$), the other measuring 7 mm in preservative ($32 \times 1.1.0.1.1$). As was noted by Alder & Hancock (1845–55), the lateral teeth resemble those of *O. inconspicua* but it may be remarked that they have affinity also with *O. depressa*. The cusp is relatively small in all these species and bears a short row of blunt lateral denticles and a spur-like lateral prominence. The marginals are rectangular with a sharp cusp at one corner. The oral armature exhibited a pattern of microscopic honeycombing.

HABITS. This species feeds on the sublittoral ectoproct polyzoans *Cellepora pumicosa* L. and *Porella concinna* (Busk). Details of the reproductive cycle are unknown.

DISTRIBUTION (Map 5). Scandinavian records of *O. sparsa* have come from the Swedish Gullmarfjord and the Skagerrak. The most northerly records are, however, from the Orkneys, while collections have been made from scattered localities all around the British Isles. One only French locality is known, near Mor Braz (Bouchet & Tardy, 1976), although Ortea (1979) has described specimens from Asturias, northern Spain.

Acanthodoris Gray, 1850
(*type Doris pilosa* Müller, 1789)

Onchidorids in which the dorsal mantle bears abundant small, soft, conical, villous tubercles; gills bipinnate or tripinnate; radula *n*.1.0.1.*n* ($n = 4$–8); penis armed with numerous hooked spines (perhaps these spines are absent in some north American species); buccal pump sessile, not pedunculate.

The world's species of *Acanthodoris* have been reviewed by Williams & Gosliner (1979).

42. *Acanthodoris pilosa* (Müller, 1789)
Doris pilosa Müller, 1789
Doris stellata Gmelin, 1791
Doris nigricans Fleming, 1820
Doris flemingii Forbes, 1838
Doris sublaevis Thompson, 1840
Doris similis Alder & Hancock, 1842
Doris subquadrata Alder & Hancock, 1845a
Doris fusca Lovén, 1846 (not Müller, 1776)

Doris vocinella Leach, 1847
Lamellidoris sparsa Mörch, 1868 (not Alder & Hancock, 1846)
Doris quadrangulata Jeffreys, 1869
Doris bifida Verrill, 1870
Doris pilosa stellata Sauvage, 1873
Acanthodoris stellata; Verrill, 1873
Acanthodoris citrina Verrill, 1879
Acanthodoris ornata Verrill, 1879
Lamellidoris inconspicua; Norman, 1890 (not Alder & Hancock, 1851)

APPEARANCE IN LIFE (Plate 14, Fig. 10a). This common intertidal and shallow sublittoral doridacean is soft to the touch and may reach 55 mm in length, although it does not usually exceed 40 mm. The colour varies from white or yellow through pale grey or brown to dark purplish brown or charcoal grey. Occasional individuals (usually juveniles) show mottling or freckling. According to some authorities, the palest individuals come from deeper habitats. That this cannot be the whole story is evidenced by the capture of a 9 mm white specimen on the shore at Sully Island (south Wales) in January 1962. The pallial papillae are soft, tall and conical. The rhinophoral tentacles are clubbed and bent rearwards; in a 10 mm specimen there were 17 lamellae on each tentacle, in a 30 mm specimen 23 or 24. The rhinophores issue from low pallial sheaths which have tuberculate rims. There are up to 9 voluminous tripinnate gills. Blunt, rounded oral tentacles are present.

ANATOMY (Plate 16b, Fig. 12e). Observations on the alimentary canal and reproductive organs have been published by Hancock & Embleton (1852), Hancock (1865), Bergh (1880), Lloyd (1952), Forrest (1953) and Morse (1968).

The penis bears numerous potash-resistant hooked cuticular spines. The oral canal is lined by a flimsy cuticle which exhibits microscopic honeycombing. The radula undergoes a striking change when a body-length of 13 mm is attained. Before that age, the first lateral teeth each have a row of strong denticulations along the cusp; older specimens exhibit more feeble scupturing and may lack these denticulations. In a 1 mm individual, the formula was $12 \times 2.1.0.1.2$; 4·5 mm, $25 \times 3.1.0.1.3$; 7 mm, $28 \times 4.1.0.1.4$; 13 mm, $29 \times 5.1.0.1.5$; duplicate 13 mm, $28 \times 5.1.0.1.5$; 20 mm, $34 \times 5.1.0.1.5$. According to Pruvot-Fol (1954), there may be as many as 6 marginal teeth in each half-row. The largest radula reported from British specimens (25 mm) was $30 \times 5.1.0.1.5$ (Colgan, 1914). We have examined a 20 mm preserved specimen which had the radular formula $29 \times 4.1.0.1.4$ (Fig. 12e).

HABITS. This species feeds upon encrusting polyzoans, especially species of *Alcyonidium* and occasionally *Flustrellidra hispida* (Fabricius). The smallest size at which spawning has been observed is 8 mm (Miller, 1958); egg-laying has been reported in British localities in every month of the year. The spawn masses were illustrated by Kress (1971). The usual pattern is for a major burst of spawning in the spring, with a second, minor burst in the autumn when the new generation may come prematurely into breeding condition. The normal life span does not greatly exceed 1 year (Miller, 1958). Each egg mass is ribbon-like and may contain up to 163,000 white ova, 64–76 µm in diameter, developing in 10 days at 10 °C to form planktotrophic veliger larvae of shell-type 1 (Thompson, 1961a; 1967).

DISTRIBUTION (Map 4). There are only rare records from the Mediterranean Sea (Sicily) but this species does occur on the Atlantic coast of Morocco (Gantès, 1956). It is common at numerous localities between the Bay of Biscay and northern Europe, always outside the Arctic Circle. It also penetrates the Baltic Sea, as far as Lübeck (Swennen, 1961b). Westwards, it is known from the Faeroes, Iceland, Greenland (Lemche, 1929), Canada (Johnson, 1934) and the eastern seaboard of America from New England to Maryland and Virginia (Franz, 1970a). Bergh (1880) gave a detailed description of specimens from the Pacific Ocean which agreed closely with European material. Known Pacific localities are around the Aleutian Islands and off the Canadian coast as far south as Vancouver Island. Specimens have been taken from depths to 170 m.

DISCUSSION. It seems wise to include *Doris subquadrata* Alder & Hancock, 1845a, erected on the basis of one sublittoral individual from Torbay, in which the mantle was somewhat reduced and oblong, with sparse conical papillae. Corroborative material has been noted by Walton (1908: one sublittorally off Holy Island, Northumberland), and by McMillan (1944: two on the shore at Greenisland, Co. Antrim), but in our opinion these four are probably all simply damaged or parasitized specimens of *A. pilosa*.

TRIOPHIDAE

Phanerobranchiate, limaciform, soft-bodied doridaceans with the mantle skirt reduced to form a tuberculated ridge continuous around the frontal margin. The tubercles are often arborescent and complex and similar excrescences are sometimes found on the flanks and dorsum. The bipinnate or tripinnate gills form an arc in front of the anal papilla; they lack protective pallial lobes, but they may be flanked by paired, clubbed cerata (*Plocamophorus*). Rhinophores lamellate, retractile into low, plain sheaths. Oral tentacles reduced to form semi-circular lobes. Feeding upon encrusting sponges and polyzoans.

The penis bears numerous hooked spines. The buccal mass does not possess a muscular pump. Radula $n.0.n$ ($n =$ up to 100), mesial teeth hooked, sometimes multicuspid.

Crimora Alder & Hancock, 1862
(type *C. papillata* Alder & Hancock, 1862)

Triophids having numerous small, arborescent processes around the mantle rim and on the flanks and dorsum, but without large, clubbed ceratal processes. Three gills, without protective flaps or tubercles. Radula $n.0.n$ ($n =$ up to 30), teeth highly differentiated.

This genus probably embraces *Triopha* Bergh, 1880 (type *T. carpenteri* Stearns, 1873).

43. *Crimora papillata* Alder & Hancock, 1862

APPEARANCE IN LIFE (Plate 17; volume I, Plate 6). This uncommon nudibranch may reach 35 mm in length. The background colour of the body is white or pale yellow, with numerous small but conspicuous epidermal excrescences coloured yellow or orange. These tubercular processes are forked in various ways and this becomes especially complex around the frontal margin of the head where simple forking gives way to arborescence. There are 3–5 orange or yellow tripinnate gills around the anal papilla. The rhinophores are similarly pigmented, and lamellate, issuing from plain, low sheaths. The pigmented oral tentacles are very short and blunt.

ANATOMY (Fig. 13b, Plate 16c). Hooked penial spines were visible in serial sections through the terminal part of the male duct. Jaws are absent. The radula is complex, with up to 42 rows of teeth, having the formula 32.0.32. The 2nd lateral tooth is strongly developed; towards the margins the teeth become long and slender, with fine denticles along the concave side of the cusp. Haefelfinger (1962a) obtained the formula 39×23–$21.0.21$–23 from Mediterranean material. This is one of the most remarkable and highly differentiated radulae possessed by any dorid nudibranch.

It is unfortunate that Pruvot-Fol (1954) illustrated the radula of *Limacia clavigera* in mistake for the present species.

HABITS. This species, until 1972 regarded as one of the most rare of the British nudibranchs, has recently been encountered in large numbers in restricted sublittoral areas (down to 30 m) off the Devon, Cornwall and Pembrokeshire coasts, feeding upon the ectoproct polyzoans *Flustra foliacea* (L.) and *Chartella papyracea* (Ellis & Solander). Spawn is yellow in colour; each mass may contain up to 5000 ova. Development to hatching occupies 11 days at 16 °C (Schmekel & Portmann, 1982).

DISTRIBUTION (Map 5). Alder & Hancock (1862) first described this species from Guernsey, but only in recent years have large numbers been obtained from south west England. The most northerly records are from the Dale Peninsula (Hunnam & Brown, 1975) and Galway Bay in Ireland. Bouchet & Tardy (1976) have provided recent records from various French Atlantic localities; furthermore, this species was found off the Mediterranean coast of France during 1960 (Haefelfinger, 1962a). Gantès (1956) collected specimens from Temara near Rabat on the

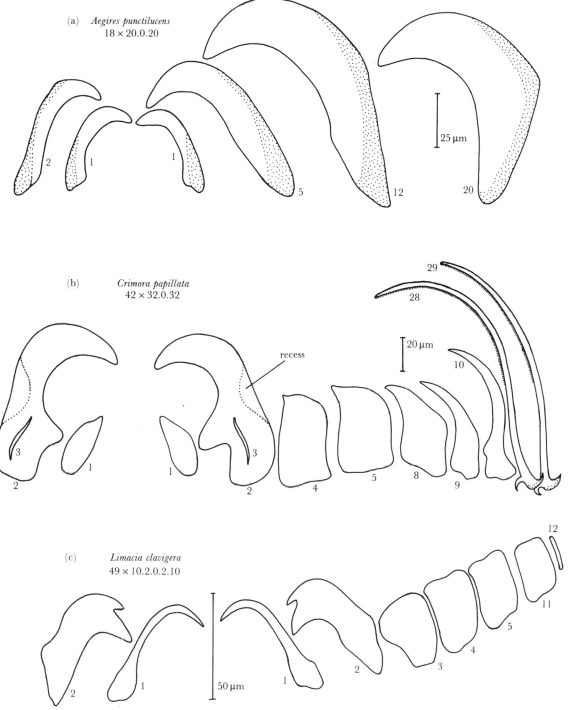

FIG. 13. Camera lucida drawings of doridacean radulae: *Limacia*, *Aegires* and *Crimora*. Teeth are numbered from the mid-line.

   (a) *Aegires punctilucens*, length 13 mm, Dale, S. Wales.
   (b) *Crimora papillata*, preserved length 9 mm, Lundy Island.
   (c) *Limacia clavigera*, preserved length 9 mm, Lundy Island.

Atlantic coast of Morocco, and these represent the most southerly records. Recently, Schmekel (1968) published a Mediterranean range-extension to the Bay of Naples, and we now know by virtue of a personal communication from Mr Bernard Picton that *C. papillata* occurs on the coasts of Galicia and Santander in northern Spain. All records are sublittoral, to 43 m.

NOTODORIDIDAE

Phanerobranchiate, limaciform, firm-bodied doridaceans with the mantle skirt reduced and indistinct. The dorsal surfaces bear short, blunt tubercles. Projecting pallial lobes protect the gills, which may be simple, bipinnate or tripinnate, arranged in an arc anterior to the anal papilla. Rhinophores smooth, issuing from tuberculated, raised sheaths (*Aegires*) or from expanded pallial flaps (*Notodoris*).

Oral tentacles reduced (*Aegires*) or absent (*Notodoris*). Feeding upon encrusting polyzoans and sponges.

The penis bears numerous hooked spines. The buccal mass does not possess a muscular pump, but includes a stout median dorsal cuticular plate or jaw in *Aegires*. Radula $n.0.n$ ($n = 10$–31); the teeth are hooked and sometimes bifid at the tip.

Aegires Lovén, 1844
(type *Polycera punctilucens* Orbigny, 1837)

Notodorids with numerous blunt, pedunculate, dorsal tubercles. Rhinophores issue from cylindrical pallial sheaths bearing tubercles around the rims. Gills protected by tuberculate lobes. Median dorsal cuticular plate or jaw is present in the buccal mass. Radular teeth entire, not bifid.

44. ***Aegires punctilucens*** (Orbigny, 1837)
    Polycera punctilucens Orbigny, 1837
    Doris maura Forbes, 1840
    Aegires leuckarti Vérany, 1853
    Polycera horrida Hesse, 1872
    Aegirus hispidus Hesse, 1872

APPEARANCE IN LIFE (Plate 17). The adults reach only 20 mm and this well-camouflaged species is probably more common than the sparse records indicate. The body is brown in colour, paler in juveniles, darker in adult Manx records than elsewhere. There are abundant white specks, forming patches or streaks on the rhinophores and gills, together with a number of dark brown oval areas, each containing an iridescent blue-green spot, resembling an ocellus. The dorsum bears abundant spiculose, clubbed tubercles, each of which has a dark red or black apical spot. There are 3 tripinnate gills anterior to the anal papilla, protected by 3 tuberculate pallial lobes. Approximately 5 tubercles ring the rim of each rhinophore sheath. The rhinophores are smooth and cylindrical; short, rounded oral tentacles are present on each side of the head. The propodium is squarish in outline.

ANATOMY (Fig. 13a). Alder & Hancock (1845–55) stressed the close resemblance between *Aegires* and *Polycera*, the main differences residing in the buccal mass, which in the former contained a median dorsal "mandible" with lateral plates consisting of countless rod-like elements. These authors say that this mandible is not unlike the jaw of *Limax*. Nothing is known about its mode of functioning. It is difficult to discern in a caustic preparation, but it is strikingly obvious in stained serial sections.

The radular formula of British specimens reaches maximally $23 \times 22.0.22$ (in a 12 mm preserved specimen from the Isle of Man). A 13 mm specimen from Japan had a smaller radula with the formula $15 \times 15.0.15$ (Baba, 1974). All the teeth are smooth, lacking denticulations; the teeth from near the mid-line have a straight neck below the hooked tip, while teeth from the margins are more smoothly curved.

The penis bears numerous elongated, conical, slightly hooked spines, 12 μm in length.

HABITS. *Aegires* is extremely well camouflaged in its natural habitat, among the encrusting sponges (*Leucosolenia* spp.) upon which it feeds. It appears to be a nudibranch of shallow waters, with rare deep records (for instance, 100 m off the Isle of Man, according to Miller (1958)).

Spawn has been noted in June (Isle of Man and Villefranche (Haefelfinger, 1968)) and July (St Andrews (M'Intosh, 1863)). Each mass may contain up to 1500 eggs laid in a white, spiralling band. Planktonic larval development, metamorphosis and early benthic development were described by Thiriot-Quiévreux (1977). Development to hatching occupies 12 days at 16 °C (Schmekel & Portmann, 1982).

DISTRIBUTION (Map 5). This species, despite its cryptic nature, is recorded from all around the British Isles. Collections from Sweden and the Skagerrak (Jaeckel, 1952) supplement Scandinavian records along the Norwegian coast north to Brønnøysund, 150 miles north of Trondheimsfjord, just outside the Arctic Circle. Bouchet & Tardy (1976) listed several French Atlantic localities, and its occurrence in the Mediterranean Sea was noted at Villefranche and Banyuls (Haefelfinger, 1968), Marseilles (Vicente, 1967) and Naples (Schmekel, 1968). Surprisingly, Baba (1974) described what appears to be the same species from the warm temperate waters around Sado Island, Japan. The deepest collection remains 100 m recorded both by Friele & Grieg (1876: Bergen) and Miller (1958: Isle of Man).

POLYCERIDAE

Phanerobranchiate, soft, limaciform doridaceans with the mantle skirt reduced to form a rim which often bears elongated papillae situated laterally and on the frontal margin. Rhinophores lamellate, sometimes arising from well-developed sheaths, more often lacking such sheaths. Head with flattened oral lobes, or grooved oral tentacles. Feeding upon erect and encrusting ectoproct polyzoans.

The penis bears numerous hooked spines. The buccal mass does not possess a muscular pump. Chitinous jaws are usually present, often exhibiting a prominent wing-like process. Radula 2–8.2.0.2.–8, the 2nd laterals especially strongly developed.

Polycera Cuvier, 1817
(type *Doris quadrilineata* Müller, 1776)

Polycerids which lack oral tentacles and rhinophore sheaths; with strong pallial tubercles which arise from the mantle rim on either side of the branchial circlet and from the frontal margin. Radula 3–5.2.0.2.3–5; jaws conspicuous, sometimes exhibiting a prominent wing-like process.

45. *Polycera faeroensis* Lemche, 1929
? *Polycera nonlineata* Thompson, 1840

APPEARANCE IN LIFE (Plate 18). This is a sublittoral species which reaches 45 mm in length, coloured white, with pigment blotches of yellow. The body has a soft appearance, and often seems to be loosely swollen. The yellow pigment covers the frontal tubercles and most of the papillae close to the gills. It can also form a narrow line on the lateral crest from the hindmost frontal papilla. The same colour covers the median keel of the metapodium and occasionally tinges the margins of the foot. Finally, it is found on the outermost half of each rhinophore and the tips of the gills. There are usually no mid-dorsal blotches of yellow pigment like those of *P. quadrilineata*. Occasionally, areas of black pigment may be present (this tendency towards sporadic melanism is characteristic of the genus). A further distinction is in the manner in which the animals react when preserved by dropping directly into alcohol: *Polycera quadrilineata* usually contracts so that the body-length is considerably shortened, whereas *P. faeroensis* is more inclined to keep its natural shape.

The head is semicircular in front, extending to form numerous (normally 8) finger-like, tapering frontal papillae. The number of these papillae will normally distinguish this species from the related *P. quadrilineata*, but in both species the number may vary. In *P. faeroensis* it may go down to 6, or up to 12, whereas *P. quadrilineata* from the vicinity of Bergen (Norway), with the same colours as *P. faeroensis*, may have 3 to 6 papillae.

Closely surrounding the distinct anal papilla are the gills, 5–8 in number. Each gill consists of a compressed stem carrying about 15 lamellae on each side, often with accessory ones in between.

ANATOMY (Fig. 14a). The radula has the formula 3.2.0.2.3 in all the specimens examined, except the type, in which there were no more than 2 marginal teeth on each side, and the specimen illustrated in Fig. 14a, in which a vestigial 4th marginal tooth was present in each right half-row (absent on the left). The number of rows of teeth varies from 12 to 16 in adult specimens. The 2nd lateral teeth tend to be less slender than the analogous teeth of *P. quadrilineata* (Lemche & Thompson, 1974). The shape of the jaws varies much. Lemche (1929) drew them with large and rounded shoulders, but Odhner (1941) indicated the shoulders to be almost absent. Of 2 specimens from Ireland (Lemche & Thompson, 1974), one was fairly close to the type in shape, whereas the other had jaws with high shoulders as in *P. quadrilineata*; only the masticatory edge was strong.

HABITS. The preferred prey of this carnivorous species includes several colonial polyzoans, especially *Crisia denticulata* Lamarck, *Cellepora pumicosa* L. and *Bugula plumosa* (Pallas). Records of feeding upon *Membranipora membranacea* (L.) may mistakenly relate to *P. quadrilineata*.

Spawning has been reported in British waters in the months of May to September. The spawn takes the form of a white band 2–3 mm broad and up to 30 mm in length if unrolled. It is attached to erect polyzoans, and forms an open coil of one and a half whorls or less. The number of ova per spawn mass was estimated in one instance to be 10,000 (Lemche & Thompson, 1974).

Polycera faeroensis occurs commonly in shallow sublittoral situations, down to 35 m in the British Isles (to 80 m at Bohuslän, Sweden, according to Odhner, 1941; to 120 m between Nolsø and Østerø, Faeroes, according to Lemche, 1929).

DISTRIBUTION (Map 6). Some older observations (e.g. Elmhirst, 1922) under the name *Polycera quadrilineata* may have related instead to *P. faeroensis*. The first authoritative report of the latter species from the British Isles came from Lemche & Thompson (1974). The most reliable recognition feature is the number of frontal tubercles, 4 (rarely up to 6) in *P. quadrilineata*, 6 or more in *P. faeroensis*; other cardinal features are mentioned under *P. quadrilineata*.

46. ***Polycera quadrilineata*** (Müller, 1776)
Doris quadrilineata Müller, 1776
Doris flava Montagu, 1804
Doris cornuta Rathke, 1806
Policere lineatus Risso, 1826
Doris ornata Orbigny, 1837
Polycera varians M. Sars, 1840
Polycera typica Thompson, 1840
? ***Thecacera capitata*** Alder & Hancock, 1854a
Polycera mediterranea Bergh, 1879a
Polycera nigropicta Ihering, 1885
Polycera nigrolineata Dautzenberg & Durouchoux, 1913
Polycera salamandra Labbé, 1931

APPEARANCE IN LIFE (Plate 18; volume I, Plate 1). This common species may reach 39 mm in length, coloured white, with pigment blotches of yellow or orange. There is usually a row of ovoid blotches down the mid-line of the back, with others on the flanks and the dorsal metapodium. Occasionally, streaks and blotches of jet black may be present (said by Haefelfinger (1960c) to be dark blue in Villefranche specimens). These may take the form of a delicate dappling, as shown in Plate 18c, but more usually they unite to form a number of longitudinal streaks (4 in Müller's type). Out of 100 individuals reported from the Isle of Man,

Miller (1958) noted that 70 exhibited black markings in some degree. Colgan (1914), however, found black pigment to be rare in the shallow waters of Co. Dublin.

On either side of the branchial circlet, the mantle rim is produced into a strong, posteriorly directed yellow or orange tipped papilla or tubercle. Similar conspicuous tubercles project from the frontal margin, usually 4 in number, but rarely 6 (Odhner, 1941). The rhinophores, propodial tentacles, oral lobes, and the simple pinnate gills are tipped with yellow.

ANATOMY (Fig. 14b, Plate 16d). Descriptions have been furnished of the nervous system (Pelseneer, 1894), reproductive organs (Pohl, 1905; Lloyd, 1952; Schmekel, 1970), wandering cells (Millott, 1937) and skin histology (Herdman & Clubb, 1892; Hecht, 1895; Thompson, 1960b). The pallial papillae are charged with large proteinaceous glands and similar glands discharge to the exterior near the bases of the gills; their secretions are presumably defensive. The skin contains numerous sausage-shaped, rough, calcareous spicules. Penial spines can be detected, each of which is hooked and becomes effective during copulation only after the extension of the penis and its consequent eversion.

The jaws are strong and exhibit microscopic striations and upper wing-like expansions. The radula of an 18 mm specimen had the formula $12 \times 5.2.0.2.5$, whereas that of a 2 mm specimen was $11 \times 3.2.0.2.3$ (Miller, 1958). This confirms Colgan's (1914) assertion that the number of rows of radular teeth does not increase appreciably with age in British material. Haefelfinger's (1960c) largest Villefranche specimen, however, measured 35 mm in length and had the formula $20 \times 5.2.0.2.5$. We re-investigated the radula of a 13 mm preserved specimen with radular formula $9 \times 5.2.0.2.5$ (Fig. 14b). The 2nd lateral tooth was more slender than the comparable tooth of *P. faeroensis*, as was noted by Lemche & Thompson (1974).

HABITS. This species is exclusively carnivorous, despite Alder & Hancock's (1845–55) claim, repeated by Garstang (1889), that it is "phytivorous". The preferred diet consists of encrusting ectoproct polyzoans, especially *Membranipora membranacea* (L.), but also *Electra pilosa* (L.), *Callopora dumerilii* (Audouin), *Celleporella hyalina* (L.) and *Tegella unicornis* (Fleming) (Elmhirst, 1922; Miller, 1961). It is especially common on the shore and in shallow water, down to 60 m (Odhner, 1939; Miller, 1958; Haefelfinger, 1960c), although in Arctic waters it may be found at greater depths, down to 160 m (Lemche, 1929).

Spawning has been reported in British waters in all months of the year except January and February (M'Intosh, 1863; Elmhirst, 1922; Miller, 1958; Allen, 1962). Growth from metamorphosis to maturity may be rapid; 6 weeks, according to Elmhirst (1922). The larval shell was first described and adequately illustrated by Mazzarelli (1904); it conforms to shell-type 1 (Thompson, 1961a). A spawn mass investigated in the Menai Straits contained 21,550 eggs, each ovum 70–90 μm in diameter before cleavage; the embryonic period was 18–20 days at 8·5–9·5 °C. In the Isle of Man, Miller (1958) concluded that the normal life span was approximately 1 year.

DISTRIBUTION (Map 5). Common all around the British Isles; Atlantic and Mediterranean France and Spain; Italy (Vicente, 1967; Schmekel, 1968; Ros, 1975; Hunnam & Brown, 1975; Ortea, 1976a; Bouchet & Tardy, 1976). Also north Britain to Denmark, Sweden, south west Norway, the Faeroes, Iceland and west Greenland (Lemche, 1941; Rasmussen, 1973).

DISCUSSION. This species may have been confused in the past with *Polycera faeroensis*, but the 2 species are only superficially similar. *Polycera quadrilineata* has only 4–6 finger-like frontal processes and an epidermal pattern which consists of numerous blotches as shown in the colour plate, while *P. faeroensis* has 8 or more (very rarely as few as 6) frontal processes and sparser epidermal blotches. Internally, there are differences in the radula. The second lateral tooth is more slender in *P. quadrilineata* and there tend to be more marginal teeth (3.2.0.2.3 to 5.2.0.2.5, compared with 2.2.0.2.2 to 3.2.0.2.3 in *P. faeroensis*).

Thecacera capitata Alder & Hancock, 1854a is considered to be a probable synonym of *P. quadrilineata*. The radulae do not appear to differ significantly, and the colour pattern which Alder & Hancock described matches well a recognizable variety of *Polycera quadrilineata*, shown in Plate 18c. We have examined the type material of *T. capitata* in the British Museum (Natural History) and can find nothing to distinguish it from *P. quadrilineata*.

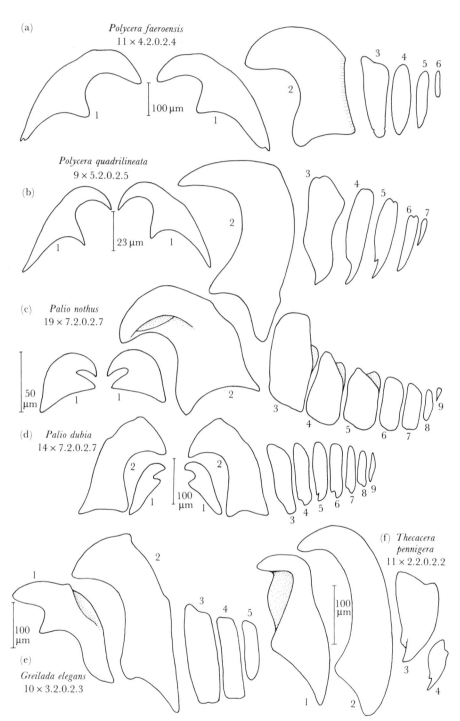

FIG. 14. Camera lucida drawings of doridacean radulae: *Polycera*, *Palio*, *Greilada* and *Thecacera*. Teeth are numbered from the mid-line.

   (a) *Polycera faeroensis*, preserved length 18 mm, Lundy Island.
   (b) *Polycera quadrilineata*, preserved length 13 mm, Cornwall.
   (c) *Palio nothus*, length 12 mm, Skerries, Portrush.
   (d) *Palio dubia*, preserved length 12·5 mm, Black Head, Ireland.
   (e) *Greilada elegans*, preserved length 16 mm, Lundy Island.
   (f) *Thecacera pennigera*, preserved length 9 mm, Ilfracombe, Devon.

Thecacera Fleming, 1828

(type *Doris pennigera* Montagu, 1815)

Polycerids which possess rhinophore sheaths, simple in some species, but complex, lobed and massive in others. The frontal margin of the body is devoid of pallial tubercles, but 1 or more prominent tubercles on either side flank the branchial circlet. Laterally, just below each rhinophore base, is a deep pit, the floor of which may be folded into lamellae. Radula 2–5.2.0.2.2–5. Jaws exhibit a prominent wing-like process.

47. ***Thecacera pennigera*** (Montagu, 1815)
   Doris pennigera Montagu, 1815
   Thecacera maculata Eliot, 1905
   Thecacera lamellata Barnard, 1933

APPEARANCE IN LIFE (Plate 19). This gaudy polycerid has a worldwide distribution and may be locally abundant. Specimens from Brazil and from Japan are not so brightly patterned as the English type. In European waters it may reach 30 mm in length, with a ground colour of white, exhibiting numerous irregularly shaped orange areas and black and yellow spots. The black spots are always much smaller than the orange areas. The upper part of each lamellate rhinophore is yellow with black speckling. The rhinophores each issue from a complex, raised sheath, which is open mesially and rises to form a substantial club behind. There are 3–5 bipinnate or tripinnate gills, white with orange and black spots, and behind the gills is a pair of club-like, dorso-lateral, glandular, defensive, ceratal processes. The front of the foot is produced into antero-lateral points, the propodial tentacles. No oral tentacles are present.

ANATOMY (Fig. 14f). Marcus (1957: Brazil), Macnae (1957: South Africa) and Willan (1976: New Zealand) have investigated features of the reproductive and other systems. Minute gizzard plates are said to be present in the midgut (we have failed to confirm this), and hooked spines (15 μm in length) have been observed, associated with the base of the penis.

The jaws of *T. pennigera* each have a wing-like process (Baba, 1960). The radula of an 18 mm specimen from Arcachon has been examined; the formula was $8 \times 2.2.0.2.2$. Up to 15 rows of teeth have been found in exotic material (Macnae, 1957). The teeth are well differentiated, but nothing is known about the functional roles of the different components. The large 1st and 2nd lateral teeth are robust and resistant to squash deformation to such an extent that microscopic preparations usually show them edge-on, as can be seen in Fig. 14f, from a 9 mm preserved specimen, formula $11 \times 2.2.0.2.2$.

HABITS. Despite reports that this species is herbivorous, feeding upon algae (Jaeckel, quoted by Swennen, 1961), it is now certain that it subsists on a diet of the polyzoan *Bugula*. Spawn has been observed at British localities between May and August. Each spawn mass is a flattened fawn to white band, attached along one edge to flat surfaces, or wound around a colony of *Bugula*.

DISTRIBUTION (Map 6). This cosmopolitan species appears to be best adapted to temperate waters, having been recorded from Britain as far north as Gairloch, west Scotland and from the North Sea (Lowestoft), the Netherlands (Swennen, 1961b), Atlantic France (Bouchet & Tardy, 1976), Spain (Santander), Sicily (Vayssière, 1913), Brazil (Marcus, 1957), South Africa (as *T. lamellata* Barnard, 1933), Pakistan (as *T. maculata* Eliot, 1905), Japan (Baba, 1960), Australia (Allan, 1957) and New Zealand (Willan, 1976). It is probable that this distribution has been assisted by shipping, which is also believed to have spread one of the prey species, *Bugula flabellata* Gray (Ryland, 1970).

DISCUSSION. The type material of the problematic *Thecacera capitata* Alder & Hancock, 1842 has been re-examined, and it is considered that these specimens are indistinguishable from a recognizable variety of *Polycera quadrilineata* (Müller). Similarly, *T. virescens* Forbes & Hanley, 1851 is probably embraced by *Palio nothus* (Johnston).

Burn (1978) commented upon the probable analogy between the lateral pits on the head of *Thecacera* and similar depressions lying dorsal to the oral tentacles of the Australian polycerid

Tambja verconis (Basedow & Hedley). The depressions appear to have a sensory function in short-distance location of live *Bugula*, and in the orientation of the head prior to feeding.

Greilada Bergh, 1894
 (type *Greilada elegans* Bergh, 1894)
 Polycerids which lack rhinophore sheaths; strong pallial tubercles arise from the frontal margin only (absent from the mantle rim on either side of the branchial circlet). Radula 3.2.0.2.3. Jaws do not exhibit a prominent wing-like process.

48. **Greilada elegans** Bergh, 1894
 Polycera messinensis Odhner, 1941
 ?**Polycera atlantica** Pruvot-Fol, 1955a

APPEARANCE IN LIFE (Plate 20). This is one of the most beautiful nudibranchs of the north Atlantic Ocean. It reaches 48 mm in length, coloured yellow to orange, with conspicuous elevated brilliant blue blotches on the dorsum and flanks. These blue areas sometimes contain small crimson specks. Rare specimens may exhibit a pair of tiny papillae behind the gills (Edmunds, 1961). The frontal margin of the head bears up to 22 finger-like processes. There are 5–7 tripinnate gills in front of the anal papilla. Each rhinophore exhibits up to 19 lamellae.

ANATOMY (Fig. 14e). Little is known concerning the structure of the nervous or reproductive systems. Odhner (1941), however, described and figured the anterior genital mass (of what he called *Polycera messinensis*) and remarked on the presence of a large spherical spermatheca connected proximally to the vaginal duct. The equivalent receptaculum in other British polycerids is an elongated sac. The thick and muscular vas deferens he followed back to a large prostate gland which had only slightly folded walls. The penis sheath bears numerous hooked spines, up to 7·5 μm long.

 The radula of a 42 mm specimen had the formula 11 × 3.2.0.2.3. The 2nd lateral teeth are the largest and most effective teeth and show most wear (Edmunds, 1961). The marginal teeth are vestigial. We have examined the radulae of 2 preserved specimens, 16 and 20 mm long, both of which had the formula 10 × 3.2.0.2.3. The large lateral teeth interlock and are extremely robust, so that they can only be represented edge-on as shown in Fig. 14e. The jaws are similar in shape to those of *Polycera quadrilineata*.

HABITS. The diet consists of the erect sublittoral polyzoans *Bugula flabellata* Gray and *B. plumosa* (Pallas), which are locally abundant in gulleys and on sublittoral cliffs, where clear oceanic water meets exposed islands and headlands.

 Spawn is produced in British waters in June and July, attached to *Bugula*. The ova are orange in colour, up to 15,000 per spawn mass.

DISTRIBUTION (Map 6). This species is locally common around south west coasts of the British Isles. The northernmost records are from Kilkieran Bay, Galway, and Skomer Island in Wales. Other British records include specimens from the coasts of Cornwall and Devon, as well as large numbers (several hundreds) from Lundy Island in the Bristol Channel. It is recorded also from Atlantic France (Bouchet & Tardy, 1976), northern Spain (Ortea, 1976a), Villefranche (Haefelfinger, 1960a), the Bay of Naples (Schmekel, 1968), Messina in Sicily (Odhner, 1941) and Rovigno in the Adriatic Sea (Pruvot-Fol, 1954).

Palio Gray, 1857
 (type *Polycera ocellata* Alder & Hancock, 1842)
 Polycerids which lack rhinophore sheaths; small pallial tubercles arise from the frontal margin, low mamillae cover the dorsal and lateral surfaces of the body, and there is, on either side of the branchial circlet, a large tubercle (lobulated in adults) which projects backwards. Radula 3–8.2.0.2.3–8.

49. **Palio dubia** (M. Sars, 1829)
    Polycera dubia M. Sars, 1829
    Polycera lessoni Orbigny, 1837
    Polycera citrina Alder, 1841
    Polycera illuminata Gould, 1841
    ?**Euplocamus holbölli** Möller, 1842
    Polycera modesta Lovén, 1844
    ?**Polycera pudica** Lovén, 1846
    Palio pallida Bergh, 1880
    ?**Polycera maculata** Pruvot-Fol, 1951
    ?**Folietta maculata**; Nordsieck, 1972

APPEARANCE IN LIFE (Plate 20). This delicate, small (up to 29 mm), sublittoral species has in the past often been confused with the closely related *P. nothus*.

The ground colour of the limaciform body is olive green with white or yellow tubercles and papillae. The lamellate rhinophores and the tripinnate gills (up to 5 in number) are translucent brown, and the other conspicuous excrescences, the dorsal and lateral tubercles and the frontal papillae, are pale yellow. Especially obvious are the whitish compound papillae situated on either side of the branchial circlet, and the median longitudinal, dorsal metapodial papillae. The pedal sole and the flattened oral lobes are paler in colour. The general texture of the body is, compared with *P. nothus*, more flaccid and less opaque.

ANATOMY (Fig. 14d). Jaws are absent. The radula of a mature specimen from North Wales had the formula 13×7–$8.2.0.2.7$–8. We examined the radula of a 12·5 mm preserved specimen from Ireland; the formula was $14 \times 7.2.0.2.7$. The teeth are longer and more slender than those of *P. nothus*.

Serial sections showed that minute spines are present on the penis.

HABITS. This nudibranch is especially common sublittorally, from 10 to 100 m. The diet consists principally, if not exclusively, of the anascan polyzoan *Eucratea loricata* (L.).

Spawning has been recorded in British waters in April, May and June. According to Miller (1958), a single spawn mass contained 4600 ova, each of which measured 77–81 μm in diameter. Roginsky (1962) records a spawn mass from the White Sea (where spawning occurs from May to July) containing approximately 10,000 eggs. In the Baltic Sea, Jaeckel (1952) found this species to survive down to a salinity of 14; it spawned in April and May. Jaeckel wrongly stated that *P. dubia* was herbivorous.

DISTRIBUTION (Map 6). Because of the confusion between the two British species of *Palio*, it is not possible to accept many distributional records uncritically. Certainly, both species occur all around the British Isles, although *P. dubia* may be less common, preferring offshore habitats. All intertidal collections of *Palio* examined by us have been *Palio nothus*, while offshore collections have usually included both species.

Both species of *Palio* have amphi-Atlantic distributions (verified from photographs of New England material kindly supplied by Dr T. M. Gosliner), and, while Gould & Binney (1870) described *P. dubia* from north America, Meyer's (1971))description appears to be embraced by *P. nothus*.

Although Lemche (1929, 1938) did not directly quote *nothus* as a synonym of *dubia*, he merged *ocellata* (= *nothus*) with *dubia*. Thus his records from Greenland, Iceland, and the Faeroes are questionable. Odhner (1907, 1939) was, however, referring to *P. dubia* explicitly when quoting distributional records from the White Sea and the whole of the Norwegian coast, so a boreo-arctic range is certain. Both species occur on the French Atlantic coast, but Mediterranean records from southern France and from the Adriatic need corroboration. Schmekel & Portmann (1982) combine *dubia* with *nothus*.

DISCUSSION. The Pacific American species *Palio zosterae* (O'Donoghue, 1924, as *Polycera*) appears to be very similar to *P. dubia*; comparative studies have not yet been undertaken.

50. **Palio nothus** (Johnston, 1838b)
   Triopa nothus Johnston, 1838b
   Polycera ocellata Alder & Hancock, 1842
   ? **Polycera plebeia** Lovén, 1846

APPEARANCE IN LIFE (Plate 20). This is a hardy species, found frequently intertidally in pools and under boulders.

The ground colour of the body is dark brown to olive green, with pale brown or white excrescences. The lamellate rhinophores and the tripinnate gills (up to 5 in number) are brownish. The frontal margin is slightly bilobed and bears a variable number of short, finger-like, pale brown papillae. On either side of the branchial circlet is a cluster of club-like, pale papillae. Over the remainder of the dorsum and flanks are numerous low pale tubercles. Rarely, these tubercles may form a line of median dorsal, metapodial papillae. The pedal sole and the flattened oral lobes are virtually colourless. The general texture of the body is, compared with *P. dubia*, more firm and less transparent.

ANATOMY (Fig. 14c). The radula of a 2 mm preserved specimen from the Isle of Man had the formula 15×5–6.1.1.0.1.1.6–5. The teeth are shorter and broader than those of *P. dubia*. We have examined the radula of a 12 mm specimen from Ireland; the formula was $19 \times 7.2.0.2.7$ (Fig. 14c).

Serial sections showed the presence of 7 μm spines on the penis.

HABITS. According to McMillan (1942), this species will feed in captivity on the polyzoan *Bowerbankia imbricata* Adams. Although most records of *P. nothus* have come from the seashore, it has on occasion been taken by aqualung divers at depths of 20–30 m.

Spawning occurs in British waters in April (Jersey), June (Greenisland, Co. Antrim) and July–September (N.E. England).

DISTRIBUTION (Map 6). While *P. nothus* has been reported from all around the British Isles, and is certainly an amphi-Atlantic, boreo-arctic species, many localities around north Atlantic coasts need to be confirmed.

DISCUSSION. The two British species of *Palio* have often been confused, but the following table of salient features should enable accurate diagnoses:

Palio dubia	*Palio nothus*
Dorsal and lateral papillae are delicate, elevated, and pointed	more broad and flattened
Rhinophores short and slender	longer and more stout
Rhinophore lamellae up to 13	up to 9
Median dorsal metapodial papillae usually prominent	rarely prominent
Body more flaccid, less opaque	more firm, less transparent
Radular teeth longer and more slender	shorter and broader
Marginal radular teeth narrow	broader

Finally, it should be noted that a detailed comparison of the two Pacific American species *Palio zosterae* (O'Donoghue, 1924) and *P. pallida* Bergh, 1880 with their British congeners, has yet to be undertaken.

Limacia Müller, 1781
   (type *Doris clavigera* Müller, 1776)
Polycerids which lack rhinophore sheaths; strong pallial tubercles arise all around the mantle rim, those from the frontal margin being minutely papillate. Mamillae project from the dorsal mantle. The oral tentacles are elongated, and grooved along the dorsal surface. Radula 9–12.2.0.2.9–12. Jaws flimsy or absent.

51. **Limacia clavigera** (Müller), 1776
   Doris clavigera Müller, 1776
   Tergipes pulcher Johnston, 1834
   Euplocamus plumosus Thompson, 1840
   Triopa lucida Stimpson, 1855

APPEARANCE IN LIFE (Plate 19). This species is a familiar one in shallow sublittoral areas on the western coasts of Europe and Africa, from Bergen to Cape Town; despite its delicate appearance it has a robust, though small body. It may reach 18 mm in length (larger in the south Atlantic, where it is commonly 20 mm long, even reaching 40 mm, according to Macnae (1957)). It is always white, with yellow or orange tips. The mantle edge is produced to form paired series of strong, finger-like ceratal processes; these are held erect. The frontal margin of the head bears similar processes and these are beset with small papillae. Numerous low papillae project from the central dorsum (one of Garstang's (1889) specimens had only a single longitudinal row), and these are heavily pigmented, as also is the rear tip of the foot. Yellow-tipped lamellate rhinophores issue from plain, low sheaths. There are 3 (rarely 4) tripinnate, yellow-tipped gills, forming an arc in front of the anal papilla. The oral tentacles are conspicuous, elongated and grooved along the upper surface. Many of these features are difficult to detect in juveniles (Miller, 1958).

ANATOMY (Fig. 13c). Macnae (1957) has illustrated the anterior genital mass, following Bergh's (1880) discovery that the penis is armed with small, hooked spines. Hecht (1895) described the complex defensive glands that fill the cerata.

Jaws are flimsy or absent. The radular formula of a 16 mm specimen was $50 \times 11.2.0.2.11$; of an 8 mm specimen $54 \times 10.2.0.2.10$; and of a 5 mm specimen $46 \times 11.2.0.2.11$. The teeth are highly differentiated. The shape in South African specimens was found to be similar (Macnae, 1957), although it is noteworthy that the radulae appeared to be significantly shorter, with formulae $25-35 \times 10-12.2.0.2.10-12$.

We recently investigated the radula of 2 specimens from Lundy Island, one of which measured 9 mm preserved (formula $49 \times 10.2.0.2.10$); the other was 16 mm alive ($42 \times 10.2.0.2.10$). Unusually in the Polyceridae, the first lateral teeth are slender and needle-like. The 2nd lateral is a formidable bicuspid erect structure, while the marginals are flattened plates of little functional importance (Fig. 13c).

HABITS. Miller (1961) found this species associated with *Callopora dumerilii* (Audouin), *Cryptosula pallasiana* (Moll), *Electra pilosa* (L.), *Membranipora membranacea* (L.), *Porella concinna* (Busk), *Schizoporella unicornis* (Johnston) and *Umbonula littoralis* Hastings, down to 20 m. *Limacia clavigera* occurs in clear, shallow waters (rarely on the seashore) down to 80 m. Spawn has been found in Britain during the month of June.

DISTRIBUTION (Map 6). This shallow-water species penetrates the Arctic Circle along the Norwegian coast to Finmarken (Lemche, 1941). It also occurs around the Faeroes and is one of the commonest nudibranchs all around the British Isles. Numerous records exist from the French Atlantic coast (Bouchet & Tardy, 1976) and the northern Spanish coast. There are, however, only rare records from the Mediterranean Sea, from Spain (Ros, 1975), Marseilles (Vicente, 1967) and the Bay of Naples (Schmekel, 1968). Gantès (1956) extended the known range to include Morocco, but it appears unlikely that there are no gaps in distribution between north Africa and the Cape of Good Hope (where it is commonly found).

Superfamily EUDORIDOIDEA (= CRYPTOBRANCHIA)

The mantle is ample. Strong jaws are absent, but the oral canal may exhibit a sculptured or patterned cuticle. The radula is usually broad. The penis may bear minute hooks or a more elaborate stylet, together with an adjacent penetrant rod. The gills are retractile into a common branchial pit in the postero-dorsal mantle (except in the tropical genus *Hexabranchus*).

CADLINIDAE

Cryptobranchiate, moderately soft, broad doridaceans with an ample mantle skirt. The dorsal mantle is glossy and relatively smooth, sometimes bearing sparse, low, soft tubercles. The rhinophores are lamellate, without conspicuous pallial sheaths. The branchial pocket lacks flap-like valves. The head has flattened oral lobes or short tentacles. The propodium is bilaminate, not notched in the mid-line. Feeding upon encrusting sponges.

The penis is armed with numerous tiny hooked chitinous spines; there is no distinct prostate gland. The buccal mass does not possess a buccal pump. Chitinous jaws are absent, but the oral canal is strongly cuticularized. Radula *n*.1.*n*, all the teeth bearing denticulate ridges on the hooked cusps.

Cadlina Bergh, 1878b

(type *Doris repanda* Alder & Hancock, 1842)

Cadlinids with a flattened, patelliform shape. The rachidian radular tooth lacks a median cusp but instead has up to 6 equal denticles, symmetrically disposed.

(This genus was validated in ICZN Opinion 812.)

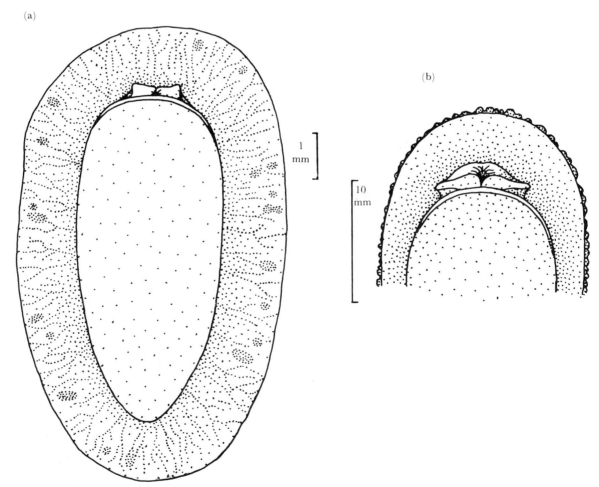

Fig. 15. Ventral views of (a) *Cadlina laevis*, and (b) *Doris sticta*.

52. **Cadlina laevis** (L., 1767)
    Doris laevis L., 1767
    Doris obvelata Müller, 1776
    Doris repanda Alder & Hancock, 1842
    ?**Doris glabra** Friele & Hansen, 1876

APPEARANCE IN LIFE (Plate 21e, Fig. 15a). The white body may reach a length of 32 mm. It is flattened, fragile, glistening and translucent. The expanded margins of the ample mantle skirt are thin and frail; the under-side of this skirt exhibits a delicate white tracery of spiculose markings. The dorsal mantle epithelium bears sparse, soft, low, conical tubercles. A characteristic feature is the presence of lemon-yellow or chalk-white multicellular defensive glands deep within the lateral mantle. In a rare northern variant, the whole body is tinged with yellow (M'Intosh, 1863); we have examined a number of examples from southern Britain in which the mantle skirt alone was yellow in colour. There are up to 5 (rarely 7) tripinnate gills. The lamellate rhinophores issue from low, crenulate, pallial sheaths. The oral tentacles are short, broad and flattened.

ANATOMY (Fig. 17a, Plate 16a). Observations on the reproductive, alimentary and nervous systems have been published by, respectively, Odhner (1939), Hancock & Embleton (1852) and Bergh (1879b). Sections showed the vas deferens to have large numbers of slender, gently curved spines, each 22·5 μm in length.

Many authors have examined the radula. Bergh (1900) counted 86 rows of teeth in the radula of a specimen 13 mm long in preservative; other authors have described shorter radulae (Meyer & Möbius, 1872; Odhner, 1907, 1939). Our observations relate to 3 specimens from the Northumberland coast, preserved and measured in alcohol; length 14 mm, formula 61 × 23.1.23; length 22 mm, formula 67 × 24.1.24; length 24 mm, formula 70 × 23.1.23. The median tooth has 3 denticulations on each side of the mid-line (only 2 in the 11·5 mm juvenile illustrated in Fig. 17a). The first few lateral teeth have denticulations on both edges, but towards the margins these denticles form a row on one edge only; there may be up to 20 such denticles in each row. SEM pictures of the radula were first published by Thompson & Hinton (1968).

The oral cuticle is highly differentiated and exhibits countless 20 μm long forked rods.

HABITS. This northern species, which bridges the gap anatomically between the temperate archidorids, discodorids and kentrodorids, on the one hand, and the warm-water chromodorids on the other, feeds upon encrusting slime-sponges (Harris, 1973), more precisely *Halisarca dujardini* Johnston (Barbour, 1979). Spawning has been seen in February and March in Northumberland but M'Intosh (1863) reported spawn near St Andrews in the month of November, and such autumnal breeding has been found also in the Baltic Sea (August and September, according to Jaeckel, quoted by Swennen (1961b). In Northumberland, the white spawn coils each contain up to 600 ova, 370–390 μm in diameter, developing in about 50 days at 10 °C to give rise to hatchlings which immediately begin benthic life. This direct development of *Cadlina* was first reported by Roginsky (1962) from the White Sea; it was later confirmed in British material. Many observations on development are detailed by Thompson (1967).

Although often found intertidally around Scotland and on the North Sea coasts, this dorid appears to be restricted to the sublittoral around southern shores.

DISTRIBUTION (Map 6). This common British species is distributed throughout the Arctic Ocean, at depths to 800 m, from Spitzbergen and the White Sea to Iceland, Greenland and North America (Lemche, 1929, 1941; Swennen, 1961). Along European coasts it is common in innumerable localities between Norway and northern Spain. A colour variety with a yellow mantle border is recorded from southern Britain to the Mediterranean Sea, although Bouchet & Moreteau (1976) questioned the conspecificity of this form. Certainly, *Cadlina laevis* is well adapted to constant subzero temperatures around Greenland (Lemche, 1941), which suggests an unusually wide environmental tolerance if the same species is indeed present in the

Mediterranean, where records include several from the French coast (Haefelfinger, 1960a) and the Adriatic Sea.

Franz (1970a) reported an American distribution extending southwards along the eastern seaboard to northern New England, but a record from the Gulf of Mexico (Lemche, 1929) is probably mistaken.

ALDISIDAE

Cryptobranchiate, broad doridaceans with an ample mantle skirt. The dorsal mantle bears abundant spiculose tubercles. Rhinophores lamellate with tuberculate pits. Branchial pocket lacking flap-like valves. Head with short oral tentacles.

The penis is armed with numerous hooked spines. The buccal mass does not possess a buccal pump. The oral cuticle is flimsy and lacks microscopic substructure. Radula *n*.0.*n*, the teeth greatly elongated, fragile, and armed with a denticulate fringe.

Aldisa Bergh, 1878
(type *Doris zetlandica* Alder & Hancock, 1854a)
(With the characters of the family.)

53. **Aldisa zetlandica** (Alder & Hancock, 1854a)
Doris zetlandica Alder & Hancock, 1854a

APPEARANCE IN LIFE (Fig. 19c). This rare species may attain a length of 35 mm. The body is white to grey-green in colour in contrast to the pale yellow gills and rhinophores. The ample mantle bears sparse, conical, pointed tubercles. The margins of the rhinophore pits are each protected by 4–5 tubercles. There are 6–8 tripinnate gills. The oral tentacles are short and rounded. The propodium is bilaminate but lacks a notch in the anterior lamina.

ANATOMY (Fig. 18a). The radula of an 11 mm preserved specimen had the formula $51 \times 100.0.100$ (Bergh, 1900). Neither Odhner (1939) nor our own observations on serial sections brought to light any evidence of genital armature. Bergh (1900), however, described and illustrated hooked spines on the penis. MacFarland (1966) claimed similar spines in the Pacific North American *Aldisa sanguinea* (Cooper). The discrepancy is at present inexplicable.

We have measured the radulae of 2 specimens, one from north east of the Shetland Islands, the other off Co. Galway. The formula of the former was $25 \times 39.0.39$. The teeth conform to an unusual type; they are elongated and concave, with minute pointed denticles in a row along one edge, continuing for some way around the distal tip of the tooth (Fig. 18a). They appear to be extremely slender and fragile, and the roots of the teeth are readily detached from the radular membrane.

HABITS. Nothing is at present known about the biology of this rare and unusual species.

DISTRIBUTION (Map 6). There are only two recent confirmed records of this species from the British Isles: Mr Bernard Picton obtained a specimen in shallow water off western Ireland, while another individual was included in collections from the oilfields 200 miles to the north east of the Shetland Islands. More frequent records exist from all along the Norwegian coast and off Sweden (Odhner, 1939) at depths approaching 280 m. Lemche (1938) dredged specimens from the north and east of Iceland and also from deep water (1900 m) to the south. Odhner (1907) provided the most southerly record, from the Azores.

ROSTANGIDAE

Cryptobranchiate, broad doridaceans with an ample mantle skirt. The dorsal mantle bears abundant, small, even, spiculose tubercles. Rhinophores lamellate, without conspicuous pallial sheaths. Branchial pocket lacking flap-like valves. Head with short oral tentacles. Propodium bilaminate, the anterior lamina notched in the mid-line. Feeding upon encrusting sponges.

The penis is unarmed, but a masculine stylet is present in *Awuka* Marcus, 1955. A distinct prostate gland is present. The buccal mass does not possess a buccal pump. Chitinous jaws are absent, but the oral canal is lined by a cuticle which often (but not in *Boreodoris* Odhner, 1939) exhibits a microscopic substructure of small rods. Radula *n.0.n*, lateral teeth numerous and hook-shaped, marginal teeth bifid, or multifid and brush-like.

Rostanga Bergh, 1879c

(type *Doris coccinea* Alder & Hancock, 1848)

Rostangids with a discrete prostate gland, but lacking masculine armature. Oral canal lined by a strong cuticle. Gills simple pinnate, occasionally bipinnate.

54. ## *Rostanga rubra* (Risso, 1818)
   Doris rubra Risso, 1818
   Doris coccinea Alder & Hancock, 1848b
   Rostanga perspicillata Bergh, 1881
   Rostanga rufescens Iredale & O'Donoghue, 1923

APPEARANCE IN LIFE (Plate 21, Fig. 16d–f). The body may reach 15 mm in length (20 mm in the Mediterranean, according to Schmekel & Portmann, 1982), bright scarlet in the typical variety, although occasional individuals vary from pink to pale orange-yellow. Scattered small black specks are present on the dorsum and there is a highly characteristic pale yellow patch

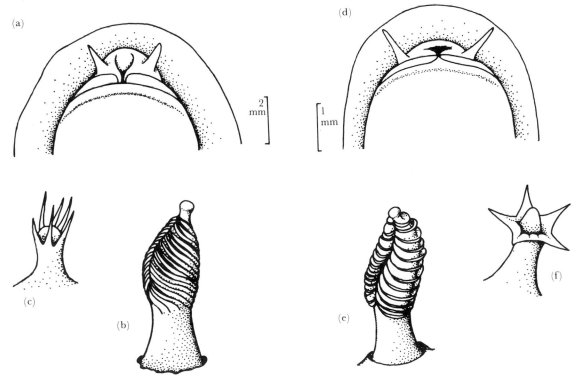

FIG. 16. Details of doridacean morphology.
   (a) *Jorunna tomentosa*, ventral view of the head.
   (b) detail of the rhinophore of the same.
   (c) detail of mantle papilla of the same.
   (d) *Rostanga rubra*, ventral view of the head.
   (e) detail of the rhinophore of the same.
   (f) detail of mantle papilla of the same.

between and around the rhinophores. When in motion, the sides of the body are more or less parallel, contrary to the oval shape of most other British cryptobranch doridaceans. All over the ample mantle are close-set, short, blunt, rather uniform, spiculose tubercles. There are up to 10 simple pinnate yellowish gills; the lamellate rhinophores are the same colour. Digitiform oral tentacles project from the sides of the head. The propodium is bilaminate, the anterior lamina notched in the mid-line.

ANATOMY (Plate 13c, d, Fig. 17f). The radula of a 7 mm specimen from Cornwall was examined; the formula was $60 \times 55.0.55$. An 8·5 mm specimen had the same formula. The first lateral tooth was very small, 14 μm in length, and bore 5–9 fine denticles along the cusp (these were missed by earlier workers). Other lateral teeth were larger but lacked such denticles. The marginal teeth were greatly enlarged with slender, bifid tips; these teeth reached 100 μm in length. *Rostanga rubra* has a simpler radula than other species of this genus (Thompson, 1975).

The largest radula on record was from an 11 mm preserved specimen found near Plymouth; the formula was $71 \times 39.31.1.0.1.31.39$. A 15 mm specimen from Lundy Island is shown in Fig. 17f, with the radular formula $58 \times 32.32.1.0.1.32.32$.

The oral cuticle is robust and exhibits microscopic honeycombing.

HABITS. This brightly coloured species feeds upon red siliceous sponges, such as *Microciona atrasanguinea* Bowerbank. It is arguable that some or all of the pigmentation of the nudibranch may be derived from its prey (Hecht, 1895; Cuénot, 1927b). Spawning occurs in June and July (Cornet & Marche-Marchad, 1951; Colgan, 1911; Cuénot, 1927). Although the spawn has been illustrated (Kress, 1971), nothing is known about the development of the egg, and the larval form remains undescribed, although the embryonic development and larval metamorphosis of *Rostanga pulchra* MacFarland from Friday Harbor, Washington, were recently described by Chia & Koss (1978).

DISTRIBUTION (Map 7). This species is well known from the Mediterranean Sea, including the coast around Villefranche (Haefelfinger, 1960a) and the Bay of Naples (Schmekel, 1968), where it is found at depths between 10 and 60 m. On Atlantic coasts it is often intertidal and has been recorded all along the French seaboard (Bouchet & Tardy, 1976) and around the British Isles, but it reaches its northern limit off southern Norway near Kristiansund (Jaeckel, 1952). A record from the Faeroes under the name *D. coccinea*, published by Mörch (1868), was probably a mistaken identification of *Jorunna tomentosa* (according to Lemche, 1929).

DISCUSSION. A review of the world's species of *Rostanga* has recently been completed (Thompson, 1975).

DORIDIDAE

Cryptobranchiate, broad doridaceans with an ample mantle skirt. The dorsal mantle bears numerous large spiculose tubercles which are sometimes pedunculate and in other cases interconnected by anastomosing ridges. Rhinophores lamellate, often guarded by pallial tubercles; similar tubercles often guard the branchial pocket. Head with short, sometimes grooved, oral tentacles. Propodium bilaminate, the anterior lamina entire, not notched. Feeding upon encrusting sponges.

The penis may be armed with hooks (*Artachaea* Bergh, 1882; *Alloiodoris* Bergh, 1904) or unarmed (*Doris* L., 1758; *Austrodoris* Odhner, 1926). No distinct prostate gland. The buccal mass does not possess a buccal pump. Chitinous jaws are absent and the oral canal is lined by a flimsy cuticle exhibiting little or no substructure. Radula *n.0.n*, all the teeth smoothly hooked, lacking subsidiary denticulations (except in *Artachaea* and *Alloiodoris*).

Doris L., 1758
(type *Doris verrucosa* L., 1758)
Doridids with smooth radular teeth; penis without hooks.

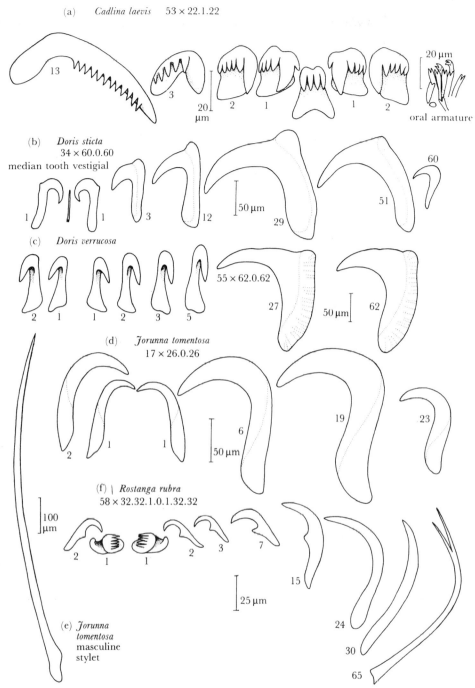

FIG. 17. Camera lucida drawings of doridacean radulae: *Cadlina*, *Doris*, *Jorunna* and *Rostanga*. Teeth are numbered from the mid-line.
  (a) *Cadlina laevis*, length 11·5 mm, Scapa Flow.
  (b) *Doris sticta*, preserved length 25 mm, Lundy Island.
  (c) *Doris verrucosa*, preserved length 19·5 mm, Arcachon, France.
  (d) *Jorunna tomentosa*, length 38 mm, Donegal (a radula from the variety *lemchei*).
  (e) *Jorunna tomentosa*, masculine stylet of an 18 mm specimen from Porlock, Somerset.
  (f) *Rostanga rubra*, length 15 mm, Lundy Island.

55. *Doris sticta* (Iredale & O'Donoghue, 1923)

Doris maculata Garstang, 1895 (not *Doris maculata* Montagu, 1804, which is a dotoid dendronotacean)
Archidoris (Staurodoris) maculata lutea Vayssière, 1919 (not *Doris lutea* Risso, 1818)
Doridigitata sticta Iredale & O'Donoghue, 1923

APPEARANCE IN LIFE (Plate 21, Fig. 15b). This beautiful and highly distinctive species has been recorded very few times, although it is both large and conspicuous, reaching 45 mm in length. The body colour is yellow with purplish grey pigment on the summit of many of the mantle tubercles. These tubercles may be up to 3 mm in diameter, interconnected by a complex network of pale cream ridges. Especially strong tubercles guard the rhinophoral pockets; typically, each pocket has 2 tubercles, one mesially and the other laterally placed. There are about 8 similar tubercles around the edge of the branchial pocket. The gills number up to 5, bipinnate or tripinnate (not simple pinnate, as Garstang (1895) claimed). The head bears a pair of short, grooved oral tentacles. The propodium is bilaminate, the anterior lamina entire, not notched.

ANATOMY (Fig. 17b). Vayssière (1919) reported a flimsy oral armature consisting of tiny cuticular rods. According to Eliot (1905), the radular formula may reach $40 \times 80.0.80$. A 4 cm preserved specimen was examined; the formula was $45 \times 70.1.70$, all the teeth being smooth hooks. There was undoubtedly a rudimentary median tooth faintly detectable in some of the transverse rows. In the 25 mm preserved specimen illustrated (radular formula $34 \times 60.0.60$), the medial tooth is vestigial. There is plainly considerable variation in this functionless feature of the radula.

The robust oral cuticle exhibited microscopic polygonal substructure, contrary to Vayssière's assertion. The penis is unarmed (Eliot, 1910).

HABITS. There is a general impression that this species feeds upon encrusting sponges, but dietary preferences remain obscure. Little is known about breeding, except that spawn masses are produced in June, July and August around south west Britain, each mass being a wide yellow band, up to 12 cm in length, attached in a spiral along one edge.

DISTRIBUTION (Map 7). This distinctive, sublittoral species has recently been collected in abundance around South Wales, Lundy and the Scilly Isles. However, the greater part of its range extends southwards from northern Brittany (Bouchet & Tardy, 1976) to Morocco (Gantès, 1956). There are also infrequent Mediterranean records from Banyuls and Marseilles (Pruvot-Fol, 1954).

DISCUSSION. Garstang (1895), in his original description of the species, chose the name *Doris maculata*, which was preoccupied by a dotoid named by Montagu (1804). Vayssière (1919) introduced *lutea* as a subspecific name for a variety found near Marseilles, but that too is preoccupied, by *Doris lutea* Risso, 1818 (although this latter species has never been traced and is *incertae sedis*). Iredale & O'Donoghue (1923) proposed a new name, *sticta*, which is acceptable.

56. *Doris verrucosa* L., 1758

Doris derelicta Fischer, 1867
Doris eubalia Fischer, 1872
Doris sepositosa Fischer, 1872
Doris biscayensis Fischer, 1872
Staurodoris januari Bergh, 1878c
Staurodoris bobretzki Gadzikiewics, 1907
Doris verrucosa mollis Eliot, 1910
Doridigitata derelicta; Iredale & O'Donoghue, 1923

APPEARANCE IN LIFE (Plate 21). This large species may reach 70 mm in length. The body is grey or yellowish in colour, often darker in the middle of the back, which is covered by rounded tubercles of various sizes, the largest reaching 4 mm in diameter. The size of the tubercles decreases towards the edge of the ample mantle skirt. The branchial pocket is guarded by 8 or more especially prominent pedunculate tubercles. A pair of tubercles guards each rhinophore pocket, one tubercle mesially and one laterally placed. There are up to 18 pinnate (usually simple) gills. The head bears a pair of short, grooved, oral tentacles. The propodium is bilaminate, the anterior lamina entire, not notched.

ANATOMY (Fig. 17c). The radula of a 6 mm specimen from Arcachon was examined; the formula was $50 \times 63.0.63$. All the teeth were smooth hooks. No median tooth could be detected, contrary to *D. sticta*. Four specimens (7·5–45·0 mm in length) examined by Franz (1970b) had shorter and narrower radulae, encompassed by the general formula $24–42 \times 25–47.0.47–25$.

The specimen illustrated in Fig. 17c was 19·5 mm long in preservative and had the radular formula $55 \times 62.0.62$. The strong cusps can measure up to $150\,\mu m$ in height. There are no vestigial marginal teeth.

The oral cuticle was robust but lacking in microscopic differentiation.

HABITS. The natural prey is sponges, but, as in *D. sticta*, nothing is known about preferences in British waters (although *Halichondria* is said by Schmekel (1968) to be preferred in the Bay of Naples). Breeding is unknown in the British Isles, but spawn has been reported from July to October in the Bassin d'Arcachon (Cuénot, 1903). Specimens are adult from 10 mm onwards. The ova measure 99–106 μm in diameter. This species is known to be a host for the degenerate parasite *Splanchnotrophus* (Cuénot, 1903).

DISTRIBUTION (Map 7). This large and conspicuous dorid appears in the English Channel to be at the limit of its range. Although frequently found in the vicinity of Arcachon and further south in the Bay of Biscay, only one specimen (of the subspecies *mollis* Eliot, 1906, from Salcombe, Devon, in 1900) can be unquestioningly accepted as a British record. However, Eliot (1910) described museum specimens labelled as coming from the Firth of Clyde, while Farran (1903) claimed a record from western Ireland. Mediterranean localities include the Adriatic, the Bay of Naples (Schmekel, 1968) and the French coast (Vayssière, 1888; Haefelfinger, 1960a; Vicente, 1967). Furthermore, the distribution is amphiatlantic, from the Bay of Biscay (Bouchet & Tardy, 1976) to the Azores and Brazil (Marcus, 1955) and extending northwards to encompass Florida (Bergh, 1894), Georgia (Marcus & Marcus, 1967), South Carolina (Eliot, 1910), Connecticut and Massachusetts (Franz, 1970b).

ARCHIDORIDIDAE

Cryptobranchiate, broad doridaceans with an ample mantle skirt. The dorsal mantle bears abundant spiculose tubercles which sometimes coalesce to form anastomosing ridges with intervening arcades of various degrees of complexity. Rhinophore sheaths and flap-like valves around the branchial pocket are sometimes present; rhinophores lamellate. Propodium bilaminate, the anterior lamina sometimes notched in the mid-line (not in *Archidoris*). Head with flattened oral lobes or finger-like oral tentacles. Feeding upon encrusting sponges. The penis is usually unarmed (but bears a spine in *Peronodoris*); no distinct prostate gland. The buccal mass does not possess a muscular pump. Chitinous jaws are absent, and the oral canal is lined by a flimsy cuticle exhibiting little or no substructure. Radula *n.0.n*, all the teeth smoothly hooked, lacking subsidiary denticulations.

Archidoris Bergh, 1878b
(type *Doris tuberculata* Müller, 1778)

Archidorids which have a tuberculate, spiculose, dorsal mantle without ridges or arcades. The propodium is bilaminate, both laminae entire, not notched in the mid-line. Penis unarmed.

57. **Archidoris pseudoargus** (Rapp, 1827)
    Doris argo Pennant, 1777, not Bohadsch, 1761
    ?**Doris tuberculata** Müller, 1778; Cuvier, 1804, not Cuvier, 1836
    Doris obvelata Fabricius, 1797, not Müller, 1776
    Doris pseudoargus Rapp, 1827
    Doris britannica Johnston, 1838a
    Doris montagui Johnston, 1838a
    Doris mera Alder & Hancock, 1844
    Doris flammea Alder & Hancock, 1844
    Doris argus; Bergh, 1871b in Mörch, 1871

APPEARANCE IN LIFE (Plates 22, 23, Fig. 20a). This is the largest and most commonly found dorid nudibranch on British shores, and is colloquially termed the "sea-lemon". It may reach 120 mm in length, and the dorsum, covered by spiculose blunt tubercles of various sizes, bears blotchy irregular markings of yellow, brown, pink, green and white pigments. The variety *flammea* is red or purple all over.

There are 8–10 tripinnate gills which are paler than the mantle; they frequently bear opaque white or sulphur-yellow pigment. The head bears vestigial blunt oral tentacles.

ANATOMY (Fig. 22b). Because of its large size, this is the most frequently studied British nudibranch (volume I, Fig. 13g). The earliest studies of the anatomy were those of Alder & Hancock (1845–55), Hancock & Embleton (1852) and Hancock (1865), followed by Bolot (1886) and Vayssière (1888). These observations were reviewed and extended by Eliot (1910). More recently, the structure of the alimentary canal was investigated by Forrest (1953) and by Rose (1971). The reproductive organs were studied by Lloyd (1952) and by Thompson (1966); the latter paper included an account of the embryonic and larval development. The histology and physiology of the blood-gland, situated above the cerebro-pleural and pedal ganglia, were carefully investigated by Schmekel & Weischer (1973), who measured high concentrations of copper in this gland, believed to relate to haemocyanin synthesis. Potts (1981) has described the anatomy and histology of the gills and epidermal glands.

The oral canal is lined by a flimsy cuticle which exhibits little or no microscopic substructure. The radula consists of large numbers of smoothly hooked teeth. The formula of a 13 mm specimen was $23 \times 38.0.38$; 14 mm: $27 \times 40.0.40$; 35 mm: $35 \times 55.0.55$; 39 mm: $39 \times 71.0.71$; 70 mm: $38 \times 85.0.85$. No median tooth is present; Runham (1963) was mistaken.

HABITS. The disruptive pigment pattern of this shallow water and sublittoral species renders it fairly inconspicuous, especially when submerged. It feeds upon encrusting siliceous sponges, chiefly *Halichondria panicea* (Pallas) and *Hymeniacidon perleve* (Montagu). Alder & Hancock (1845–55) state that it will take the calcareous sponge *Grantia compressa* (Fabricius) and Miller (1961) includes also *Tethya aurantia* (Pallas), but these reports are unconfirmed. In the Isle of Man (Thompson, 1966), *Archidoris* has an annual life cycle, growth occurring during the autumn and winter months, then a lengthy spawning season occupies the spring, followed by death. Juveniles begin to appear again in the late summer and the cycle is repeated. More detailed observations on the life cycle and the histology of the ovotestis were presented in volume I (p. 93 *et seq.*).

Each spawn mass (illustrated by Thompson, 1966, and by Kress, 1971) is a large cream or white coil attached to a hard substratum; it may contain up to 300,000 ova in Welsh or Manx material (up to 645,000 in Ireland, according to Colgan, 1914). The ova vary from 140 to 170 μm in diameter, and develop in about 28 days at 10 °C to give planktotrophic larvae of shell-type 1 (Thompson, 1961a). The veliger larva is illustrated in volume I (Fig. 40a–e). Breeding activities have been reported at various times of the year (every month except November) (Sutherland, 1890; Hecht, 1895; Cuénot, 1903; Storrow, 1911; Renouf, 1915; Elmhirst, 1922; McMillan, 1944; Marine Biological Association, 1957; Allen, 1962; Miller, 1962; Thompson, 1966).

The ectosymbiont *Lichomolgus agilis* (Leydig) has been reported on the gills (Leigh-Sharpe, 1935). *Archidoris* was at one time used as fish-bait in Shetland (Jeffreys, 1867).

DISTRIBUTION (Map 7). This well-documented species is known from all around the British Isles and along the Norwegian coast to Varanger Fjord east of Nordkapp (Odhner, 1939). Lemche (1929, 1938) identified specimens from Iceland and the Faeroes.

Archidoris is tolerant of lowered salinity and penetrates the Baltic Sea to the Bay of Kiel (Swennen, 1961b). South of Britain *A. pseudoargus* has usually been referred to as *A. tuberculata* and these records demonstrate a distribution which includes the Atlantic coasts of France (Bouchet & Tardy, 1976), Spain (personal communication from Mr Bernard Picton), Portugal (Oliveira, 1895) and numerous Mediterranean localities (Pruvot-Fol, 1954).

While this species is well adapted to the rigours of intertidal existence, sublittoral populations are not uncommon and specimens have been dredged from depths between 250 and 300 m in the fjords of Norway. The red variety *flammea* is encountered north of Donegal and the Isle of Man, and is a common component of Scandinavian populations.

Bergh's (1894) records from Bare I. in the northern Pacific and from lower California refer to closely related but distinct species. Similarly, a record from the Gulf of Mexico quoted by Lemche (1929) was probably a misidentification.

Atagema Gray, 1850

(type *Doris carinata* Quoy & Gaimard, 1832)

Archidorids which have a uniform covering of small spiculose tubercles over the dorsal mantle, and a median longitudinal, elevated, zig-zag ridge; in some exotic species this ridge has deep arcades on either side, together with a few larger spiculose excrescences. The propodium is bilaminate, the anterior lamina usually notched in the mid-line. Penis unarmed.

58. *Atagema gibba* Pruvot-Fol, 1951

APPEARANCE IN LIFE (Plate 23, Fig. 21). This highly distinctive species has been found in only one British locality, at Porthkerris Point, Cornwall, under 12 m of water. Five specimens from 7–68 mm in length have been found. The general aspect of the body is flattened, but there is a conspicuous elevated zig-zag dorsal pallial ridge from between the bases of the rhinophores to the gills. The edge of the mantle skirt is usually loosely crenulated; some of these crenulations persist permanently even when the animal is actively creeping. The dorsal mantle surface is stiff and velvety to the human touch, covered with small, digitiform, spiculose tubercles which are not uniformly distributed but form a pattern of papillose, anastomosing strands surrounding smooth areas. The papillae are whitish, while the smooth areas are dark brown. Papillae predominate in areas of the mantle near the median dorsal line, while smooth brown areas predominate near to the mantle edge. In this way the characteristic pattern is produced.

The trumpet-shaped rhinophore sheaths bear characteristic tubercles and encircle the mottled brown and white rhinophoral tentacles; each rhinophore bears 19 or 20 lamellae. The branchial pocket has a crenulated rim which forms 5 flanges, each overlying a mottled tripinnate or quadripinnate gill plume; the gills are directed posteriorly rather than dorsally (contrary to other British cryptobranchiate dorids).

The underside of the mantle is mottled brown and white while the pedal sole is a more uniform pale brown. The propodium is transversely grooved and notched in the mid-line anteriorly. The head bears a pair of finger-like oral tentacles.

ANATOMY (Fig. 22a). This was studied by Thompson & Brown (1974). Paired systems of gill retractor and buccal mass retractor muscles are conspicuous (like those of *Archidoris*), and a substantial blood gland overlies the ganglia of the central nervous system. The stomach is largely free from the digestive gland, and possesses a small but distinct gastric caecum projecting ventrally. The penis is long and slender and bears no armature. The radular formula of a 68 mm specimen was $23 \times 30.0.30$ (a 45 mm individual exhibited $18 \times 28.0.28$). Each tooth had a dark base and a smooth, erect, hook-like cusp. In each tooth row the components were smaller near the mid-line and near the periphery. The teeth have a smaller basal attachment than those of *Archidoris*.

HABITS. Nothing is known about the diet or breeding biology of this rare species. *Atagema* is principally a tropical genus and *A. gibba* has elaborate branchial and rhinophoral protective devices and a rapidity of response and retraction unusual in the British doridaceans.

DISTRIBUTION (Map 7). Only one British locality is known for this species, near Porthkerris Point, Cornwall. Elsewhere it is known only from Banyuls-sur-Mer, on the Mediterranean coast of France, although a questionable record exists from Brittany (Pruvot-Fol, 1951) (see Bouchet & Moreteau, 1976).

DISCUSSION. There can be little doubt that the Porthkerris specimens correspond with Pruvot-Fol's *A. gibba*, although her original description was gravely deficient in that no mention was made of dimensions (other than "plus grand que *A. rugosa*"), colour (other than "brune . . . avec . . . taches presques noires"), date of collection, depth nor anatomy.

There are 3 known species in the eastern north Atlantic: *Atagema rugosa* Pruvot-Fol, 1951 (single specimen from Banyuls), *A. gibba* Pruvot-Fol, 1951 (two specimens from Banyuls) and *A. africana* Pruvot-Fol, 1953 (single specimen from Dakar, north Africa). Later, Pruvot-Fol (1954) recorded a dubious specimen in the collections of the Roscoff Biological Laboratory, but was unable to do more than give it the generic name *Atagema*.

DISCODORIDIDAE

Cryptobranchiate, tough, broad doridaceans with an ample mantle skirt. The dorsal mantle usually bears abundant, small, spiculose tubercles which have a granular appearance and texture; autotomy of the edges of the mantle occurs. Defensive acid secretion through the skin of the dorsal mantle occurs in many species. Rhinophores lamellate. Head with finger-like oral tentacles, sometimes longitudinally grooved. Propodium bilaminate, the anterior lamina notched in the mid-line in many cases. Feeding upon encrusting sponges. Swimming escape movements have been observed.

The penis is usually unarmed; prostate gland gathered into a greatly swollen region of the vas deferens. The buccal mass does not possess a muscular pump. Chitinous jaws are absent, but the oral canal cuticle is resistant and often exhibits microscopic honeycomb substructure. Radula *n.0.n*, all the teeth hooked and usually smooth, but extreme marginals may be denticulate or pectinate.

Discodoris Bergh, 1877
(type *Discodoris boholiensis* Bergh, 1877)

Discodorids which have a tuberculate, spiculose, dorsal mantle, granular in appearance. Low, smooth rhinophore sheaths and flap-like valves around the branchial pocket are sometimes present.

The penis is unarmed. The oral cuticle usually exhibits microscopic substructure which is ridged or resembles a honeycomb. Radula *n.0.n*, all the teeth smoothly hooked, but the marginals are sometimes denticulated.

59. *Discodoris millegrana* (Alder & Hancock, 1854a)
Doris millegrana Alder & Hancock, 1854a
Aporodoris millegrana; Ihering, 1886

APPEARANCE IN LIFE (Fig. 19a, b). This species has not been seen alive, and the description that follows is based upon 2 Alder & Hancock syntypes in the British Museum (Natural History). These are the same specimens that were described by Eliot (1910); they remain beautifully preserved (up to 28 mm long in preservative). The ample mantle is covered by innumerable, tiny, rather uniform, conical, spiculose papillae. It is impossible to guess at the colour of the body in the living state. The lamellate rhinophores and the tripinnate gills are retracted into pits which are not especially well armoured. Curved digitiform oral tentacles are present on the head. The propodium is bilaminate, the anterior lamina notched in the mid-line.

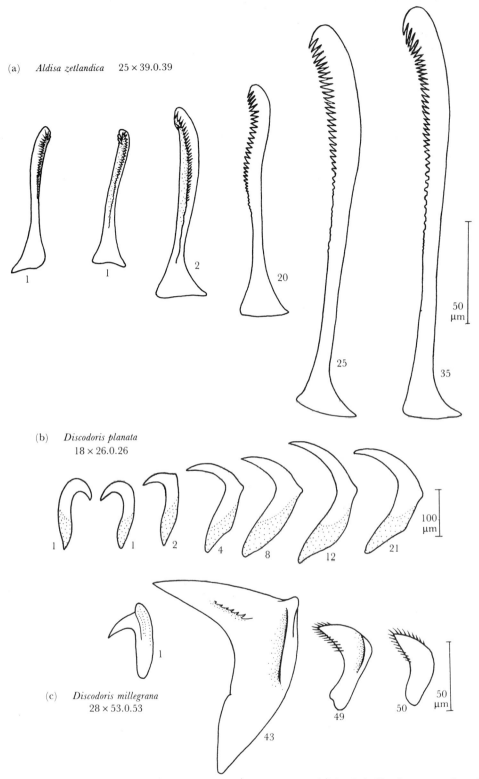

(a) *Aldisa zetlandica*   25 × 39.0.39

(b) *Discodoris planata*
18 × 26.0.26

(c) *Discodoris millegrana*
28 × 53.0.53

FIG. 18. Camera lucida drawings of doridacean radulae: *Aldisa* and *Discodoris*. Teeth are numbered from the mid-line.

(a) *Aldisa zetlandica*, preserved length 9 mm, 200 miles NE of the Shetland Islands.
(b) *Discodoris planata*, preserved length 24 mm, Helford, Cornwall.
(c) *Discodoris millegrana*, preserved length 25 mm, Torbay, Devon (BMNH 1980101).

ANATOMY (Fig. 18c). The larger preserved specimen, 28 mm long, was dissected. The anterior genital mass was well developed, the receptaculum seminis was swollen with allosperms as was the vesicula seminalis with autosperms; there was a swollen, pedunculate, prostate gland. After dissection from its sheath, the penis was found to be short and conical in shape. Under the microscope no cuticular armature was detectable on the vas deferens or the penis.

The oral canal of the larger specimen was examined after treatment with caustic soda. A substantial cuticle survived this and showed fine ridging under the optical microscope. The radula of the smaller preserved specimen, 25 mm long, had been dissected out by Alder & Hancock, and examined by them and by Eliot (1910). Of the published descriptions, Eliot's was the less accurate.

The radular formula is 28–29 × 53.0.53; the teeth all bear hooked cusps. The smallest teeth are near the mid-line of the radula and at the margins. The 4 or 5 extreme marginals are degraded; the next 4–6 have a denticulate ridge consisting of up to 8 tiny denticles. The degraded marginals bear faint denticulations which in favourable cases show up as a tuft of bristles replacing the cusp. Eliot stated that all the teeth were denticulate, but this is untrue.

HABITS. Nothing is known about the life history of this species. Pruvot-Fol's (1954) statement that it can be host to the ectoparasitic copepod *Lichomolgus* may be a mistaken attribution.

DISTRIBUTION (Map 6). If *D. millegrana* does not embrace "*Thordisa*" *dubia* Bergh, 1894 (see below), the only records of this species come from Torbay and the French coast near Wimereux (Bouchet & Tardy, 1976).

DISCUSSION. Ihering (1886) introduced an unnecessary genus, *Aporodoris*, to embrace this species, but *millegrana* is a straightforward member of the widespread older genus *Discodoris*.

According to Marcus (1955), Odhner (*in litt.*) was of the opinion that *millegrana* included "*Thordisa*" *dubia* Bergh, 1894 from Brazil. This is unlikely because Bergh's species had denticulate central lateral teeth in the radula, smooth in *millegrana*.

60. ***Discodoris planata*** (Alder & Hancock, 1846)
  Doris planata Alder & Hancock, 1846
  Doris testudinaria Alder & Hancock, 1862
  Archidoris stellifera Vayssière, 1904

APPEARANCE IN LIFE (Plates 23, 24, Fig. 20b). This is the best known northern European member of a familiar group of tropical doridaceans. This British species shows the habit, prevalent among discodorids, of secreting defensive acid fluids through the dorsal mantle (but not from the head or the foot). Numerous (up to 12) pale, star-shaped markings indicate the presence of patches of especially strongly developed acid-secreting epithelia. Each star radiates from a white papilla. The rest of the dorsal mantle is mottled brown or purple, generally less variegated and colourful than *Archidoris pseudoargus*. *Discodoris planata* is a smaller species, reaching only 65 mm in length, while *A. pseudoargus* may reach 120 mm. The dorsum is covered by spiculose tubercles of generally smaller size than those of *Archidoris*. Finally, the head in *D. planata* bears conspicuous finger-like oral tentacles, lacking in *Archidoris*. The propodium is bilaminate, the anterior lamina notched in the mid-line.

Juveniles are usually paler in colour and more flattened.

ANATOMY (Fig. 18b). The oral canal is lined by a cuticle which exhibits microscopic honeycombing, in which small rod-like elements may be discerned in favourable individuals. The radula consists of numerous hook-like teeth, becoming degenerate at the margins of the radular ribbon. The formula of a 15 mm specimen was 20 × 31.0.31; 24 mm (in preservative): 18 × 26.0.26; 31 mm (in preservative): 18 × 30.0.30 (Eliot, 1906); 40 mm (in preservative): 18 × 57.0.57. Schmekel & Portmann (1982) describe a radula in which the marginal teeth bear minute serrations.

The anterior genital mass was briefly described by Schmekel (1970) under the name *Anisodoris stellifera*.

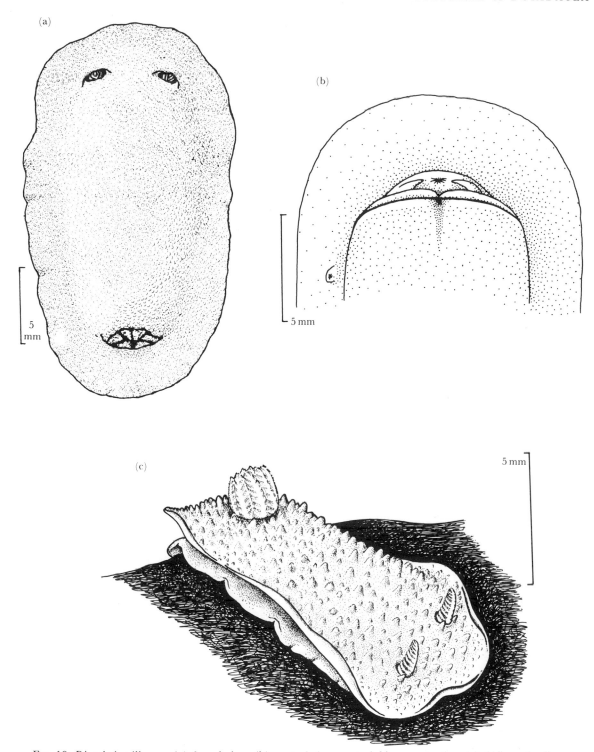

FIG. 19. *Discodoris millegrana* (a) dorsal view, (b) ventral view; material in preservative; (c) *Aldisa zetlandica*, latero-dorsal view.

HABITS. This shallow water species feeds upon sponges, for instance, *Hemimycale columella* (Bowerbank) (Forrest, 1953). It is drab and has undoubtedly been confused in the past with *Archidoris pseudoargus*. Little is known about its biology. Colgan (1914) estimated that a single spawn mass contained 480,000 ova, but the extent of the breeding season is unknown.

89

DISTRIBUTION (Map 7). This species has only been found recently around west coasts of the British Isles, although there are records from southern Norway and the North Sea (McKay & Smith, 1979). It is common along the French Atlantic coast (Bouchet & Tardy, 1976) and in the Mediterranean. In the Bay of Naples it is frequently encountered between 10 and 45 m.

DISCUSSION. Alder & Hancock described *Doris planata* in 1846 and *D. testudinaria* in 1862. Later, Alder stated (in Jeffreys, 1867) that he had come to the conclusion that the former was the juvenile form of the latter. Changes of shape and coloration frequently occur during early life in discodorids, and we accept Alder's judgement.

(a)

(b)

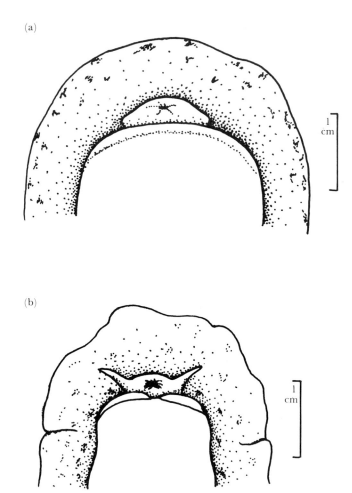

FIG. 20. Ventral views of the head of (a) *Archidoris pseudoargus* and (b) *Discodoris planata*.

KENTRODORIDIDAE

Cryptobranchiate, broad doridaceans with an ample mantle skirt. The dorsal mantle bears abundant small, even, spiculose tubercles. Rhinophores lamellate, without conspicuous pallial sheaths. Branchial pocket lacking flap-like valves. Head with short oral tentacles. Propodium bilaminate, the anterior lamina notched in the mid-line. Feeding upon encrusting sponges.

The penis is unarmed, but a masculine stylet is present in *Jorunna* Bergh, 1876. A distinct prostate gland is present. The buccal mass does not possess a buccal pump. Chitinous jaws are absent but the oral canal is lined by a cuticle which often exhibits a microscopic substructure of small rods. Radula *n*.0.*n*, all the teeth hooked and smooth, but extreme marginals are occasionally denticulate or pectinate.

Jorunna Bergh, 1876

(type *Doris johnstoni* Alder & Hancock, 1845b)

Mantle surface velvety, covered with small, uniform, spiculose tubercles, each of which has a retractile central projecting finger. A masculine stylet is present. Radular teeth smooth.

The world's species of *Jorunna* have been reviewed by Marcus (1976).

61. **Jorunna tomentosa** (Cuvier, 1804)
Doris tomentosa Cuvier, 1804
Doris obvelata Johnston, 1838a, not Müller, 1776
Doris johnstoni Alder & Hancock, 1845b
Doris johnstoni alba Bergh, 1881
Jorunna lemchei Marcus, 1976
?**Jorunna luisae** Marcus, 1976

APPEARANCE IN LIFE (Plate 21, Fig. 16a–c). This species is common and widespread on the shores and in the shallow waters of the British Isles. It may reach 55 mm in length and is soft and velvety to the touch. The ample mantle is covered with small, glistening, uniform, spiculose tubercles of the caryophyllidia type, each of which has a characteristic, retractile, central, digitiform papilla. The colour of the mantle gives good camouflage while on or near encrusting sponges; it is sandy brown, yellow, orange or grey-white. A fairly constant feature is the presence of an approximately paired series of dark brown blotches down the sides of the dorsum. Specimens from the west coast of Ireland often lack these spots (variety *lemchei* Marcus, 1976). The margins of the rhinophore pits are slightly raised and crenulated. There are up to 17 tripinnate gills; the gill pocket, when the gills are fully extended, is elevated and forms a short, cylindrical, vase-like base for the gill circlet. The head bears slender, digitiform oral tentacles. The propodium is bilaminate, the anterior lamina notched in the mid-line.

ANATOMY (Fig. 17d, e). An important paper on the functioning of the alimentary canal of this species was published by Millott (1937). Other aspects of the anatomy and histology were touched upon by Hancock & Embleton (1852), Bergh (1880) and Odhner (1939). The anterior genital mass possesses a remarkable chitinous masculine stylet. This is non-tubular and is housed at the mouth of a blind-ending masculine gland; it is presumed to effect the arousal of a mate. In an 18 mm specimen from the Somerset coast, the stylet was 1·1 mm long. Under the highest power of the optical microscope there was a longitudinal fibrous appearance.

A number of radulae were examined, from Cornwall or the Isle of Man; length of body 5·5 mm, formula 15 × 14.0.14; length 10 mm, 16 × 16.0.16 + 5 degraded marginals each side; length 12 mm, 18 × 17.0.17 + 3–4 degraded marginals each side; length 15 mm, 15 × 20.0.20; length 21 mm, 16 × 16.0.16 + 5 degraded marginals each side; length 28 mm, 22 × 25.0.25; length 30 mm, 20 × 27.0.27; length 38 mm, 17 × 26.0.26; length 54 mm, 27 × 33.0.33. The teeth are all smoothly hooked. The oral cuticle exhibited fine microscopic honeycombing.

HABITS. On British shores this species feeds chiefly upon the encrusting siliceous sponges *Halichondria panicea* (Pallas) and *Haliclona* spp. (Millott, 1937; Miller, 1961; Thompson, 1964).

The smallest individual known to spawn measured 19 mm in length (Miller, 1958); egg-laying has been recorded from February to August (Miller, 1958; Allen, 1962), but at Arcachon on the French Biscay coast we have found it to spawn in September. Each spiral spawn mass contains up to 145,000 ova, 69–90 μm in diameter, which develop in 23 days at 9–10 °C to planktotrophic veliger larvae of type 1 (Thompson, 1961a, 1967). The newly hatched larva is illustrated in volume I (Fig. 41a).

DISTRIBUTION (Map 7). At one time it was thought that this species had a cosmopolitan, world-wide distribution, but material from Tanzania (Edmunds, 1971) and Hawaii (Kay & Young, 1969) has recently been allocated by Marcus (1976) to two new species, *J. malcolmi* and *J. alisoni* respectively. The distribution of true *J. tomentosa* appears to be from 65 °N on the Norwegian coast to the Faeroes, all coasts of Britain, France, Portugal, Morocco and the Adriatic Sea (Marcus, 1976). Schmekel's material from the Bay of Naples was also re-

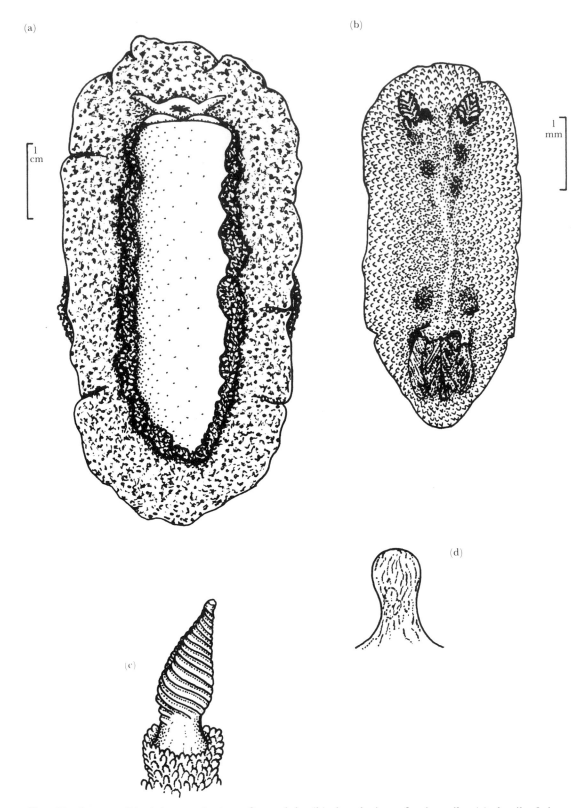

FIG. 21. *Atagema gibba* (a) ventral view of an adult, (b) dorsal view of a juvenile, (c) detail of the rhinophore, (d) detail of pallial papilla.

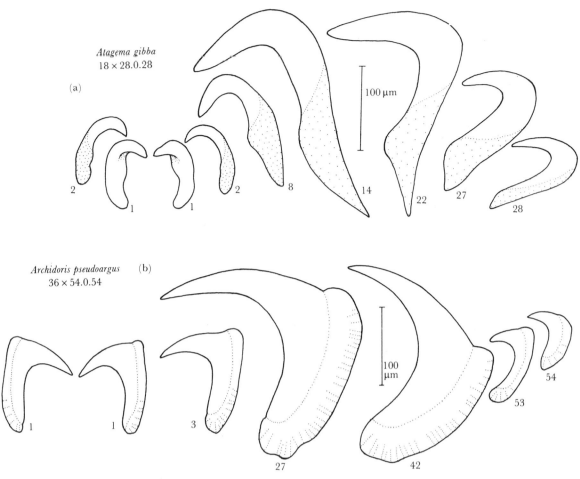

FIG. 22. Camera lucida drawings of doridacean radulae: *Atagema* and *Archidoris*. Teeth are numbered from the mid-line.
   (a) *Atagema gibba*, length 45 mm, Porthkerris Point, Cornwall.
   (b) *Archidoris pseudoargus*, preserved length 25 mm, Plymouth, Devon.

investigated by Marcus (1976) who concluded that it was separate and distinct, with the name *J. luisae* Marcus, 1976. We are not convinced of the need for this.

   We have been able to examine specimens of *Jorunna lemchei* Marcus, 1976 from western Ireland; these lack the dark dorsal spots of typical *J. tomentosa* but are otherwise indistinguishable in habits, external morphology or internal anatomy. Accordingly, we have merged these two species.

Suborder III ARMINACEA

This is the most difficult suborder to define and is certainly difficult to recognize from external features alone. The 9 families (3 of them British) currently contained here will not fit into any other suborder, and, to a certain extent, this explains the conviction that a separate suborder is needed for them. Species of *Armina* found in the north Atlantic have, externally, some of the features of a primitive doridacean, such as *Phyllidia* or *Corambe*, with external pallial leaflets forming a respiratory series beneath the mantle skirt, but internally *Armina* proves to be very unlike a dorid for it has a much-divided digestive gland, a laterally placed anal papilla and strong jaws. Similarly, the arminaceans *Hero, Janolus* and *Proctonotus* have a strong resemblance to the aeolidacean nudibranchs, but their cerata lack cnidosacs, a fact which instantly separates them from the aeolids.

Arminacean nudibranchs have some significant features in common. Their rhinophoral tentacles lack external protective pallial sheaths. There is a tendency in both Arminidae and Janolidae towards the development of a median accessory sensory caruncle close to the rhinophoral bases. The anal papilla is usually situated rather far forward, either dorsally, or laterally, on the right side. The radula may be multiseriate or triseriate, but it is never uniseriate. Oral tentacles are usually lacking.

The diets of arminaceans are varied. The tropical *Pinufius* takes the scleractinian *Porites*, but *Doridomorpha* prefers the octocoral *Heliopora* (Rudman, 1982). *Hero formosa* certainly feeds upon gymnoblastic and calyptoblastic hydroids, and *Armina* is known to attack pennatulacean cnidarians. But *Janolus* and *Proctonotus* feed upon encrusting and erect polyzoans, while the north American Pacific *Dirona albolineata* is reported (Robilliard, 1971) to devour a wide variety of shelled molluscs and other invertebrates. The larval shell is of veliger type 1 (Thompson, 1961a).

For an authoritative discussion of the subdivisions of the Arminacea, see Odhner (1939).

KEY TO THE BRITISH SPECIES OF ARMINACEA

1. Body with a row of plate-like gills beneath the mantle
rim   *Armina loveni* (p. 96)
Such gills absent   **2**

2. Sensory pleated caruncle present between the
rhinophores   **3**
No caruncle   **4**

3. Cerata smooth, with a bluish superficial sheen   *Janolus cristatus* (p. 97)
Cerata with small epidermal warts, brownish   *Janolus hyalinus* (p. 98)

4. Cerata warty but not branched   *Proctonotus mucroniferus* (p. 100)
Cerata branched, arborescent   *Hero formosa* (p. 100)

Superfamily EUARMINOIDEA

Arminaceans which possess lamellate rhinophoral tentacles, each retractile into a pallial emargination or pit. The mantle is ample and entire, lacking prominent excrescences; under the rim there is a symmetrical series of branchial processes (not in *Heterodoris* or *Doridomorpha*). The anus is lateral.

ARMINIDAE

Euarminoid arminaceans in which the mantle is ample and contains numerous pustular lateral glands; beneath its rim on either side is a symmetrical series of branchial projections, the gills and the lateral lamellae. Large jaws are present. Radula broad, with well-developed

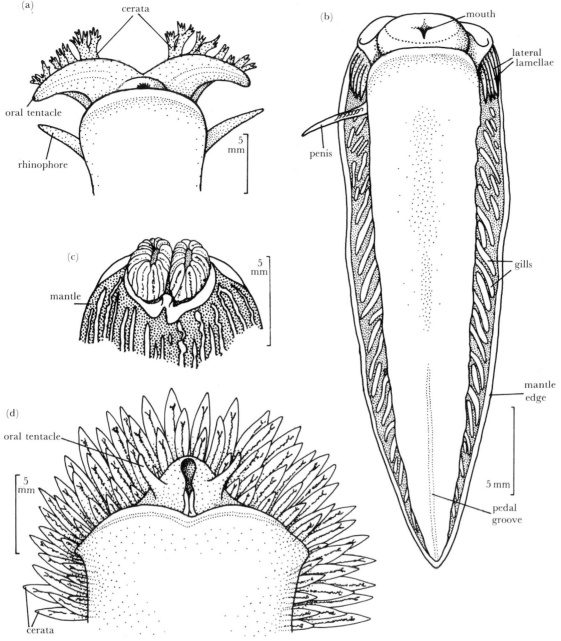

FIG. 23. Details of arminacean morphology.
(a) *Hero formosa*, ventral view of the head.
(b) *Armina loveni*, ventral view.
(c) detail of rhinophores of the same, dorsal view.
(d) *Janolus cristatus*, ventral view of the head.

denticulate rhachidian teeth. The digestive gland is much-divided, and its lobes penetrate the branchiae.

Armina Rafinesque, 1814

(type *Armina tigrina* Rafinesque, 1814)

Arminids with a wrinkled sensory caruncle just in front of the rhinophores. The upper surface of the mantle is ridged longitudinally.

62. *Armina loveni* (Bergh, 1860)

Pleurophyllidia loveni Bergh, 1860
Pleurophyllidia lineata "Otto, 1820" Bergh, 1866 (not Otto, 1820)
Pleurophyllidia henneguyi Labbé, 1922

APPEARANCE IN LIFE (Plate 25, Fig. 23b, c). This is a highly distinctive species, quite unlike any other British nudibranch. Despite this, there have been relatively few reported finds, probably because it burrows into soft substrata in search of its cnidarian prey. Adults may reach 40 mm in length. The ample mantle is ovate when viewed from above, and brick-red in colour with up to 50 prominent white longitudinal ridges. These ridges are irregularly wavy and may in places break up into pustules, especially towards the pale edges of the mantle. The frontal margin of the mantle is indented to accommodate and protect the longitudinally ridged, pale rhinophores, which are united at the base. A swollen sensory caruncle is present just in front of the rhinophore bases. The mantle rim contains numerous large defensive glands.

Ventrally, a number of characteristic structures can be seen. To the hyponotum is attached a symmetrical series of gills and lateral lamellae, all probably respiratory in function. The coarse lateral lamellae comprise up to 20 separate, oblique folds on either side, while the delicate plicate gills may contain up to 33 subunits on each side. The foot is white, tinged with red; it often exhibits a median longitudinal groove of unknown significance (Fig. 23b). The head is large and flattened, produced laterally into blunt tentacles.

ANATOMY (Fig. 25a). The radula of an 8 mm specimen (measured in preservative) from the northern Irish Sea was examined. The formula was 24 × 20.1.20. The large median tooth has a prominent cusp with 4–8 irregular denticles on each side. The first lateral on either side is small and somewhat different in shape from the others. Those 6 or 7 lateral teeth nearest the mid-line have a denticulate ridge along the cusp; the marginal teeth are smooth. Eliot (1910) described the radula of a larger specimen, 35 mm long alive, which had the formula 25 × 30.1.30. A 12 mm specimen from the Isle of Man had the formula 29 × 25.1.25. The longest radula on record was from a Plymouth specimen, 17 mm preserved, with the formula 33 × 27.1.27 (Fig. 25a). Substantial jaws are present, having masticatory edges which exhibit a characteristic substructure consisting of a great number of pointed thorn-like denticles. The penis is a simple blunt-tipped cylindrical structure, often extruded on death.

HABITS. This species is always found among beds of *Virgularia mirabilis* Müller, in mud in sheltered localities. It is presumed to feed upon the cnidarian just as *Armina californica* preys upon the related *Renilla koellikeri* in the Pacific Ocean (Bertsch, 1968), but no corroborative observations have been reported. Nothing is known about its reproduction.

DISTRIBUTION (Map 7). The published records of *A. loveni* indicate an unusually restricted distribution, between southern Norway (Lemche, 1929) and the French Biscay coast (Bouchet & Tardy, 1976). The majority of the records have come from Scandinavian coasts, particularly around the entrance to the Baltic Sea, and as far north as the Shetland Islands. Most of the recent British collections are from sheltered northern localities, and from north west Ireland, at depths between 10 and 75 m.

DISCUSSION. *A. loveni* closely resembles *A. neapolitana* (Chiaje, 1824, as *Pleurophyllidia*), common in the Mediterranean and the Bay of Biscay. The 2 species can be separated by examination of the radula: the lateral teeth are smooth hooks in *A. loveni*, but are flattened and divided into irregular cusps in *A. neapolitana*.

Superfamily METARMINOIDEA

Arminaceans which possess smooth, papillate, or lamellate, non-retractile rhinophoral tentacles. Numerous digitiform or arborescent latero-dorsal ceratal processes are present, resembling those of many aeolidaceans, but lacking cnidosacs. The mantle rim is indetectable and has no branchial processes. The anus is dorsal or lateral.

JANOLIDAE

Metarminoid arminaceans in which the digitiform latero-dorsal ceratal processes continue around the frontal margin of the head and contain tributaries of the digestive gland (except in *Bonisa*). The radular formula is *n*.1.*n* (*n* = 20 or more). The anal papilla is postero-dorsal. A sensory caruncle between the rhinophore bases may be present (*Janolus*) or absent (*Proctonotus*).

Janolus Bergh, 1884

(type *Janolus australis* Bergh, 1884)

Janolids in which the rhinophores are lamellate; a sensory caruncle is present between the rhinophore bases. The cerata each contain a tributary of the digestive gland which may subdivide near the apex.

The world's species of *Janolus* have been reviewed by Gosliner (1981a).

63. *Janolus cristatus* (Chiaje, 1841)
Eolis cristata Chiaje, 1841
Janus spinolae Vérany, 1846
Antiopa splendida Alder & Hancock, 1848a
?Janolus flagellatus Eliot, 1906

APPEARANCE IN LIFE (Plate 26, Fig. 23d). This magnificent and most colourful British arminacean may reach 75 mm in length; it is oval in outline and somewhat flattened dorso-ventrally compared with the aeolidaceans which it otherwise resembles. The body is pale brown or cream in colour but the rhinophores and the median sensory caruncle are darker, tending towards amber. The latero-dorsal cerata are numerous, finger-like, inflated and smooth; they form a lateral series on either side and unite around the frontal margin in front of the rhinophores. Through the translucent skin of each ceras can be discerned a slender tributary of the digestive gland. Such tributaries are dark brown, swollen, and often bifurcated or further subdivided at the tip. The apices of the cerata are peppered externally with white guanine pigment and have an iridescent blue quality. Similar white pigment forms a line or blotches down the bare central dorsum, on the metapodium, and on and around the lamellate rhinophores. The bases of the two rhinophores are united by the swollen sensory caruncle. The anal papilla is on the dorsal metapodium.

Ventrally, the head gives rise to a pair of slender digitiform oral tentacles; the frontal margin of the bilaminate propodium is produced into obtuse lateral points.

ANATOMY (Fig. 36c). A general account of the anatomy was given by Alder & Hancock (1845–55), with details later added by Trinchese (1881), Pelseneer (1891), Henneguy (1925) and Schmekel (1970, 1971). The early authors established that the digestive gland branched within the body as well as in the cerata; they claimed with less evidence that each ceras had a terminal pore, but this lacks corroboration. Henneguy described a substantial anal "defensive" gland, the function of which is not known.

The radula of a 15 mm preserved specimen was examined; the formula was 23 × 34.1.34. The median tooth was small and slender while the laterals, with the exception of a few degraded marginals, were elongated and robust, with a slender cusp and a subterminal shoulder or spur. None of the teeth exhibited any subsidiary denticulations. Stout jaws were present.

HABITS. This large species feeds upon erect ectoproct polyzoans, such as *Bugula turbinata* Alder (Tardy, 1962) and possibly *Cellaria* (Graham, 1955).

Spawn has been reported in Britain in the months of April, May, June, July, August, September and December (personal observations and Marine Biological Association, 1957). The egg mass contains a linear series of bead-like white ovoids. Hecht (1895) thought that each of these was an ovum, but it is now known that each is instead a capsule containing up to 250 separate individual spherical ova. This means that a single spawn spiral may contain up to 160,000 ova (Pelseneer, 1935). Development to hatching takes 23 days at 10–12 °C (Kress, 1972). The planktotrophic larva on hatching has a veliger shell of type 1 (Trinchese, 1881).

97

The ectoparasitic copepod *Lichomolgus agilis* (Leydig) has been found on the skin (Pruvot-Fol, 1954); *Janolus cristatus* is also a common host for the internal parasite *Splanchnotrophus*.

DISTRIBUTION (Map 7). This common British sublittoral species reaches its northern distributional limit around the Shetland Islands and southern Norway (Bergen). Further south in the Atlantic Ocean, it occurs along the French coast as far south as Morocco (Gantès, 1956). It is well known in the Mediterranean Sea, both from southern France (Pruvot-Fol, 1955b) and the Bay of Naples (Schmekel, 1968). The recorded maximal depth is 40 m; it cannot survive in turbulent waters because it is fragile, and therefore is normally restricted to hard substrata under clean, shallow, calm water.

64. *Janolus hyalinus* (Alder & Hancock, 1854a)
Antiopa hyalina Alder & Hancock, 1854a

APPEARANCE IN LIFE (Plate 26). This is one of the rarest British nudibranchs. It may reach 26 mm and is highly distinctive. The body is cream coloured with dark brown blotches and a few white patches on the central dorsum. The cerata are numerous, elongated and finger-like, forming a pair of latero-dorsal series which are continuous around the frontal margin anterior to the lamellate rhinophores. Each ceras exhibits numerous, small, wart-like, epidermal papillae which are streaked and spotted with white pigment; between the papillae the skin bears tiny dark brown or orange flecks. Within the cerata on the sides of the body can be discerned a central tributary of the digestive gland extending only half-way up the cerata; no such tributaries can be seen in the anteriorly placed cerata. The bases of the dark speckled rhinophores are united by the swollen, wrinkled, sensory caruncle. The anal papilla is on the dorsal metapodium.

Ventrally, the head exhibits a pair of slender, digitiform, oral tentacles; the frontal margin of the bilaminate propodium is smoothly rounded.

ANATOMY (Fig. 25d). A general description of the reproductive organs was given by Schmekel (1970).

The buccal mass of a Lundy specimen, 13·5 mm in preserved length was prepared for the microscope. The formula was $21 \times 32.1.32$, much larger than previously described (up to $15 \times 13.1.13$, according to Eliot, 1910); the median tooth was rudimentary, lacking an erect cusp, and absent in the newest part of the radula. The lateral teeth were smoothly curved, each cusp ending in a blunt tip (in contrast to the slender, sharp cusps of *J. cristatus*); no subsidiary denticulations were detectable. The denticles reported by Farran (1909), Eliot (1910) and Schmekel & Portmann (1982) are evidently juvenile features.

The jaws are stout and consist of two large wing-like elements joined by a third, median element of rounder shape.

HABITS. This species has been found on *Bugula flabellata* (Thompson) and associated with various species of *Bugula* in Lough Ine, Ireland. In September 1980 a large number of Lough Ine specimens included spawning individuals up to 26 mm, as well as juveniles only 4 mm long. The white egg masses were similar in appearance to those of *J. cristatus*, and uncleaved ova were between 75 and 85 μm in diameter.

The ectoparasitic copepod *Lichomolgus agilis* (Leydig) has been found on the skin.

DISTRIBUTION (Map 8). The most northerly records of this species are from the Isle of Man and Galway Bay, west Ireland. Although not found in the North Sea, it is reported from Plymouth and from the Normandy and Brittany coasts of France (Bouchet & Tardy, 1976). There are rare Mediterranean records from the Bay of Naples (Schmekel, 1968) and Banyuls (Pruvot-Fol, 1954). Gantès (1956) extended the known range to include the Atlantic coast of Morocco. An interesting specimen was identified as *J. hyalinus* from Victoria, Australia (Burn, 1958). There are no other records which might suggest a cosmopolitan distribution, but the prey *Bugula flabellata* has probably reached Australia.

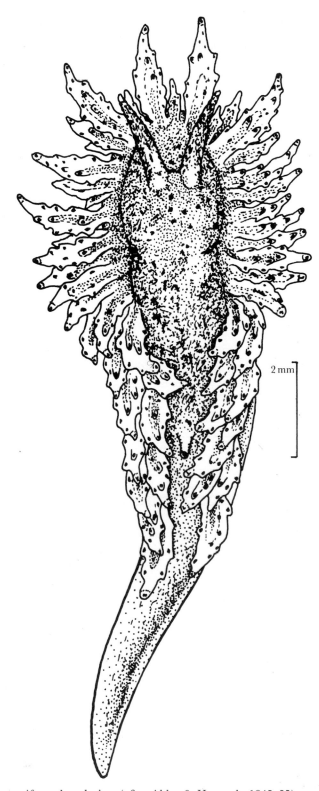

2 mm

FIG. 24. *Proctonotus mucroniferus*, dorsal view (after Alder & Hancock, 1845–55).

Proctonotus Alder, 1844
(type *Venilia mucronifera* Alder & Hancock, 1844)
Janolids in which the rhinophores are wrinkled and papillate, not lamellate; no sensory caruncle is present between the rhinophore bases. The cerata each contain an undivided tributary of the digestive gland.

65. **Proctonotus mucroniferus** (Alder & Hancock, 1844)
   Venilia mucronifera Alder & Hancock, 1844
   Zephyrina pilosa Quatrefages, 1844
   Zephyrina mucronifera Iredale & O'Donoghue, 1923

APPEARANCE IN LIFE (Fig. 24). This is one of the most rare British nudibranchs; it has been seen only 6 or 7 times, always in northern localities, chiefly Irish. In some respects, it resembles *Janolus hyalinus* but it lacks the sensory caruncle so characteristic of that species.

Proctonotus mucroniferus may reach a length of 11 mm, pale brown, marbled with darker brown. The cerata are digitiform and inflated, with numerous epidermal warts or tubercles. They form latero-dorsal series united around the frontal margin. These frontal cerata apparently lack the yellow digestive gland tributaries evident elsewhere. The ceratal tips are curiously truncated and button-like. The rhinophores lack even, serial lamellae, but exhibit irregular corrugations. The anal papilla is on the dorsal metapodium, behind the bare central area of the back.

The head bears short digitiform oral tentacles. The bilaminate propodium is smoothly rounded.

ANATOMY. Alder & Hancock (1844) could not make out many details of the radula. No central tooth was present and the laterals were "numerous, plain, simply hooked". The jaws were examined and, according to these authors, consisted of a pair of lateral plates with a third, smaller plate at the anterior angle.

HABITS. The diet is unknown. Hecht (1895) found spawn in the month of August, remarking that it took the form of a spiral like that of the aeolidacean *Facelina*. The same author also recorded the ectoparasitic copepod *Lichomolgus agilis* (Leydig) on the skin.

DISTRIBUTION (Map 8). Since Alder & Hancock's description of the first specimen, collected in Malahide Bay, near Dublin, there have been two records from western Ireland and one from the Clyde Sea area. All the published reports are over 50 years old, including the only collections outside British waters, off the coast of Brittany, by Quatrefages (1844) and Hecht (1895).

HEROIDAE

Metarminoid arminaceans in which the arborescent latero-dorsal ceratal processes contain subdivided tributaries of the digestive gland; one pair of cerata lies anterior to the smooth rhinophores. The radular formula is 1.1.1. The anal papilla is lateral.

Hero Alder & Hancock, 1855
(type *Cloelia formosa* Lovén, 1841)
With the characters of the family.

66. **Hero formosa** (Lovén, 1841)
   Cloelia formosa Lovén, 1841
   Cloelia trilineata M. Sars, 1850
   Hero formosa arborescens Eliot, 1905

APPEARANCE IN LIFE (Plate 25, Fig. 23a). The body may reach 38 mm in length, white or pink in colour, with 3 opaque lines down the back. On each side of the back is a row of up to 8

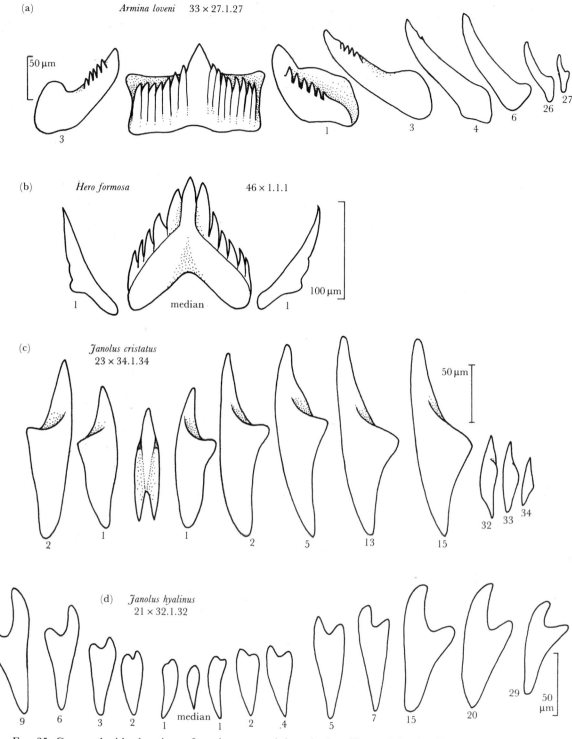

(a)   *Armina loveni*   33 × 27.1.27

(b)   *Hero formosa*   46 × 1.1.1

(c)   *Janolus cristatus*
23 × 34.1.34

(d)   *Janolus hyalinus*
21 × 32.1.32

FIG. 25. Camera lucida drawings of arminacean radulae: *Armina*, *Hero* and *Janolus*. Teeth are numbered from the mid-line.

(a) *Armina loveni*, preserved length 17 mm, Plymouth, Devon.
(b) *Hero formosa*, preserved length 13 mm, Isle of Man.
(c) *Janolus cristatus*, preserved length 15 mm, Aberdeen, Scotland.
(d) *Janolus hyalinus*, preserved length 13·5 mm, Lundy Island.

pedunculate, dichotomously divided, yellow cerata, giving an arborescent appearance to the dorsum. The first pair of cerata lies anterior to the rhinophores. Branches of the digestive gland subdivide within the cerata; these branches are grey or brown at the base of each ceras, but may be reddish at the tips. The rhinophores are smooth. The anal papilla is lateral, on the right side, behind the 2nd ceras; the genital openings lie close to the base of the right rhinophore. A very conspicuous feature is the presence on the head of a pair of greatly enlarged, curved oral tentacles.

ANATOMY (Fig. 25b). The radulae of a number of specimens from the Isle of Man were examined. They were all long and slender, each row always exhibiting the formula 1.1.1. At 5 mm the number of rows was 55; 6 mm, 48; 15 mm, 51; 26 mm, 47. The median tooth has a prominent cusp and a row of 5–6 denticles on either side. Each lateral tooth has a smooth, slender spine, and a clubbed base. A strong pair of denticulate jaws is present.

HABITS. There are few published records of this species, although it may be locally common, for instance off the Isle of Man, feeding on calyptoblastic (*Abietinaria abietina* (L.) and *Hydrallmania falcata* (L.)) and gymnoblastic (*Tubularia indivisa* L. and *T. larynx* Ellis & Solander) hydroids (Miller, 1961; Thompson, 1964). Specimens found in the Sound of Mull were feeding on *Nemertesia ramosa* (Lamouroux) and lacked the deep pink coloration of animals taken from *Tubularia*.

Spirals of spawn are deposited from April to August, up to 45,000 ova in each spiral (Miller, 1958). The ova measure 75–89 µm in diameter and develop to hatching in 17–19 days at 10 °C to give veliger larvae of type 1 (Thompson, 1961a, 1967). A normal table of embryonic development is given in volume I (p. 85) and the newly hatched veliger is illustrated in volume I (Fig. 41(f), p. 83).

DISTRIBUTION (Map 8). Despite early records from Plymouth, south west Ireland and Anglesey, recent collections suggest that this species becomes common only to the north of the Isle of Man, around the Scottish Isles, and along the Norwegian coast to Nordkapp, within the Arctic Circle (Lemche, 1941). Unlike other cold-water nudibranchs of the north Atlantic, it has not been reported from Iceland or the Faeroes.

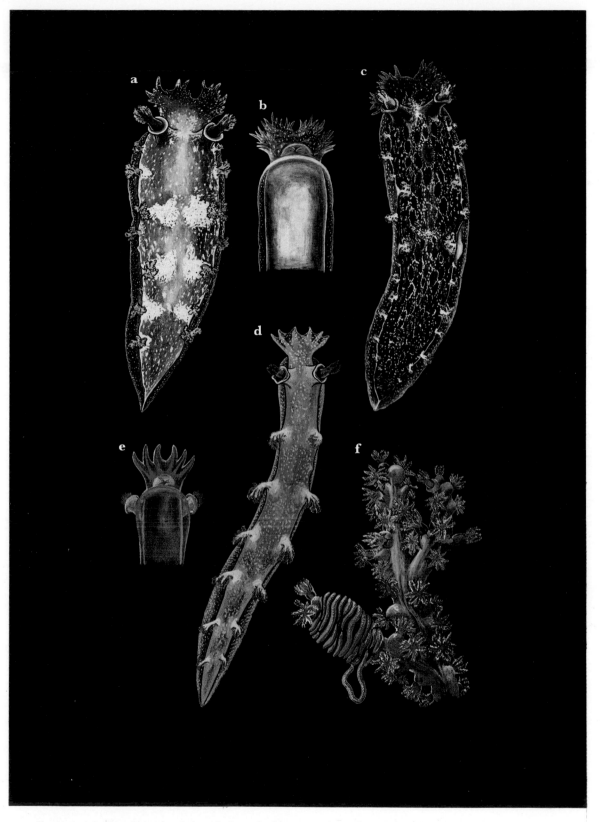

PLATE 1. (a) *Tritonia hombergi*, dorsal view of a 36 mm specimen from Menai Bridge, N. Wales, (b) ventral view of the head of the same, (c) dorsal view of a 48 mm specimen from Mull, Scotland.
(d) *Tritonia nilsodhneri*, dorsal view of a 34 mm specimen from Lundy Island, (e) ventral view of the head of the same, (f) spawn mass of the same, attached to the gorgonian *Eunicella*.

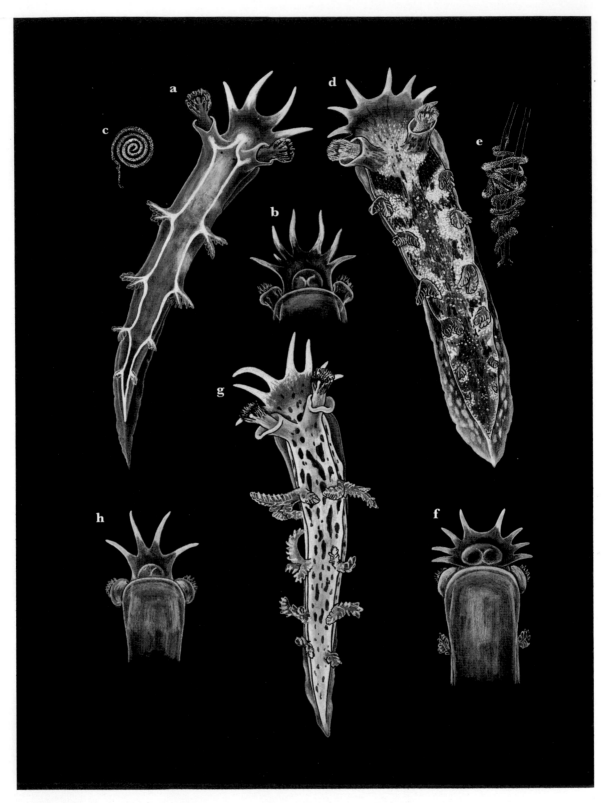

PLATE 2. (a) *Tritonia lineata*, dorsal view of an 18 mm specimen from Skomer Island, S. Wales, (b) ventral view of the head of the same, (c) spawn mass of the same.

(d) *Tritonia plebeia*, dorsal view of a 16 mm specimen from Mull, Scotland, (e) spawn mass of the same, (f) ventral view of the head of the same.

(g) *Tritonia manicata*, dorsal view of an 11 mm specimen from Lundy Island, (h) ventral view of the head of the same.

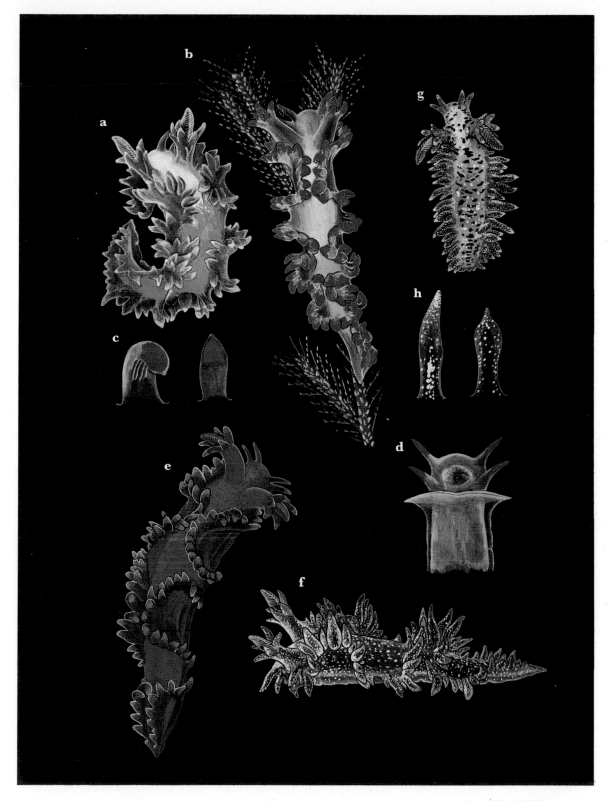

PLATE 3. (a) *Lomanotus genei*, dorsal view of a swimming specimen from Anglesey, N. Wales, 24 mm in length, (b) dorsal view of a 35 mm specimen from Strangford Lough, Northern Ireland, (c) pallial papillae of the same, (d) ventral view of the head of the same, (e) dorsal view of a 60 mm specimen from the Sound of Mull, Scotland.

(f) *Lomanotus marmoratus*, left lateral view of a 39 mm specimen from Donegal, Ireland, (g) an 8 mm juvenile from Lundy Island, (h) pallial papillae of the Irish specimen.

PLATE 4. Nudibranch radulae (Dendronotacea); scanning electron micrographs.
(a) *Tritonia hombergi*
(b) *Tritonia lineata*
(c) *Lomanotus genei*
(d) *Dendronotus frondosus.*

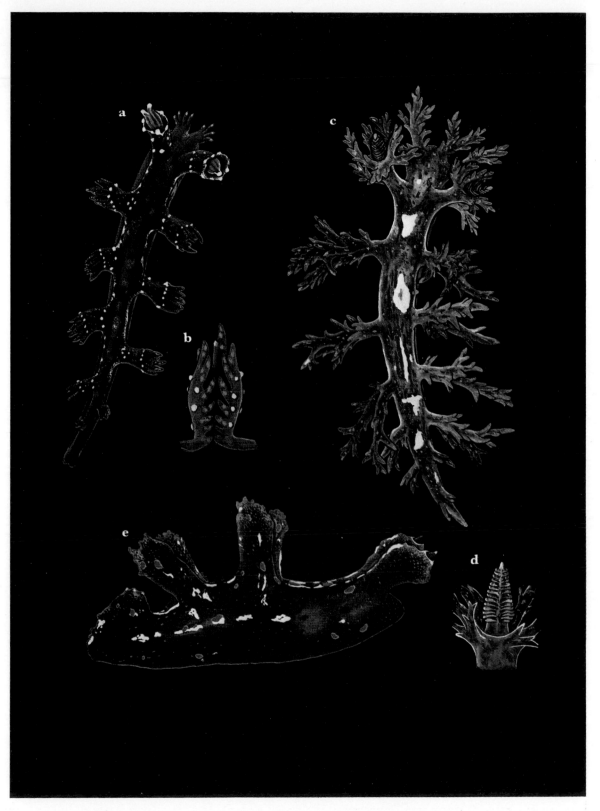

PLATE 5. (a) *Hancockia uncinata*, dorsal view of a 9 mm specimen from Sidmouth, Devon, (b) lateral view of a ceras of the same.

(c) *Dendronotus frondosus*, dorsal view of a 12 mm specimen from Menai Bridge, N. Wales, (d) rhinophore and sheath of the same.

(e) *Scyllaea pelagica*, right lateral view of a 13 mm specimen from Corpus Christi, Texas, U.S.A.

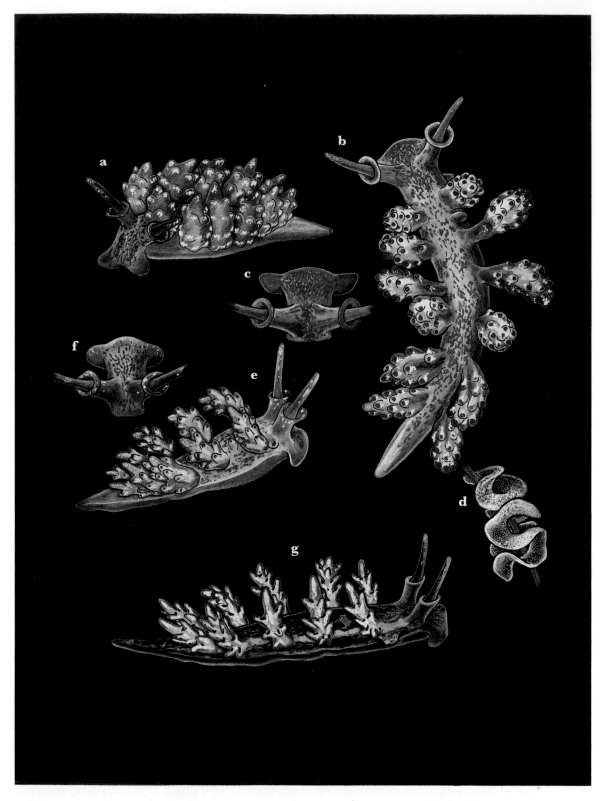

PLATE 6. (a) *Doto lemchei*, left lateral view of a 10 mm specimen from Lundy Island.
(b) *Doto coronata*, dorsal view of an 18 mm specimen from Mull, Scotland, (c) dorsal view of the head of the same, (d) spawn mass of the same.
(e) *Doto maculata*, right lateral view of a 6 mm specimen from Galway Bay, Ireland, (f) dorsal view of the head of the same.
(g) *Doto koenneckeri*, right lateral view of a 10 mm specimen from Galway Bay, Ireland.

PLATE 7. (a) *Doto fragilis*, dorsal view of a 26 mm specimen from Skomer Island, S. Wales, (b) *Doto fragilis*, dorsal view of a 34 mm specimen from Lundy Island, (c) spawn mass of the same.
    (d) *Doto cuspidata*, dorsal view of a 14 mm specimen from Lundy Island.
    (e) *Doto pinnatifida*, right lateral view of a 24 mm specimen from Lundy Island.

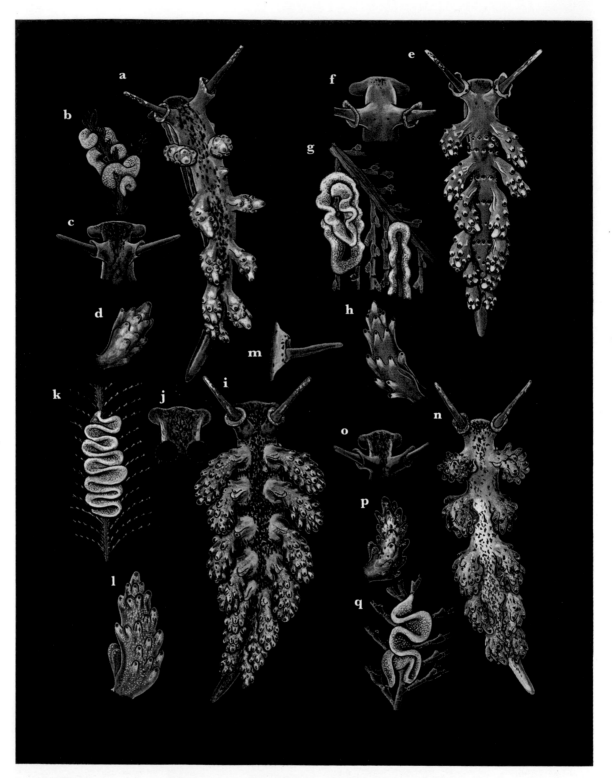

PLATE 8. (a) *Doto eireana*, dorsal view of a 7 mm specimen from Lundy Island, (b) spawn mass of the same, (c) dorsal view of the head of the same, (d) a single ceras of the same.

(e) *Doto tuberculata*, dorsal view of a 10 mm specimen from Lundy Island, (f) dorsal view of the head of the same, (g) spawn masses of the same, (h) a single ceras of the same.

(i) *Doto dunnei*, dorsal view of a 25 mm specimen from Lundy Island, (j) dorsal view of the head of the same, (k) spawn mass of the same, (l) a single ceras of the same, (m) extruded penis of the same.

(n) *Doto millbayana*, dorsal view of an 11 mm specimen from Lundy Island, (o) dorsal view of the head of the same, (p) a single ceras of the same, (q) spawn mass of the same.

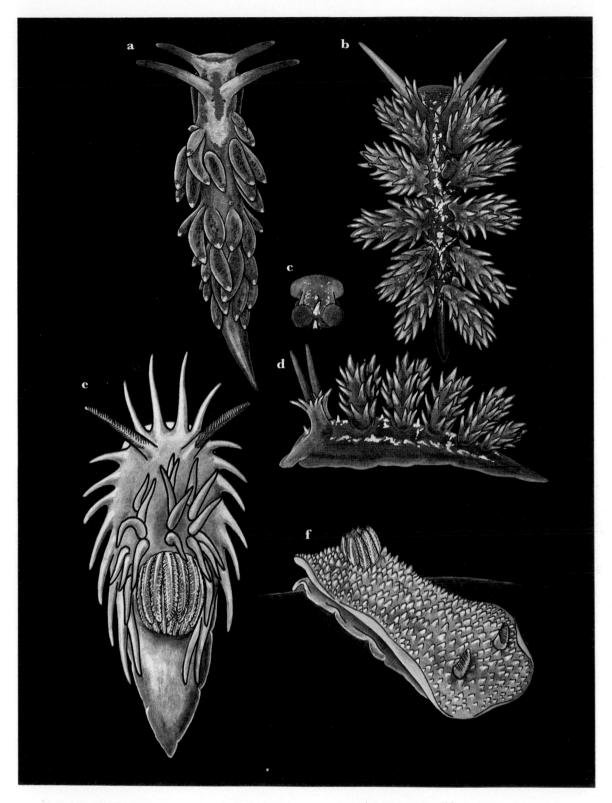

PLATE 9. (a) *Cuthona genovae*, dorsal view of a 6 mm specimen from Lough Ine, Ireland.
  (b) *Doto hystrix*, dorsal view of a 12 mm specimen from Sherkin Island, (c) dorsal view of the head of the same, (d) left lateral view of the same.
  (e) *Okenia leachi*, dorsal view of a 16 mm preserved specimen from the Celtic Sea, 40 miles south of Co. Cork, Ireland.
  (f) *Aldisa zetlandica*, dorsal view of a 15 mm specimen from Galway Bay, Ireland.

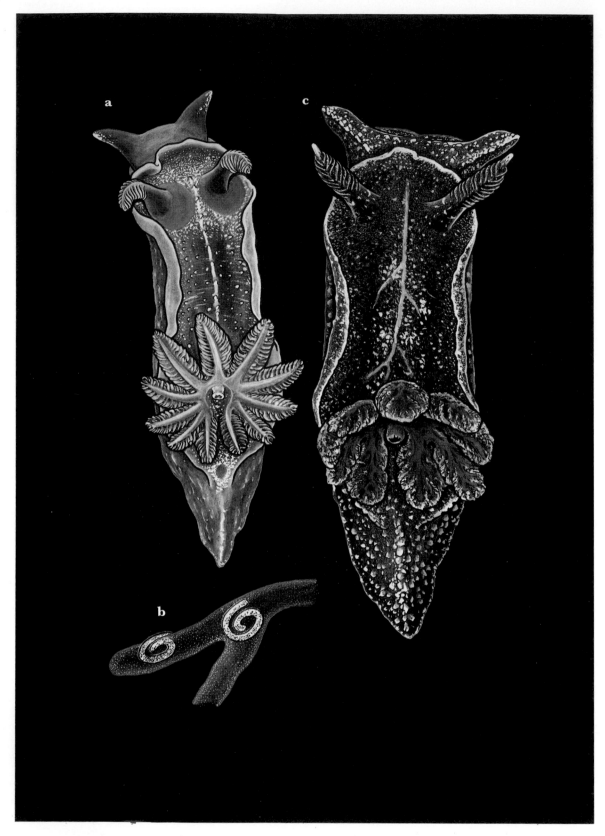

PLATE 10. (a) *Goniodoris nodosa*, dorsal view of a 22 mm specimen from Lundy Island, (b) spawn mass of the same, attached to the polyzoan *Alcyonidium*.
(c) *Goniodoris castanea*, dorsal view of a 40 mm specimen from the Sound of Mull, Scotland.

PLATE 11. (a) *Okenia elegans*, ventral view of the head of a 46 mm specimen from Lundy Island, (b) dorsal view of the same, (c) typical view of a specimen preying upon the tunicate *Polycarpa*, (d) spawn mass. (e) *Okenia pulchella*, left lateral view of a 9 mm specimen from Lundy Island.

PLATE 12. (a) *Ancula gibbosa*, dorsal view of an 18 mm specimen from Martinshaven, S. Wales, (b) ventral view of the head of the same, (c) left lateral view of the same, (d) spawn mass of the same.

(e) *Trapania maculata*, dorsal view of a 17 mm specimen from Portland Bill, Dorset, (f) ventral view of the head of the same.

(g) *Trapania pallida*, dorsal view of a 13 mm specimen from Lundy Island.

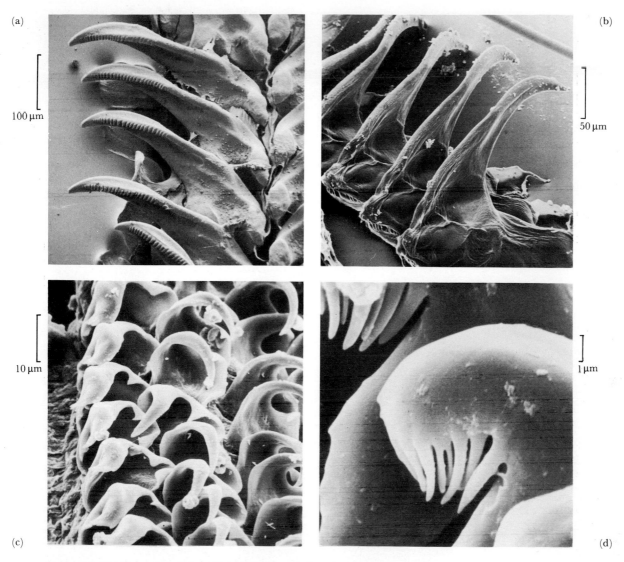

PLATE 13. Nudibranch radulae (Doridacea); scanning electron micrographs.
   (a) *Goniodoris nodosa.*
   (b) *Goniodoris castanea.*
   (c) *Rostanga rubra*, mid-lateral teeth.
   (d) *Rostanga rubra*, first lateral tooth.

PLATE 14. (a) *Acanthodoris pilosa*, dorsal view of an 18 mm specimen from Oban, (b) spawn mass of the same.

(c) *Adalaria proxima*, dorsal view of an 11 mm specimen from Menai Bridge, N. Wales.

(d) *Adalaria loveni*, dorsal view of an 8 mm specimen from Kerrera, Scotland, (e) spawn mass of the same.

(f) *Onchidoris bilamellata*, dorsal view of a 14 mm specimen from Kerrera, Scotland.

(g) *Onchidoris muricata*, dorsal view of a 14 mm specimen from Porlock, Somerset.

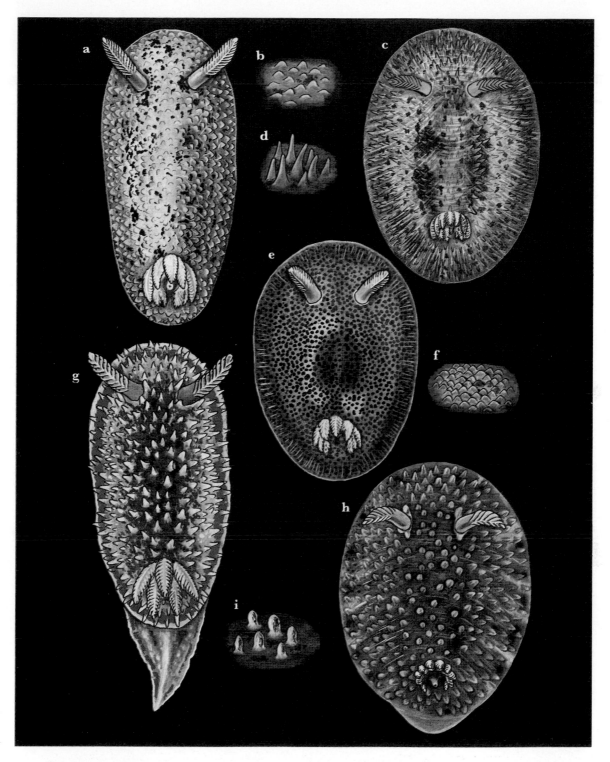

PLATE 15. (a) *Onchidoris oblonga*, dorsal view of an 8 mm specimen from Lundy Island, (b) mantle papillae of the same.

(c) *Onchidoris depressa*, dorsal view of a 7 mm specimen from Scapa Flow, Orkney, (d) mantle papillae of the same.

(e) *Onchidoris pusilla*, dorsal view of a 5 mm specimen from Lundy Island, (f) mantle papillae of the same.

(g) *Onchidoris luteocincta*, dorsal view of an 11 mm specimen from Lundy Island.

(h) *Onchidoris sparsa*, dorsal view of an 8 mm specimen from the Orkney Islands, (i) mantle papillae of the same.

PLATE 16. Nudibranch radulae (Doridacea); scanning electron micrographs.

(a) *Cadlina laevis.*
(b) *Acanthodoris pilosa.*
(c) *Crimora papillata.*
(d) *Polycera quadrilineata.*

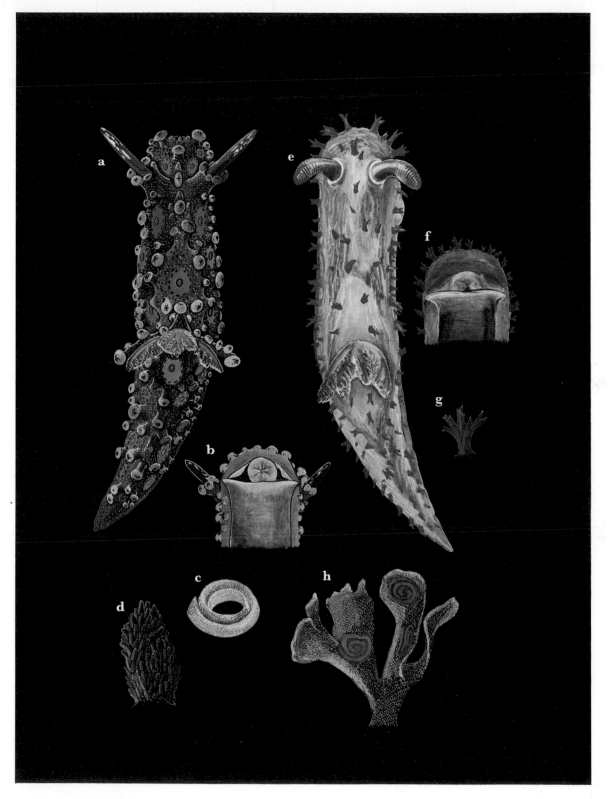

PLATE 17. (a) *Aegires punctilucens*, dorsal view of a 15 mm specimen from Martinshaven, S. Wales, (b) ventral view of the head of the same, (c) spawn mass of the same, (d) a clump of the preferred sponge, the calcareous *Leucosolenia*.

(e) *Crimora papillata*, dorsal view of a 23 mm specimen from Skomer Island, S. Wales, (f) ventral view of the head of the same, (g) pallial process from the anterior margin, (h) spawn mass attached to the polyzoan *Flustra*.

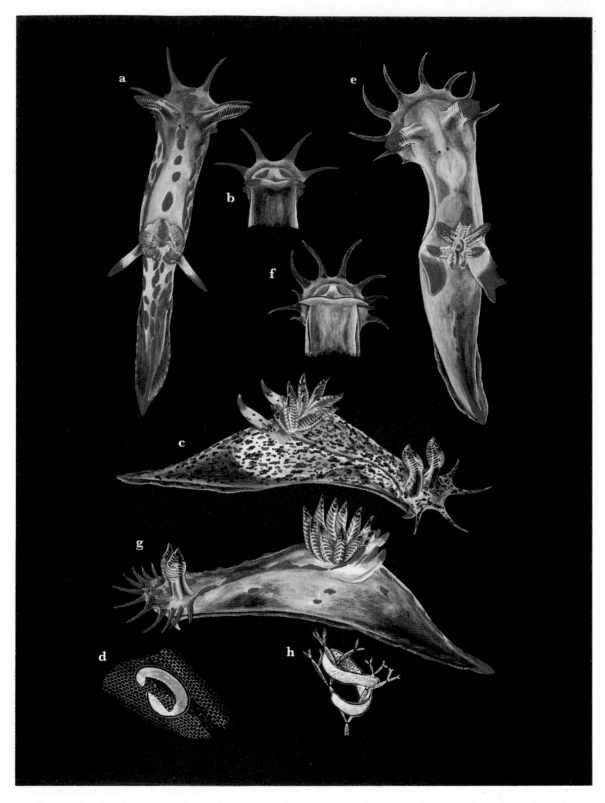

PLATE 18. (a) *Polycera quadrilineata*, dorsal view of an 18 mm specimen from Skomer Island, S. Wales, (b) ventral view of the head of the same, (c) right lateral view of a speckled individual from Lundy Island, 12 mm in length, (d) spawn mass attached to the polyzoan *Membranipora*.

(e) *Polycera faeroensis*, dorsal view of a 32 mm specimen from Skomer Island, S. Wales, (f) ventral view of the head of the same, (g) left lateral view of a 38 mm specimen from the Isle of Man, with extra body pigment and multiple anal papillae, (h) spawn mass attached to the polyzoan *Crisia*.

PLATE 19. (a) *Limacia clavigera*, dorsal view of a 14 mm specimen from Lundy Island, (b) spawn mass of the same, (c) ventral view of the head of the same.

(d) *Thecacera pennigera*, dorsal view of a 16 mm specimen from Lundy Island, (e) right lateral view of an 11 mm specimen from Watermouth Cove, Devon, (f) ventral view of the head of the same, (g) spawn mass of the same.

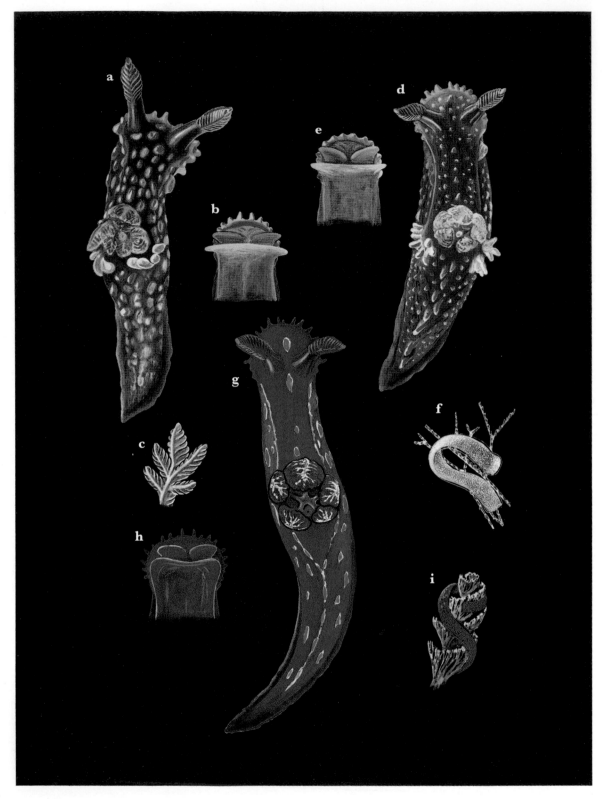

PLATE 20. (a) *Palio nothus*, dorsal view of an 8 mm specimen from Porlock, Somerset, (b) ventral view of the head of the same, (c) bipinnate gill of the same.

(d) *Palio dubia*, dorsal view of a 14 mm specimen from Strangford Lough, Northern Ireland, (e) ventral view of the head of the same, (f) spawn mass.

(g) *Greilada elegans*, dorsal view of a 29 mm specimen from Lundy Island, (h) ventral view of the head of the same, (i) spawn mass attached to the polyzoan *Bugula*.

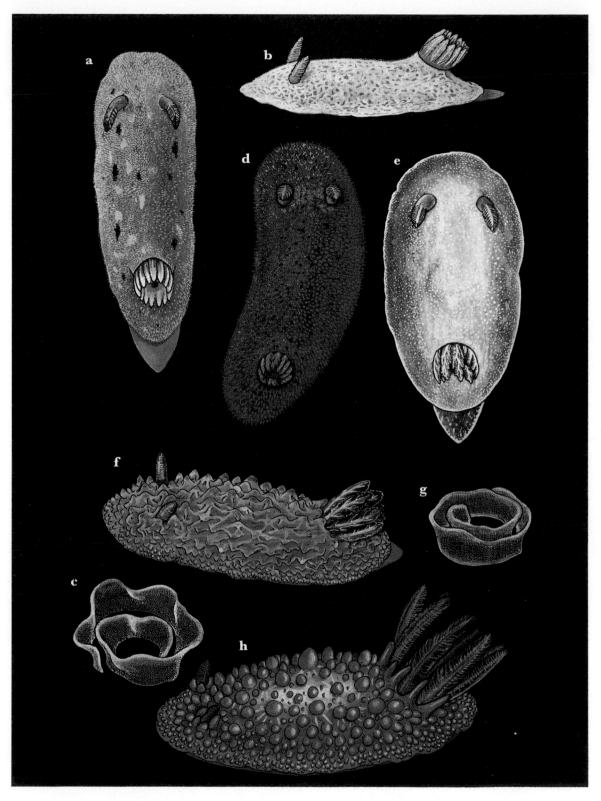

PLATE 21. (a) *Jorunna tomentosa*, dorsal view of a 32 mm specimen from Skomer Island, S. Wales, (b) left lateral view of a 34 mm specimen of *Jorunna tomentosa* from Donegal, Ireland, (c) spawn mass of the same.
 (d) *Rostanga rubra*, dorsal view of a 15 mm specimen from Skomer Island, S. Wales.
 (e) *Cadlina laevis*, dorsal view of a 16 mm specimen from Lundy Island.
 (f) *Doris sticta*, left lateral view of a 34 mm specimen from Lundy Island, (g) spawn mass of the same.
 (h) *Doris verrucosa*, left lateral view of a 34 mm specimen from Arcachon, France.

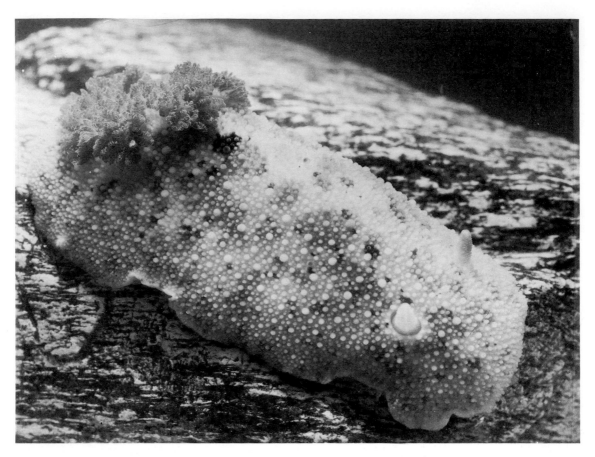

PLATE 22. *Archidoris pseudoargus*. (Photographed at Dale, S. Wales, by Heather Angel.)

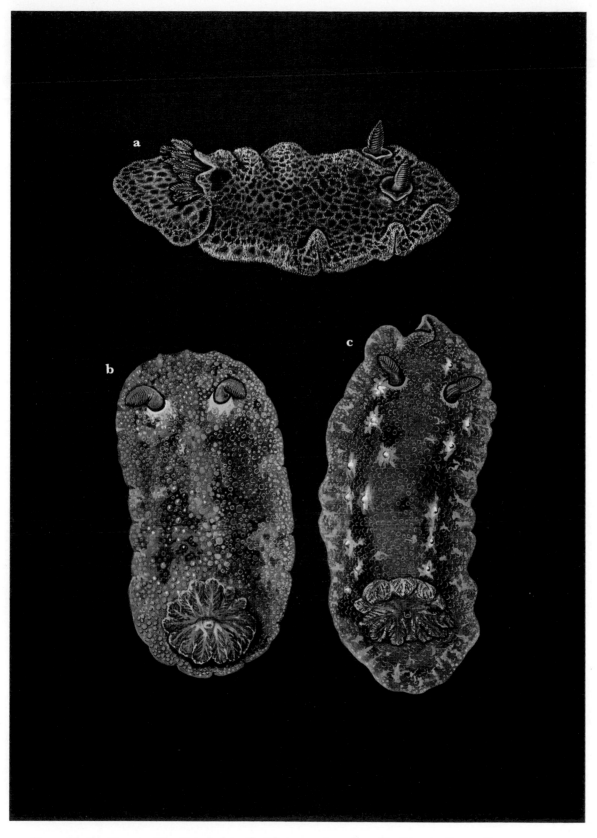

PLATE 23. (a) *Atagema gibba*, dorsal view of a 68 mm specimen from Porthkerris, Cornwall.
(b) *Archidoris pseudoargus*, dorsal view of a 45 mm specimen from Port Erin, Isle of Man.
(c) *Discodoris planata*, dorsal view of a 60 mm specimen from Donegal, Ireland.

PLATE 24. *Discodoris planata*. (Photographed on the Channel Island of Jersey by Heather Angel.)

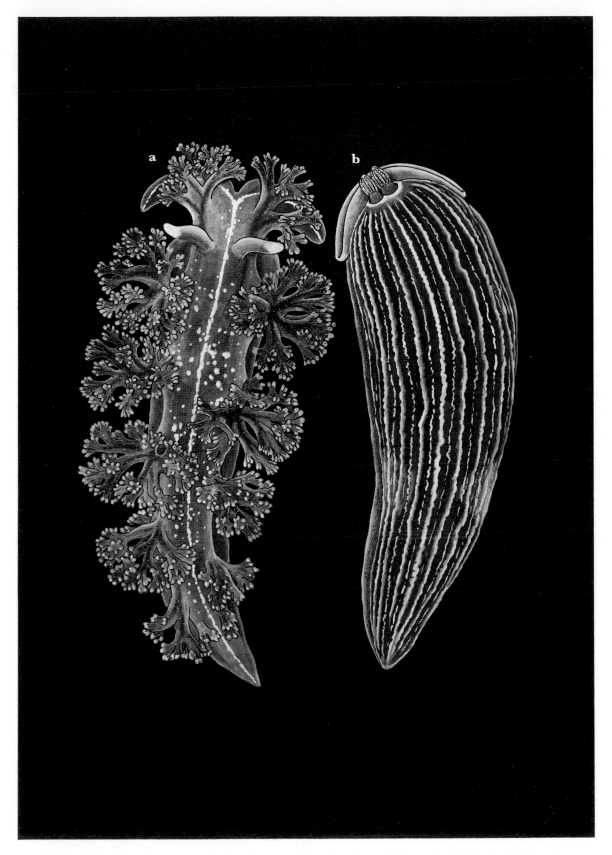

PLATE 25. (a) *Hero formosa*, dorsal view of a 32 mm specimen from the Sound of Mull, Scotland.
(b) *Armina loveni*, dorsal view of a 32 mm specimen from Fife Ness, Scotland.

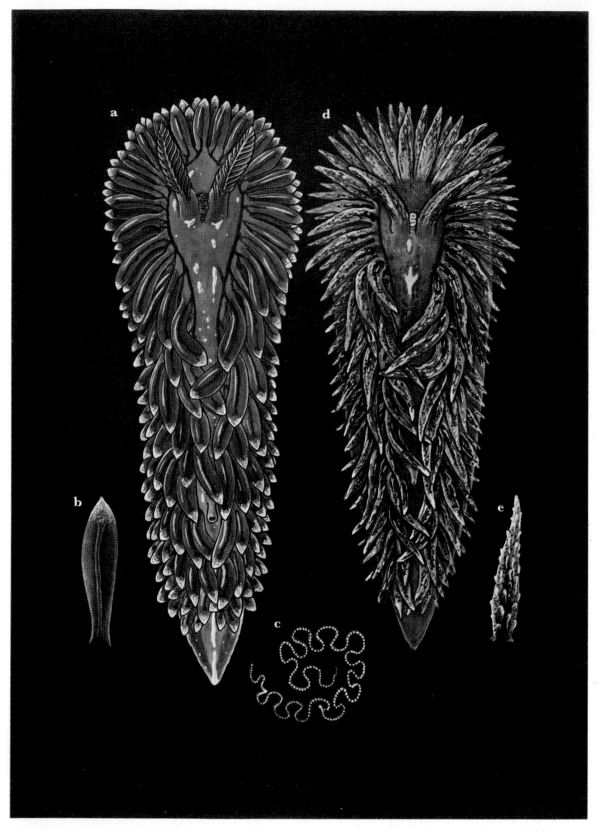

PLATE 26. (a) *Janolus cristatus*, dorsal view of a 39 mm specimen from Lundy Island, (b) a ceras of the same, (c) a spawn mass of the same (26 mm in diameter).
　　(d) *Janolus hyalinus*, dorsal view of a 24 mm specimen from Lundy Island, (e) a ceras of the same.

PLATE 27. (a) *Coryphella pellucida*, dorsal view of a 23 mm specimen from Scapa Flow, Orkney. (b) *Coryphella lineata*, dorsal view of a 30 mm specimen from Lundy Island, (c) cerata from different specimens of *C. lineata*, showing variation in the amount of white pigment, (d) ventral view of the head of *C. lineata*.

PLATE 28. (a) *Coryphella pedata*, dorsal view of a 33 mm specimen from the Lizard Peninsula, Cornwall.
(b) *Coryphella browni*, dorsal view of a 42 mm specimen from Lundy Island, (c) spawn mass of the same.
(d) *Coryphella verrucosa*, dorsal view of a 23 mm specimen from Scapa Flow, Orkney.
(e) *Coryphella gracilis*, dorsal view of a 14 mm specimen from Watermouth Cove, Devon.

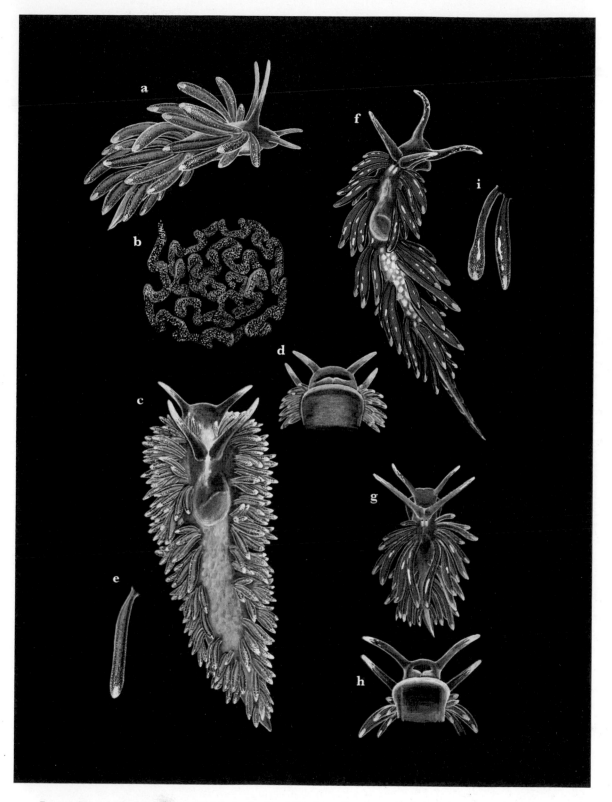

PLATE 29. (a) *Cuthona nana*, right lateral view of a juvenile, 4 mm in length, from Helford Passage, Cornwall, (b) spawn mass of the same, (c) dorsal view of a 22 mm specimen from Helford Passage, (d) ventral view of the head of the same, (e) a single ceras of the same.

(f) *Cuthona pustulata*, dorsal view of an 18 mm specimen from Skomer Island, S. Wales, (g) dorsal view of a 4 mm specimen from Lundy Island, (h) ventral view of the head of the same, (i) two cerata of the same, including one of a swollen variety.

PLATE 30. (a) *Cuthona caerulea*, dorsal view of an 11 mm specimen from Lundy Island, (b) dorsal view of a 13 mm specimen from Port Erin, Isle of Man, (c) dorsal view of a 14 mm specimen from Skomer Island, S. Wales, (d) dorsal view of an 8 mm specimen from the Scilly Isles, (e) a single ceras of a specimen from Skomer Island, S. Wales.

(f) *Cuthona viridis*, dorsal view of an 18 mm specimen from the Sound of Mull, Scotland (g) ventral view of the head of the same, (h) spawn mass of the same.

(i) *Cuthona foliata*, dorsal view of a 7 mm specimen from Lundy Island, (j) spawn mass of the same.

PLATE 31. (a) *Cuthona rubescens*, dorsal view of a 10 mm specimen from Skomer Island, S. Wales.
(b) *Cuthona amoena*, dorsal view of a 10 mm specimen from Oxwich Bay, S. Wales.
(c) *Cuthona gymnota*, dorsal view of a 14 mm specimen from Lundy Island, (d) spawn mass of the same.
(e) *Cuthona concinna*, dorsal view of a 7 mm specimen from Strangford Lough, Northern Ireland,
(f) spawn mass of the same.
(g) *Tenellia adspersa*, dorsal view of a 6 mm specimen from the Fleet, Dorset, (h) spawn mass of the same.
(i) *Tergipes tergipes*, right lateral view of a 7 mm specimen from Lundy Island.

PLATE 32. (a) *Eubranchus tricolor*, dorsal view of a 30 mm specimen from Skomer Island, S. Wales, (b) spawn mass of the same.

(c) *Eubranchus farrani*, dorsal view of a 12 mm specimen from Plymouth, Devon, (d) dorsal view of a 14 mm specimen from the Fleet, Dorset, (e) spawn mass deposited by a Fleet specimen, (f) dorsal view of a 9 mm specimen from Skomer Island, S. Wales, (g) right lateral view of an 18 mm specimen from Skomer Island, S. Wales, (h) right lateral view of an 18 mm specimen from Skomer Island, S. Wales.

PLATE 33. *Eubranchus pallidus*, dorsal view of a 19 mm specimen from Salcombe, Devon.

PLATE 34. (a) *Eubranchus cingulatus*, dorsal view of a 14 mm specimen from Scabbacombe Head, S. Devon, (b) dorsal view of a 6 mm specimen from the Lizard Peninsula, Cornwall, (c) spawn mass of the same. (d) *Eubranchus exiguus*, dorsal view of a 6 mm specimen from Lundy Island. (e) *Eubranchus doriae*, left lateral view of a 5 mm specimen from Lundy Island. (f) *Eubranchus vittatus*, right lateral view of a 29 mm specimen from Salcombe, Devon.

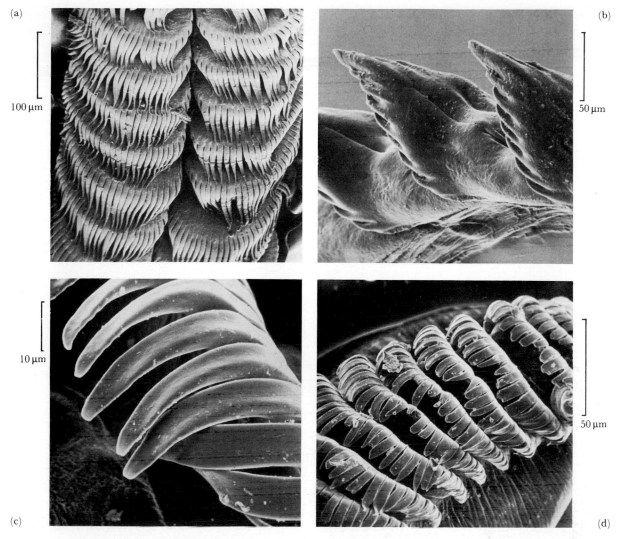

(a)

(b)

(c)

(d)

PLATE 35. Nudibranch radulae (Aeolidacea); scanning electron micrographs.
  (a) *Aeolidiella glauca*.
  (b) *Facelina coronata*.
  (c) *Aeolidia papillosa*, detail of denticulations.
  (d) *Aeolidia papillosa*, lower magnification.

PLATE 36. (a) *Calma glaucoides*, dorsal view of a 23 mm specimen from Penzance, Cornwall, (b) *Calma glaucoides*, dorsal view of a 9 mm specimen from Hope Cove, S. Devon.

(c) *Fiona pinnata*, dorsal view of a 25 mm specimen from New South Wales, Australia, (d) spawn mass of the same, (e) a single ceras of the same.

(f) *Embletonia pulchra*, dorsal view of a 4 mm specimen from Weymouth, Dorset.

(g) *Pseudovermis boadeni*, dorsal view of a 3 mm specimen from Traeth Bychan, N. Wales.

PLATE 37. (a) *Facelina coronata*, dorsal view of a 31 mm specimen from Martinshaven, S. Wales, (b) ventral view of the head of the same, (c) feeding upon *Tubularia*, (d) spawn mass of *F. coronata*.

(e) *Facelina bostoniensis*, dorsal view of a 26 mm specimen from Lundy Island, (f) ventral view of the head of the same.

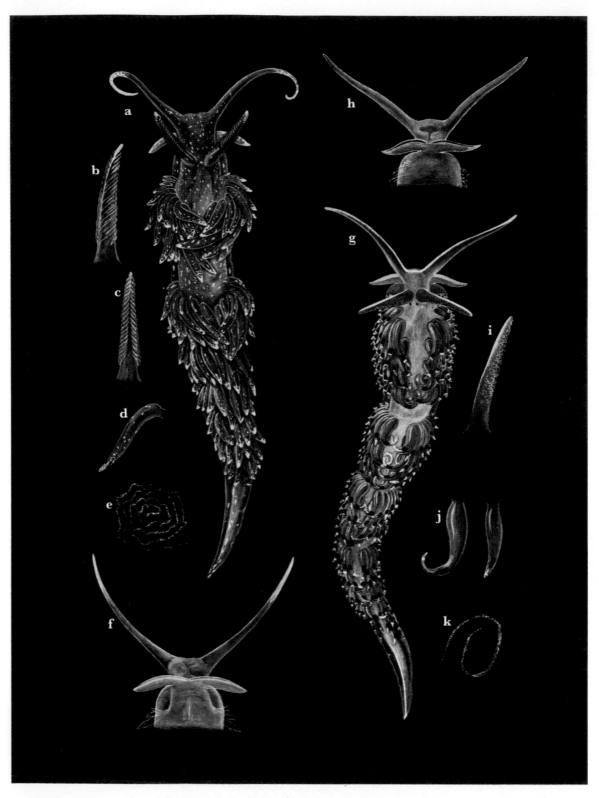

PLATE 38. (a) *Facelina annulicornis*, dorsal view of a 38 mm specimen from Lundy Island, (b) postero-lateral view of a rhinophore of the same, (c) anterior view of a rhinophore of the same, (d) a single ceras of the same, (e) spawn mass of the same, (f) ventral view of the head of the same.

    (g) *Caloria elegans*, dorsal view of a 36 mm specimen from Lundy Island, (h) ventral view of the head of the same, (i) lateral view of a rhinophore of the same, (j) two cerata of the same, (k) spawn mass of the same.

PLATE 39. (a) *Facelina dubia*, dorsal view of a 38 mm specimen from Lough Ine, Ireland, (b) dorsal view of a 17 mm specimen from the same locality, (c) spawn mass of the same.

(d) *Dicata odhneri*, dorsal view of a 13 mm specimen from Lough Ine, Ireland, (e) spawn mass of the same.

PLATE 40. (a) *Favorinus branchialis*, dorsal view of a 15 mm specimen from Watermouth Cove, N. Devon, (b) dorsal view of a 12 mm specimen from Lundy Island, (c) a single ceras from a heavily pigmented individual, (d) a single ceras from an unusual individual which was lightly pigmented overall and had orange/pink digestive gland diverticula, (e) ventral view of the head of *F. branchialis*, (f) spawn mass of the same.

(g) *Favorinus blianus*, dorsal view of a 23 mm specimen from Watermouth Cove, N. Devon, (h) ventral view of the head of the same.

PLATE 41. (a) *Aeolidia papillosa*, dorsal view of a 29 mm specimen from Dale, S. Wales, (b) spawn mass of the same.

(c) *Aeolidiella alderi*, dorsal view of a 24 mm specimen from Donegal, Ireland, (d) spawn mass of the same.

(e) *Aeolidiella sanguinea*, dorsal view of a 36 mm specimen from Donegal, Ireland, (f) spawn mass of the same.

(g) *Aeolidiella glauca*, dorsal view of a 29 mm specimen from Skomer Island, S. Wales, (h) spawn mass of the same.

Suborder IV AEOLIDACEA

This large suborder contains 21 families of nudibranchs (12 of them British), all bearing clusters, groups or rows of elongated finger-like smooth latero-dorsal ceratal processes. Primitively these cerata arise from a prominent notal ridge (which in the aeolidaceans is probably always a pedal structure, contrary to the notal or pallial rim of the doridaceans and dendronotaceans), but in advanced aeolidaceans this ridge is difficult to discern. Aeolidaceans exhibit a lateral anal position, shifted dorsally in advanced families. Propodial, oral and rhinophoral tentacles may all be present. Rhinophore sheaths are lacking.

Aeolidaceans feed upon hydrozoan, scyphozoan and anthozoan coelenterates, harvesting the nematocysts of the prey unscathed for use as defensive weapons. The nematocysts are transferred through special hepatic channels within the cerata to the cnidosacs situated near the ceratal tips. Some aeolidaceans attack other opisthobranchs, either taking the adults (e.g. the aggressive Pacific American *Phidiana pugnax* Lance, 1962) or the spawn (e.g. the British *Favorinus branchialis* (Rathke)). *Favorinus ghanensis* Edmunds feeds, however, on the bryozoan *Zoobotryon* in western Africa. *Fiona pinnata* (Eschscholtz) is a very unusual aeolidacean of wide geographical distribution, attacking stalked barnacles such as *Lepas*.

Jaws are usually well developed. In the superfamily Protoaeolidoidea the radula may be as broad as 2.5.1.5.2, but in the more advanced families which constitute the superfamily Euaeolidoidea (see volume I, p. 22) the radular formula is either triseriate or uniseriate. The larval shell may be of veliger types 1 or 2 (Thompson, 1961a).

NOTES ON THE CLASSIFICATION OF THE AEOLIDACEA

We are a long way from the attainment of a stable familial and superfamilial classification of the nudibranch suborder Aeolidacea. The aeolids have been a magnet for incompetent taxonomists who have found it easier to propose new names than to evaluate those already existing in the literature. As a result, we have probably ten times more generic and higher categories than are necessary. Few specialists would disagree that a great task of pruning and refinement awaits us, but it will be difficult to achieve this, because supposedly objective specialists are often surprisingly dogmatic and traditionalist about nomenclatural changes which may be proposed by another school.

No lasting scheme can ignore the fact that the larval shells of some families (for example, Embletoniidae, Tergipedidae, Eubranchidae, Calmidae and Fionidae) are of the inflated ovoid veliger type 2, whereas those of the remainder are of the small coiled veliger type 1 (Thompson, 1961a). No intermediates exist, and it is clear that this distinction reflects a phylogenetic dichotomy. Other morphological features which are useful in super-generic taxonomy of the Aeolidacea are the arrangement of the liver lobes from which the cerata sprout, the jaws and the radula (triseriate in the Coryphellidae, uniseriate in the Aeolidiidae, to take two examples), and the detailed arrangement of the anterior genital mass.

Odhner (1939) laid considerable emphasis on the position of the anus in relation to the ceratal pattern, and proposed to divide the aeolidacean families into three tribes, characterized as follows: Tribe 1, Pleuroprocta, in which the anus is on the right side of the body, lateral to the row or rows of cerata, e.g. Notaeolidiidac. Tribe 2, Acleioprocta, in which the anus has moved dorsally into the gap or space between those cerata emanating from the right lobe of the digestive gland and those from the (much larger) left lobe. Tribe 3, Cleioprocta, in which the anus has become surrounded by the cerata emanating from the anterior part of the left lobe of the digestive gland (this is the most diverse tribe).

In a thoughtful review, Miller (1971) disagrees with this tribal system because their use would cut across several aeolidacean families of good standing: at least two of the three tribal conditions occur in, for example, the Tergipedidae, Facelinidae or the Aeolidiidae. Gosliner (1980) lists these exceptions, for example, *Selva* is a cuthonid (Acleioprocta) but it has the anus in the clieoproct position (Edmunds, 1964). Nonetheless, the branching of the digestive gland and the position of the anus remain important and reliable features for delineating the genera. Furthermore, Miller considers that the most important features for separating aeolids at the species level are colour, pattern, and penial armature and glands.

In the hope of encouraging progress in the higher classification of this suborder, and in explanation of the conservative scheme adhered to in the present volume, we shall discuss in turn several of the key issues.

The status of the families Facelinidae and Favorinidae

Odhner (1939) first raised the group of species around *Facelina* to familial rank within his Cleioprocta, in which he also placed the Aeolidiidae. Odhner divided his Facelinidae into the subfamilies Facelininae (to include *Facelina*) and Rizzoliinae (to include *Favorinus*). The main distinction between these two subfamilies was that the right anterior hepatic lobe gave rise to rows of cerata in *Facelina*, but to an arch of cerata in *Favorinus*. Marcus (1958) proposed the elevation of the Facelininae and Rizzoliinae to familial rank. This meant that the Facelinidae accommodated only Odhner's Facelininae, and the Rizzoliinae became the Favorinidae. This latter substitution was deemed necessary because *Rizzolia* Trinchese, 1877 was unavailable (and so, therefore, was a familial or subfamilial name derived from it) because it was an objective junior synonym of *Cratena* Bergh, 1864 (in that both had *Doris peregrina* Gmelin, 1791 as the type species). *Rizzolia* was in due course rejected from the I.C.Z.N. Official List in Opinion number 776, despite a plea from Lemche that it was *Cratena* that should be suppressed, not *Rizzolia*, because of past confusion of *Cratena* with some tergipedid genera.

This recognition of the familial unit Favorinidae by Marcus has been accepted by Burn (1962, 1966), Schmekel (1968) and Thompson in volume I of the present monograph (1976). On the other hand, Edmunds (1970) has cast doubt upon the validity of the Favorinidae and Facelinidae by indicating the close similarities which exist between *Moridilla* Bergh, 1888, *Palisa* Edmunds, 1964 (both facelinids) and *Noumeaella* Risbec, 1937 (a favorinid). This last exhibits the combination of a favorinid ceratal arrangement with a primitive anterior genital mass like some facelinids. But the important distinction holds firm between cerata which are in rows in the one case, but form an arch in the other. Our view is that Odhner was generally correct to stress in his classification of the aeolidacean families the disposition of the hepatic lobes, and we believe that other features (penial armature, number and positioning of receptacula, presence or absence of propodial tentacles or jaw denticulation, smooth, lamellate or tuberculate rhinophores) are usually of merely generic or specific importance. This principle, coupled with growing knowledge concerning the veliger larval shells, has guided our grouping and positioning of aeolidacean genera in the present scheme of classification. It appears to us, nevertheless, that Miller (1974) carried this precept to an unacceptable extreme when he proposed the merging of the Facelinidae, Favorinidae, Babainidae, Pteraeolidiidae, Cratenidae, Caloriidae, Phidianidae, Myrrhinidae, Herviellidae and Glaucidae. This proposal is unattractive in two respects. First, the name that Miller chose for his enlarged family was Glaucidae (on grounds of temporal precedence), an atypical group containing hitherto only *Glaucus* and *Glaucilla*, with a handful of pleustonic species which uniquely exhibit inverted dorso-ventrality. Second, Miller achieved little genuine simplification with his proposal, because most of the groupings were retained as subfamilies. In our opinion, the gain by this radical proposal would be insufficient to balance the inconvenience of the upheaval. We agree, however, that the Caloriidae and Phidianidae should be merged into the Facelinidae, although it is probably still necessary to maintain *Caloria* Trinchese, 1888, *Phidiana* Gray, 1850 and *Facelina* Alder & Hancock, 1855 as distinct genera.

The status of some of the genera of the family Tergipedidae

Many of the genera that have been used to accommodate members of the Tergipedidae must be set aside, for reasons fully discussed by Miller (1977); these include *Cavolina* Cuvier, 1805,

Montagua Fleming, 1822, *Amphorina* Quatrefages, 1844, *Cratena* Bergh, 1864, *Diaphoreolis* Iredale & O'Donoghue, 1923, and *Cratenopsis* Lemche, 1936.

Brown (1980) argued that the differences between *Cuthona* Alder & Hancock, 1855, *Trinchesia* Ihering, 1879, and *Catriona* Winckworth, 1941 were vague, leading to the proposal that *Trinchesia* and *Catriona* should be relegated to junior status. Subsequently Gosliner (1981b) has accepted the merging of *Trinchesia* and *Cuthona*, but has retained an independent *Catriona*, characterized by a recessed median cusp on each radular tooth, and the presence of bristles or rodlets on the cutting edges of the jaws. These features are diagnostic for the type species *Catriona gymnota* but they are troublesome in, for example, *Catriona lonca* Marcus, 1965, which possesses a radula of the catrionid type but lacks the jaw-bristles. In consequence, we prefer to include *Catriona* with *Trinchesia* in an enlarged *Cuthona*, notwithstanding the contrary opinions of Williams & Gosliner (1979) and Gosliner (1981b).

Studies of the hepatic system have indicated an evolutionary progression through the tergipedids (Odhner, 1939; Miller, 1977). Among the British species, the most primitive condition is exemplified by *Cuthona nana*, which develops branched digestive ducts within the body in later life. Other *Cuthona* species retain simple digestive ducts giving rise to single ceratal rows even when fully mature. This simplification of hepatic branching and the associated reduction of ceratal numbers is taken to an extreme in *Tergipes tergipes* which has only a single ceras from each branch of the hepatic duct. It appears that selective pressures have favoured a reduction in body size and shorter life cycles, with a commensurate simplification of the hepatic system.

KEY TO THE BRITISH SPECIES OF AEOLIDACEA

1. Cerata vestigial, body vermiform, burrowing in shell gravel   *Pseudovermis boadeni* (p. 141)
   Cerata well developed; epifaunal habit   **2**

2. Cerata each bear a wavy membrane on the mesial face   *Fiona pinnata* (p. 139)
   Cerata without such a membrane   **3**

3. Rhinophores with numerous conspicuous transverse lamellae   **4**
   Rhinophores may be wrinkled, tuberculate or swollen, but without numerous conspicuous transverse lamellae   **6**

4. Blue iridescence on cerata   *Facelina coronata* (p. 150)
   No blue iridescence   **5**

5. Body pink-brown, with white spots generally distributed over dorsal surfaces   *Facelina annulicornis* (p. 147)
   Body translucent white without white speckling but with a longitudinal white bar between the rhinophores   *Facelina bostoniensis* (p. 148)

6. Rhinophores with small, irregularly distributed tubercles on the posterior surfaces   *Caloria elegans* (p. 153)
   Rhinophores not as above   **7**

7. Rhinophores bearing 1–3 bulb-like swellings   **8**
   Rhinophores without such swellings   **9**

8. Three shelf-like swellings on each rhinophore   *Favorinus blianus* (p. 154)
   One (rarely two) smooth bulb-like swellings on each rhinophore   *Favorinus branchialis* (p. 155)

9. Propodium produced to form conspicuous antero-lateral tentacles   **10**
Propodium rounded, dilated, or angular, not tentaculate   **18**

10. Muscular cerata responsible for swimming   *Cumanotus beaumonti* (p. 144)
Cerata without such ability   **11**

11. Associated with fish eggs; body flattened; anal opening lacking   *Calma glaucoides* (p. 138)
Active predators upon hydroids and anemones; body not extremely flattened; anal papilla detectable   **12**

12. Radula uniseriate   **13**
Radula triseriate   **25**

13. Radular teeth narrow, having sparse lateral denticles on either side of a prominent median cusp   **14**
Radular teeth broad with numerous lateral denticles and the median cusp inconspicuous or absent   **15**

14. Body translucent white; cerata covered with opaque white pigment   *Dicata odhneri* (p. 156)
Body with an orange or salmon-pink tinge; cerata with scattered white blotches   *Facelina dubia* (p. 151)

15. Radular teeth each form a smoothly curved arch bearing regularly graded denticles, not emarginated in the mid-line   *Aeolidia papillosa* (p. 158)
Radular teeth each form a curved denticulate arch which is markedly emarginate in the mid-line   **16**

16. Anterior one or two rows of cerata pale, forming a conspicuous collar behind the rhinophores   *Aeolidiella alderi* (p. 161)
Anterior cerata without such a collar   **17**

17. Body and ceratal colour uniformly yellow, orange or red   *Aeolidiella sanguinea* (p. 163)
Body pale yellow, cerata not uniformly coloured   *Aeolidiella glauca* (p. 161)

18. Cerata inflated, club-like, knobbly, or many-banded; radula triseriate   **19**
Cerata finger-like, rarely inflated; radula uniseriate   **30**

19. Cerata knobbly   *Eubranchus doriae* (p. 132)
Cerata smooth   **20**

20. Cerata each exhibit a golden yellow subterminal ring   **21**
Cerata without such a ring   **23**

21. Dorsum, flanks and cerata mottled with red or brown   *Eubranchus pallidus* (p. 134)
Body without red or brown mottling   **22**

22. Rhinophores and oral tentacles with bold yellow-orange pigment   *Eubranchus farrani* (p. 134)
Tentacles without such yellow-orange pigment   *Eubranchus tricolor* (p. 135)

23. Cerata inflated, urn-shaped   *Eubranchus exiguus* (p. 132)
Cerata slender, finger-like   **24**

24. Ceratal pigmentation predominantly greenish-brown   *Eubranchus cingulatus* (p. 130)
Cerata pale brown with 2 or 3 rust-coloured bands and white or pale yellow tips   *Eubranchus vittatus* (p. 137)

25. Body with a purple tinge   *Coryphella pedata* (p. 112)
Body translucent white   **26**

26. Opaque white lines present on body   **27**
Body translucent white but no such opaque white lines   **28**

27. Opaque white lines present on oral tentacles, dorsum, flanks and on cerata   *Coryphella lineata* (p. 111)
Single opaque white line down centre of dorsum; narrow white ring at tip of each ceras   *Coryphella verrucosa* (p. 114)

28. Body length 15 mm or less; narrow white ring at tip of each ceras   *Coryphella gracilis* (p. 109)
Body length up to 40 or 50 mm; tip of each ceras is white or exhibits a broad white ring   **29**

29. Distinct notal ridge connects ceratal bases between clusters; subterminal broad white rings present on cerata   *Coryphella browni* (p. 109)
Notal ridge between ceratal clusters indetectable; cerata on short peduncles; cerata capped by white pigment   *Coryphella pellucida* (p. 113)

30. Oral tentacles absent   *Embletonia pulchra* (p. 142)
Oral tentacles present, distinct   **31**

31. Head rounded, shield-like, bearing short, posteriorly directed oral tentacles   *Tenellia adspersa* (p. 128)
Head not shield-like; oral tentacles directed laterally or anteriorly   **32**

32. Cerata sparse, a single ceras arising from each hepatic branch   *Tergipes tergipes* (p. 127)
Cerata in obliquely transverse rows   **33**

33. Cerata with vivid blue areas   *Cuthona caerulea* (p. 120)
Cerata without vivid blue areas   **34**

34. Usually found on *Hydractinia echinata* on hermit crab shells   *Cuthona nana* (p. 116)
Habit not as above   **35**

35. Ceratal hepatic lobules greenish, speckled black; cnidosacs swollen and opaque, filling the tip of the ceras   *Cuthona viridis* (p. 126)
Ceratal lobules not greenish; cnidosacs not as above   **36**

36. Body and cerata very pale; no red, brown or orange markings   **37**
Red, brown and orange markings present   **38**

37. White bands present superficially on or near the tentacular and ceratal tips; penis unarmed   *Cuthona concinna* (p. 121)
White speckles all over the cerata; penis armed   *Cuthona pustulata* (p. 124)

38. Rhinophores orange or brown, darkening distally   *Cuthona gymnota* (p. 123)
Rhinophores not as above   **39**

39. Longitudinal white streaks on all tentacles; cerata with red bases   *Cuthona rubescens* (p. 125)
Tentacles having pale orange or brown rings, not as above; cerata not as above   **40**

40. Superficial orange or red markings on and behind the head; tentacles with pale orange band   **41**
Markings not as above; tentacles with olive green or brown bands   *Cuthona amoena* (p. 118)

41. Orange-red stripe traverses the head between the oral tentacles; longitudinal yellow streak present between the rhinophores   *Cuthona genovae* (p. 122)

Orange-red zone extends from each rhinophore base to the first ceratal group on each side   *Cuthona foliata* (p. 122)

FLABELLINIDAE

Active aeolidaceans having a pleuroproctic (occasionally acleioproctic) anal position; cerata in even or irregular rows, sometimes clustered into groups set upon lobes or peduncles. There may be a distinct notal ridge separating the dorsum from the flanks. Distinct propodial tentacles and long and mobile oral tentacles are conspicuous; the rhinophores may be smooth, papillate or lamellate. The penis is unarmed; there is a coiled tubular ampulla acting as a seminal vesicle; a bursa copulatrix is present, usually located distally, opening into the female atrium, but in some species it is double and situated close to the fertilization chamber.

The cutting edges of the jaws are denticulate. The radula is triseriate; the lateral teeth are usually denticulate, rarely smooth. The larval shell is of type 1.

Coryphella Gray, 1850

(type *Eolis rufibranchialis* Johnston, 1832; validated in I.C.Z.N. Opinion number 781)

Flabellinids with non-lamellate rhinophores; cerata are not borne upon raised peduncles (as they are in the related genus *Flabellina* Voigt).

We propose to retain the genus *Coryphella*, while accepting that there is some weight in the arguments in favour of merging *Coryphella* with *Flabellina* recently advanced by Miller (1971) and by Gosliner & Griffiths (1981).

Many of the N. Atlantic species of *Coryphella* have been reviewed by Kuzirian (1979). Differences between the 6 British species are tabulated by Picton (1980).

DIAGNOSTIC FEATURES OF THE BRITISH SPECIES OF *CORYPHELLA*
(modified after Picton, 1980)

	Maximal length	Radular rows	Coloration	Position of anus	Notal ridge
Coryphella browni	50 mm	11–14	broad white rings on cerata	beneath 2nd row of 2nd ceratal group	detectable
Coryphella gracilis	15 mm	11–13	narrow white rings on cerata	between 1st and 2nd ceratal group; behind renal pore	well developed
Coryphella verrucosa	35 mm	16–20	narrow white rings on cerata; rhinophores pinkish; white lines on tail and central dorsum	beneath 2nd row of 2nd ceratal group	detectable
Coryphella lineata	50 mm	11–14	white lines on oral tentacles, back and sides of body, and on cerata	beneath 2nd row of 2nd ceratal group	detectable
Coryphella pedata	48 mm	27–32	purple hue	between 1st and 2nd ceratal group; below renal pore	none, cerata on short peduncles
Coryphella pellucida	40 mm	27–35	cerata with white capping the entire tips	between 1st and 2nd ceratal group	none, cerata on short peduncles

67. **Coryphella browni** Picton, 1980
    Coryphella verrucosa verrucosa Thompson & Brown, 1976
    not **Eolidia verrucosa** M. Sars, 1829

APPEARANCE IN LIFE (Plate 28). Mature specimens may reach 50 mm in length. The body is pellucid white, with the opaque white lobes of the ovotestis visible through the skin behind the pericardium. Interrupted chalk white pigment forms a dorsal streak from the posterior cerata to the metapodial tip. Both the wrinkled rhinophores and the long, mobile oral tentacles are streaked with white; they lack the pinkish tinge found in *C. verrucosa*.

Up to 8 groups of cerata arise from the dorsal side of a distinct notal ridge. These clusters comprise numerous obliquely transverse rows of cerata, maximally 7 rows in the pre-cardial clusters and 15 rows further back. Each row may contain up to 5 cerata, on either side of the median plane. Within each ceras the digestive gland may vary in colour from bright red to dark brown; red seems characteristic of individuals which have recently fed. A sub-terminal broad band of superficial white pigment encircles each ceras.

The genital openings lie below the first ceratal cluster on the right side of the body; the nephroproct lies close to the pericardium in the right inter-hepatic space, while the anal papilla is situated ventral to the second row of the second ceratal cluster on the right.

ANATOMY (Fig. 26a). The triseriate radula consists of between 11 and 14 rows in the adult. The median tooth exhibits a recess into which the cusp of the preceding tooth can be slotted when the radular ribbon is retracted and at rest. Each median tooth has a stout central cusp and 6–9 denticles on either side. The lateral teeth are strongly developed and broad, each bearing a series of 8–14 denticles. The cutting edge of the jaw is roughened by several rows of rounded nodules. The penis is long, cylindrical and devoid of armature.

HABITS. This species forms dense populations, mixed with *C. lineata* (Lovén), feeding on the gymnoblastic hydroid *Tubularia indivisa* L. The foot is well adapted for clinging to the stems of *Tubularia* in strong currents. The terminal polyps are ingested whole.

The egg mass is an undulating spiral when deposited on a flat surface, but in the field it is more commonly looped around the hydroid prey. The ova are singly encapsulated and measure 50–75 µm in diameter. Planktotrophic veligers hatch after about two weeks. Spawn is produced between May and early August.

DISTRIBUTION (Map 8). Personal collections have been made from the Orkneys, west Scotland, the Irish Sea and the west coasts of Ireland. This is the commonest coryphellid encountered by divers off the south west coasts of Britain. On the other hand, only a single specimen has been recorded from the North Sea (near Eyemouth). It penetrates the English Channel as far as Littlehampton and ranges southwards along the French coast to Brittany (Picton, 1980).

DISCUSSION. This species has only recently been delimited and characterized after a long history of confusion with the smaller *Coryphella gracilis* (Alder & Hancock) and *C. rufibranchialis* (Johnston) (= *C. verrucosa* (M. Sars)). It is possible that the southern limit will be shown to extend as far as the western Mediterranean but the necessary reappraisal of the Iberian and Mediterranean coryphellids is lacking.

The only *Coryphella* with which the present species may be confused is *C. lineata* (Lovén, 1846), but the colour patterns are distinct. The only white pigment on the epidermis of *C. browni* consists of patches on the oral tentacles, rhinophores and metapodial tip, together with broad distal rings on the cerata.

68. **Coryphella gracilis** (Alder & Hancock, 1844a)
    Eolis gracilis Alder & Hancock, 1844a
    Eolis smaragdina Alder & Hancock, 1851
    Eolis stellata Stimpson, 1854
    Coryphella gracilis; Alder & Hancock, 1855
    Coryphella smaragdina; Alder & Hancock, 1855
    Coryphella frigida Grieg, 1907

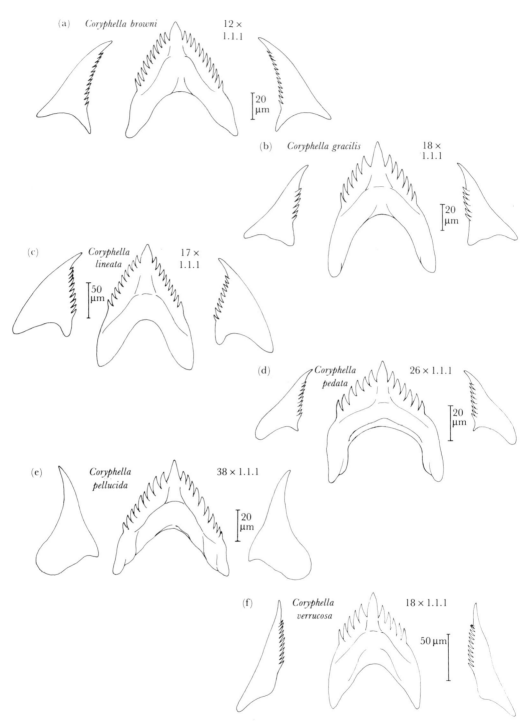

FIG. 26. Camera lucida drawings of aeolidacean radulae: *Coryphella*.
  (a) *Coryphella browni*, preserved length 35 mm, Lundy Island.
  (b) *Coryphella gracilis*, preserved length 7 mm, Manacles Rocks, Lizard Peninsula, Cornwall.
  (c) *Coryphella lineata*, preserved length 18 mm, Lundy Island.
  (d) *Coryphella pedata*, preserved length 27 mm, Lundy Island.
  (e) *Coryphella pellucida*, preserved length 28 mm, Scapa Flow, Orkney.
  (f) *Coryphella verrucosa*, preserved length 15 mm, Scapa Flow, Orkney.

Coryphella rufibranchialis gracilis*;* Odhner, 1929
Coryphella verrucosa gracilis*;* Bruce, Colman & Jones, 1963
Coryphella verrucosa smaragdina*;* Thompson & Brown, 1976

APPEARANCE IN LIFE (Plate 28). This delicate species rarely attains 15 mm in length. The body may be translucent or a more opaque white when packed with developing ova. The slender foot tapers to a finely pointed metapodium. The oral tentacles are moderately long, tapering to blunt tips. The rhinophores are smooth, thinly tapering and of a similar length to the oral tentacles. The oral tentacles, rhinophores and metapodium are all streaked with white pigment. The anterior edge of the foot is extended to form distinct, pointed, mobile propodial tentacles, equal in length to the breadth of the foot.

The ceratal contents are usually bright red but may range through various shades of red-brown to a rich grass-green. A fine band of white pigment (thinner than in *C. browni*) encircles each ceras sub-apically. The cerata are clustered in up to 6 (usually 5) groups, all originating from the distinct notal ridge. The genital openings are located below the posterior rows of the first ceratal cluster on the right, just below the notal ridge. The anal opening is situated below the first row of the second ceratal cluster, while the nephroproct lies within the inter-hepatic space.

ANATOMY (Fig. 26b). Between 12 and 20 rows of teeth make up the tapering radular ribbon. The rachidian teeth are broad, with a stout median cusp flanked by 5–9 denticles on either side. The lateral teeth are triangular plates, but the terminal spine is curved medially; the serrated inner surface bears about 8 spines, extending almost to the distal tip. The strong jaw plates display 3 or 4 rows of denticles along the prominent masticatory margin. The penis is long and cylindrical, with an unarmed conical tip.

HABITS. Around Britain, *C. gracilis* has been seen feeding upon *Eudendrium ramosum* (L.), and Kuzirian (1979) has noted a similar preference for species of *Eudendrium* in the Gulf of Maine, New England. Kuzirian observed, however, that in deeper water (25–33 m) *Eudendrium* is scarce and *Coryphella gracilis* then feeds on *Halecium articulosum* Clark and *Clytia johnstoni* Alder. The 50–60 μm ova develop to hatching as planktotrophic veliger larvae in 12–15 days at 12 °C (Kuzirian, 1979).

DISTRIBUTION (Map 8). This species has been reliably reported from the Atlantic coast of France, all around the British Isles, Denmark, Faeroes, Iceland, and along the eastern American coast from Newfoundland and Nova Scotia to New England and Cape Cod (Kuzirian, 1979).

DISCUSSION. *Coryphella gracilis* has been confused in the European literature with *C. verrucosa* and other coryphellids. A comprehensive and authoritative review of the taxonomy of this species has been published by Kuzirian (1979).

69. ***Coryphella lineata*** (Lovén, 1846)
    Aeolis lineata Lovén, 1846
    Eolis lineata Alder & Hancock, 1846
    Eolida argenteolineata Costa, 1866

APPEARANCE IN LIFE (Plate 27). This species may reach 50 mm in extended length. Long, mobile oral tentacles arise from the sides of the head, while the rugose rhinophores are as long or slightly longer. Each tentacle is colourless except for a thin white stripe. The stripes from the oral tentacles unite to form a median dorsal line which extends down the body to the tip of the pointed metapodium. Similar white lines run along each flank, starting in front of the rhinophores, to the metapodium, where they unite with the median line.

The grooved anterior margin of the foot is notched medially; the mobile propodial tentacles are at least equivalent in length to the foot-width. The cerata are clustered in 5–8 groups on an expanded notal ridge which runs down either side of the dorsum. Some white pigment is always present on the cerata, although it is not always the full pattern of two stripes (one frontal, one posterior) descending from a sub-apical ring. Poorly developed ceratal pigment can give an appearance similar to *Coryphella browni*, but the long body stripes are absolutely diagnostic of *C. lineata*. The hepatic contents vary in colour from dark brown through reddish brown to crimson, with rare examples exhibiting green cerata (perhaps a sign of starvation).

ANATOMY (Fig. 26c). The triseriate radula consists of up to 17 rows with a broad rachidian tooth and denticulate laterals. Each lateral tooth has a sharp terminal spine which bends inwards and possesses up to 16 denticles along the inner face. The masticatory margins of the large oval jaws are roughened by several rows of nodules.

The penis is long, tubular and unarmed. The anterior genital mass was described by Odhner (1939), while observations on the buccal mass, nervous system and renal and hepatic histology were presented by Trinchese (1881).

HABITS. Around the British Isles, the preferred diet consists of *Tubularia indivisa* L., competing with the large populations of *Coryphella browni*, especially around the south west coasts of Britain. Miller's (1962) study led him to conclude that there were two or more generations each year, with a normal life span of 8–12 months. This author inferred that predation occurred not only upon *Tubularia*, but also on *Sarsia eximia* (Allman), *Hydrallmania falcata* (L.) and *Sertularia argentea* (Miller, 1961). The diet is *Eudendrium* in the Mediterranean (Schmekel & Portmann, 1982).

The egg mass is a white or pink undulating spiral, from which planktotrophic veligers hatch after about 14 days under normal conditions. The spawning period is concentrated in the months from April to August around Britain; specimens may reach maturity when 10 mm in length (Miller, 1958). In the Mediterranean the 75 μm ova develop to hatching in 6–7 days at 16 °C (Schmekel & Portmann, 1982).

DISTRIBUTION (Map 8). *Coryphella lineata* has been reported in considerable numbers from the western Mediterranean Sea, but elsewhere it appears to be restricted to the European coasts of the Atlantic Ocean, northwards from Britain to Sweden (Jeffreys, 1867) and along the Norwegian coast to the Arctic Circle (Odhner, 1939). Roginsky (1962) was uncertain of the identification of specimens from the White Sea, although they bore a superficial resemblance to *C. lineata*.

Although inter-tidal reports have been rare, abundant populations can be found between 20 and 40 m in depth; a maximum depth of 360 m is on record.

70. *Coryphella pedata* (Montagu, 1815)
Doris pedata Montagu, 1815
Eolis landsburgi Alder & Hancock, 1846

APPEARANCE IN LIFE (Plate 28). Specimens up to 48 mm in length have been collected around Britain, where this is the only aeolidacean with violet or purple body pigmentation (although *Flabellina affinis* (Gmelin, 1791) has similar coloration in the Mediterranean Sea, where it co-exists with *C. pedata*, distinguished most obviously by the lamellate rhinophores of the former species).

The oral tentacles and the rugose or wrinkled rhinophores of *C. pedata* are of similar length in mature specimens; the propodial tentacles are only as long as the width of the narrow foot.

Ceratal numbers tend to be lower than in other British coryphellids and they are arranged in distinct clusters arising from slight expansions of the notal ridge. All tentacles and the cerata are tipped with superficial white pigment which enhances the beauty of this delicate and unmistakable species.

ANATOMY (Fig. 26d). The triseriate radula possesses between 14 and 28 rows of teeth. The number of denticulations on either side of the prominent median cusp tends to be fewer than in othe British coryphellids and rarely exceeds 6. Between 5 and 7 denticles are situated along the inner surfaces of the slender triangular lateral plates; the terminal spine curves inwards. The penis has a bulbous, inflated shape. Kuzirian (1977) noticed that Schmekel (1970) had found two proximal receptacula in the anterior genital mass of Mediterranean specimens, whereas

Odhner (1939) had described only one in specimens from the north Atlantic. Together with other inconsistencies, this led Kuzirian to suggest that the two populations are in fact two species. Our examination of British material has proved that there is no need to make such a separation, because there are two receptacula (opening via a common duct into the oviduct) in both Atlantic and Mediterranean samples, contrary to Odhner's assertion.

HABITS. The spawn mass is a thin white spiral, usually coiled among the *Eudendrium ramosum* (L.) growths on which the aeolid usually feeds. We cannot confirm the claim by Miller (1961) that *Tubularia* spp., *Abietinaria abietina* (L.) and *Hydrallmania falcata* (L.) are sometimes taken. Garstang (1890), however, found small specimens on *Obelia geniculata* (L.) and *Sertularella gayi* (Lamouroux), and it is possible that these juveniles indicate that there is a change during early life from calyptoblastic to gymnoblastic hydroids.

Eggs are singly encapsulated and hatch as planktotrophic larvae in 11–15 days in British waters; the spawning period extends from late May to September. In the Mediterranean the 80 µm ova are white and develop to hatching in 6 days at 16 °C (Schmekel & Portmann, 1982).

DISTRIBUTION (Map 8). *C. pedata* is reliably and extensively recorded at innumerable sites between Norway and the western Mediterranean, while we have ourselves collected it from as far east as Naples and the northern Adriatic Sea. Reports from the French Atlantic coast are tabulated by Bouchet & Tardy (1976).

71. ***Coryphella pellucida*** (Alder & Hancock, 1843)
   Eolis pellucida Alder & Hancock, 1843
   Coryphella rutila Verrill, 1879
   Coryphella landsburgi; Trinchese, 1881 (part)
   Coryphella rufibranchialis; Bergh, 1885 (part)
   Coryphella verrucosa verrucosa; Thompson & Brown, 1976 (part)

APPEARANCE IN LIFE (Plate 27). Fully mature specimens may reach 40 mm in length. The slender body is translucent, while the long narrow cerata have crimson hepatic ducts. The whole tip of each ceras bears white pigment, compared with the sub-terminal ring of pigment seen in most British coryphellids. Long, tapering oral tentacles originate dorso-laterally from the head. The rhinophores are a similar or greater length, but slightly more slender. Both pairs of tentacles have opaque white tips and often exhibit irregular streaks of white pigment running down towards their bases. The small head is almost transparent, allowing the thickened labial surfaces to be discerned. The foot is as narrow as the upper body, but produced anteriorly to form tentacles equal to or wider than the foot. The anterior surface of the labial groove, running transversely across the anterior foot margin, is notched in the mid-line. The cerata are clustered in oblique rows of up to 7 cerata per row. Up to 6 rows are included in the first cluster, but the more posterior groupings include only 1 or 2 rows. The bases originate from lateral extensions of the notal ridge.

The male and female openings are positioned separately, the male opening being slightly more anterior and situated below the centre of the first ceratal cluster on the right side. The pleuroproctic anus is situated within the interhepatic space, below the notal ridge.

ANATOMY (Fig. 26e). The triseriate radula usually consists of between 20 and 30 rows. The median plate has a strong, erect median cusp, flanked on either side by 8–11 denticles. Kuzirian (1979) noted that the denticles are often bicuspid in N. American specimens. The lateral plates are characteristically non-denticulate, but taper to a sharp terminal spine. The jaw plates are well developed, with a denticulate masticatory border. The penis is conical and unarmed. There is a distal bursa copulatrix and a proximal receptaculum seminis. The hepatic and genital systems were investigated by Odhner (1939).

HABITS. The preferred diet in British waters is *Tubularia indivisa* L., although *C. pellucida* will readily devour *Eudendrium ramosum* (L.) in laboratory vessels. Kuzirian (1979) confirmed this dietary preference in American populations. The egg mass is a thin, undulating coil, with singly encapsulated eggs measuring 70–80 µm in diameter. Planktotrophic larvae hatch in 12–14 days at 8–10 °C (Kuzirian, 1979). Miller (1958) observed spawning in the laboratory during the months of February and March.

The deepest records are from 120 m (Odhner, 1907).

DISTRIBUTION (Map 8). Nearly all recent British records have come from northern localities. While collections around Scotland and in the North Sea have been plentiful, rare specimens have been reported from the Isle of Man and from north west Ireland. We have never encountered *C. pellucida* off the south west coasts of Britain, although there are some records from the last century. A report from the Spanish Mediterranean coast (Ros, 1975) requires confirmation, in the light of Kuzirian's (1979) revision. Lemche (1929) also claimed a Mediterranean distribution, perhaps referring to a questionable record by O'Donoghue (1929).

Kuzirian (1979) established synonymy between *C. pellucida* and *C. rutila* Verrill, 1879 from the eastern American coast. The known range is therefore extended from Britain, Norway and the Faeroes to the Gulf of Maine, Massachusetts and Cape Cod.

72. **Coryphella verrucosa** (M. Sars, 1829)
   Eolidia verrucosa M. Sars, 1829
   Eolis rufibranchialis Johnston, 1832
   Eolidia embletoni Johnston, 1835
   Eolis diversa Couthouy, 1839
   Coryphella rufibranchialis; M. Gray, 1850
   Eolis mananensis Stimpson, 1854
   Aeolis bostoniensis; Mörch, 1857
   Coryphella robusta Trinchese, 1874
   Coryphella verrucosa; Friele & Hansen, 1876
   Aeolis bostoniensis approximans Mörch, 1877
   Coryphella landsburgi; Trinchese, 1881 (part)
   Coryphella diversa; Bergh, 1885
   Coryphella salmonacea; Bergh, 1886, Walton, 1908
   Coryphella rufibranchialis chocolata Balch, 1909
   Coryphella rufibranchialis mananensis Balch, 1909
   Coryphella gracilis bostoniensis Lemche, 1936
   Coryphella verrucosa rufibranchialis Odhner, 1939
   Coryphella verrucosa verrucosa Odhner, 1939

APPEARANCE IN LIFE (Plate 28). Adults may reach 35 mm in length. The elongated body is translucent white, allowing the opaque ovotestis to be seen posterior to the pericardium. The oral tentacles originate from the upper sides of the head and taper from the stout bases to the blunt tips. The rhinophores tend to be shorter than the oral tentacles and have a rugose surface. Both pairs of tentacles are streaked with white pigment. The anterior margin of the foot is transversely grooved and medially notched, the groove continuing on to the propodial tentacles which project about half the width of the foot on either side. The metapodium tapers to a fine tip from which a white stripe runs dorsally to the pericardium. The cerata are arranged in 5–7 groups inserted in diagonal rows on the dorsum above an expanded notal ridge; no ridge is evident between the clusters. The spacing of the clusters is very even compared with other British coryphellids. Odhner (1922) disagreed about this and denied that the cerata were clustered, but we consider that a degree of clustering is nearly always recognizable, especially after preservation. There are numerous small rotund cerata in the most posterior rows.

The hepatic lobes may be light brown, red-brown, maroon or crimson. The ceratal tips are covered with superficial white pigment.

The gonopore lies below the first ceratal cluster on the right side, while the pleuroproctic anus is below the third ceratal row of the second cluster.

The most reliable diagnostic features are the white dorsal stripe from the pericardium to the tip of the metapodium, and the tightly packed, even arrangement of the cerata.

ANATOMY (Fig. 26f). The triseriate radula consists of between 13 and 20 rows. The lateral plates are thin and triangular, with the distal spine usually curved outwards. Between 4 and 8 denticles flank the prominent median cusp of the rachidian tooth while the inner surface of each lateral tooth bears 7–12 denticles, stopping well short of the tip. The jaws have a prominent masticatory border bearing three or four rows of small denticulations. The penis is large and cylindrical, with a flared terminal disc with a fluted rim (Kuzirian, 1979). There is a distal bursa copulatrix and a proximal receptaculum seminis.

HABITS. The usual diet in British waters is *Tubularia indivisa* L., which is also favoured by Scandinavian populations of *C. verrucosa*. Around the coast of New England, however, Kuzirian (1979) described a life cycle which involved juveniles overwintering on a diet of thecate campanulariid and sertulariid hydroids. A shortage of these seemed to stimulate a switch to feeding upon *Hydractinia echinata* (Fleming), *Clava leptostyla* Agassiz and even the compound tunicate *Botryllus schlosseri* (Pallas). Around the Isle of Man Miller (1961) recorded predation on *Tubularia larynx* Ellis & Solander, *Sarsia eximia* (Allman), *Clytia johnstoni* (Alder), *Laomedea flexuosa* Hincks, *Dynamena pumila* (L.) and *Hydrallmania falcata* (L.).

The egg mass takes the form of a thin, non-undulating spiral on flat surfaces, or evenly spaced loops around hydroid stems (Kuzirian, 1979). Several thousand singly encapsulated ova are laid at a time and measure 85–100 µm in diameter. Development to the release of planktotrophic larvae (development-type 1 of Thompson, 1967) takes between 7 and 10 days at 13 °C (Kuzirian, 1979). The spawning period around the British Isles is from April to June and may begin when the animal has reached only 5 mm in length.

DISTRIBUTION (Map 8). This boreo-arctic species has been frequently recorded, under various synonyms, from southern New England through the Gulf of Maine to east and west Greenland, Iceland, Spitzbergen, the Barents and Kara Seas, the Faeroes, Norway and Britain (Lemche, 1941; Kuzirian, 1979). It has also penetrated to the Pacific Ocean via the Bering Sea (Lemche, 1929) and extended along the north Russian coast (Roginsky, 1964) to the Sea of Japan (Volodchenko, 1955). Specimens have been obtained from depths to 450 m (Odhner, 1907).

Recent collections around the British Isles have failed to confirm older records from south west coasts. Claims that the distribution extends along the French Atlantic coast to the Mediterranean Sea require confirmation, in the light of Kuzirian's (1979) taxonomic revision.

TERGIPEDIDAE

Generally small aeolidaceans having an acleioproctic anal position; cerata in even rows, often reduced in number, fusiform or clavate. Propodial tentacles rudimentary or absent. Oral and rhinophoral tentacles smooth, the latter $1\frac{1}{2}$ to twice as long as the former; oral tentacles reduced or lost in *Tenellia*. The penis sometimes has a tubular chitinous stylet; the vas deferens is prostatic and there is a small accessory penial gland which opens separately into the penis sheath; a bursa copulatrix is present, opening into the female atrium.

The cutting edges of the jaws are thin and often lack denticulations. The radula is uniseriate; the median cusp is sometimes dwarfed by lateral denticles, but more usually projects beyond them. The larval shell is of type 2.

Cuthona Alder & Hancock, 1855
  (type *Eolis nana* Alder & Hancock, 1842)
  = *Trinchesia* Ihering, 1879
    (type *Doris caerulea* Montagu, 1804)
  = *Precuthona* Odhner, 1929
    (type *Eolis peachi* Alder & Hancock, 1848a)
  = *Xenocratena* Odhner, 1940
    (type *X. suecica* Odhner, 1940)

    = **Catriona** Winckworth, 1941
      (type *Eolis aurantia* Alder & Hancock, 1842)
    = **Njurja** Marcus & Marcus, 1960
      (type *N. netsica* Marcus & Marcus, 1960)
    = **Narraeolidia** Burn, 1961
      (type *N. colmani* Burn, 1961)
    = **Toorna** Burn, 1964
      (type *T. thelmae* Burn, 1964)
    = **Selva** Edmunds, 1964
      (type *S. rubra* Edmunds, 1964)

Tergipedids with distinct oral tentacles and rows of cerata (not single cerata) in a series along the upper sides of the body.

73. **Cuthona nana** (Alder & Hancock, 1842)
    Eolis nana Alder & Hancock, 1842
    Eolis peachi Alder & Hancock, 1848a
    Cuthona nana; Alder & Hancock, 1855
    Montagua hirsuta Bergh, 1860
    Cuthona peachi; Alder, 1869
    Cratena nana; G. O. Sars, 1878
    Precuthona peachi; Odhner, 1929

APPEARANCE IN LIFE (Plate 29). The translucent white body may reach 28 mm in length around Britain; slightly larger specimens have been reported from Arctic waters. The body may be suffused with pink over the post-cardiac region, while in mature specimens the ovotestis is visible in the form of white or pink lobules. The broad head bears short digitiform oral tentacles arising dorsally in juveniles (less than 10 mm) but nearer the antero-lateral extremities in larger individuals. The smooth, blunt-tipped rhinophores and oral tentacles may whiten distally. Each club-shaped ceras contains a reddish brown diverticulum. Animals taken from female colonies of *Hydractinia* are darker than those from male colonies in which the male gonozoids are colourless (Swennen, 1961b; Christensen, 1977). Starved specimens become browner as the pink component fades. Epidermal glands are normally visible as minute white spots, becoming more concentrated near the ceratal tips.

The foot is expanded anteriorly, with smoothly rounded extremities. The metapodium tapers to a blunt tip just behind the rearmost cerata. On either side of the body there may be over 20 oblique and irregular rows of cerata arranged in up to 16 clusters. A total of 279 cerata were counted in an 18 mm preserved specimen. A short pre-anal row of cerata is common in fully grown specimens but is not a usual feature of animals less than 10 mm.

ANATOMY (Fig. 27f). Up to 30 teeth have been found in the tapering uniseriate radular ribbon, ranging in width from 75 to 160 μm. Alder & Hancock (1845–55) provided drawings which indicated some differences between the radulae of *C. nana* and the synonymous *Eolis peachi*, but Brown (1980) has shown that these differences lie within the range of form encountered in radulae from specimens in his collections of British *C. nana*.

The central cusp of each radular tooth is more prominent in *C. nana* than in other British species of the genus. In this respect the tooth resembles closely the median tooth of the triseriate radula of *Coryphella*, perhaps indicating a closer kinship than has been supposed.

The jaw plates are thin with a rough masticatory border. The roughness can be resolved into rounded tubercles or indistinct bumps, rather than regular denticles. Odhner (1929) described a smooth border or (1939) "two to three rows of small tubercles, irregularly crowded".

The penis is unarmed; the duct from a large penial gland unites with the vas deferens almost at the tip of the penis instead of in the basal junction found in most tergipedids. In virtually all histological details the genital organs agree with MacFarland's (1966) description of *Cuthona rosea* (= *Precuthona divae* Marcus, 1961). Odhner's (1939) account of the female organs missed the small receptaculum entering the vagina near the entrance of the oviduct.

HABITS. *Cuthona nana* feeds predominantly on *Hydractinia echinata* (Fleming) attached to hermit crab shells. *Hydractinia* will occasionally attach to other crabs or to inanimate objects; Walton (1908) obtained *C. nana* from this hydroid growing on the spider crab *Hyas coarctatus* Leach and Harris *et al.* (1975) found colonies of *Hydractinia* up to 30 cm in diameter attached to wood and rock in an area subject to fast currents. Around Britain the hydroid-affected hermit crabs are often to be found in sheltered shallow waters having full salinity, such as the Helford inlet in Cornwall.

Wright (1859) identified the nematocysts of *Clytia johnstoni* (Alder) in the cnidosacs of *C. nana*. Rivest (1978) observed predation only on *Hydractinia* in the field although his animals would take *Clava* in laboratory tests. Both Rivest and Christensen (1977) confirmed the rejection of *Tubularia* by *C. nana* and it is probable that reports suggesting that *Tubularia indivisa* L. is a diet species were mistaken (perhaps based on the related *C. gymnota*).

The reproductive season continues throughout the year, but with spawning maxima from February to May. Alder & Hancock described the egg mass of *C. nana* as similar to that of *C. foliata*, a broad, semicircular coil. This has never been confirmed, and we believe that this observation was atypical because American, Scandinavian and British populations normally lay eggs in a convoluted spiral. Independent studies of veliger development on both sides of the Atlantic have revealed some surprising differences. In Swedish waters (Christensen, 1977) the spawn mass contains 1600–5000 ova with an average diameter of 110 µm. The eggs are deposited on *Hydractinia* but hatch after about 3 weeks as planktotrophic larvae which will not metamorphose, even in the presence of *Hydractinia*, for at least 2 weeks. New England specimens sometimes produce planktotrophic larvae but Rivest (1978) studied a population producing lecithotrophic larvae, which completed metamorphosis after only 2 days, even in the *absence* of *Hydractinia*. The spawn masses produced by this N. American population contained only 450–1500 ova with an average diameter of 160 µm. British populations exhibit planktotrophic development (Brown, 1980), in agreement with Christensen's observations. Eyster (1979) has shown that pelagic and non-pelagic development occur within a single population of the tergipedid *Tenellia adspersa* on the American Atlantic coast and it seems possible that similar variability is a feature of the reproduction of *C. nana*.

DISTRIBUTION (Map 11). *Cuthona nana* has a boreo-arctic range encompassing the coasts of Brittany and Normandy in France, various localities to 30 m depth all around Britain and Ireland, the Netherlands, Sweden, Norway, Spitzbergen, W. Greenland (Odhner, 1939; Swennen, 1961b) and the north east seaboard of America as far south as New Hampshire (Harris, Wright & Rivest, 1975). *Cuthona divae* (Marcus, 1961) from the Pacific coast of America is the only other aeolidacean in the northern hemisphere which prefers a diet of *Hydractinia*. Specific distinctions given by Marcus when describing *C. divae* concerned jaw and radular morphology, but we could not confirm these in material received recently from Waldron Island, Washington. Specific divergence between Atlantic and Pacific populations may well be in progress, but it is either proceeding slowly, or has recently begun. A record of *C. nana* from the Bering Sea (Krause, 1885) could refer to the Pacific species.

DISCUSSION. The taxonomy of *C. nana* has recently been reviewed by Brown (1980). The variation displayed by this species was not appreciated by the early workers so that Alder & Hancock described juvenile and adult as distinct species, later separated into different genera by Odhner (1929).

The large number of cerata is considered to be a primitive character within the tergipedids (Miller, 1977) and distinguishes adults of this species from all other British members of the family.

74. **Cuthona amoena** (Alder & Hancock, 1845a)
    Eolis amoena Alder & Hancock, 1845a
    Cratena amoena; Bergh, 1864
    Cavolina amoena; Alder, 1869

Cuthona amoena; Eliot, 1910
Cratenopsis amoena; Lemche, 1936
Trinchesia amoena; Winckworth, 1951
not ***Aeolidia amoena*** Risbec, 1928

APPEARANCE IN LIFE (Plate 31). Mature specimens measure between 5 and 10 mm in extended length alive. Small cream-white spots are scattered over the dorsum and may be raised on tubercles on the head, between the rhinophores, and over the prominent pericardial swelling. Indistinct brown or olive-green pigment mottles much of the body surface, while the short oral tentacles are encircled by a reddish brown or olive band below an opaque white tip. White spots of surface pigment interrupt the band dorsally and may continue on to the head as an irregular streak towards the posterior surfaces of the rhinophores. Below opaque white tips another pair of reddish brown or olive bands encircle the rhinophores. Rarely these are split into two bands. The cerata are usually set in three rows anterior to the pericardial swelling with five posterior rows on either side. The ground colour of the hepatic lobes within the cerata varies between sandy brown and olive green, and is often darkest near the bases. Small reddish brown dots are scattered over the ceratal epidermis together with white spots. The reddish brown pigment is usually concentrated around the base of the ceras. A small cnidosac is visible terminally and there is often a subterminal ring of red-brown pigment spots. The metapodium is acutely pointed and streaked dorsally with white.

ANATOMY (Fig. 27a). The uniseriate radular ribbon of a 10 mm specimen tapered from a width of 70 µm to 45 µm, and possessed 17 teeth, each having 5 denticles on either side of a median cusp, all of about the same length. The radulae of other individuals had from 15 to 20 teeth. The penis is armed with a hook-shaped spine, and the vas deferens has a well-developed prostatic region.

HABITS. Sublittorally, the diet species is usually *Halecium halecinum* (L.), although Miller (1961) recorded predation on *H. beani* (Johnston). Specimens have been cast ashore on *Sertularia argentea* (L.) and *Diphasia rosacea* (L.).

The spawn mass is a small spiral usually attached to the diet hydroid. Miller (1962) showed that the life cycle can be completed in 2–6 months and he inferred that there are at least two generations per year.

DISTRIBUTION (Map 12). This is a small and well-camouflaged species, but has been recorded surprisingly often around Britain and Ireland (although some misidentifications of *Cuthona rubescens* may have been included). It is reliably recorded from Morocco, the western Mediterranean, and the Atlantic coasts of Europe as far north as the Orkneys (Pruvot-Fol, 1954; Brown, 1980). Specimens have been dredged from the Skagerak, but there are no records from the mainlands of Scandinavia.

FIG. 27. Camera lucida drawings of aeolidacean radulae: *Cuthona*, *Tenellia* and *Tergipes*.
  (a) *Cuthona amoena*, preserved length 8 mm, Skomer Island, S. Wales.
  (b) *Cuthona rubescens*, preserved length 8 mm, Skomer Island, S. Wales.
  (c) *Cuthona concinna*, preserved length 4 mm, Strangford Lough, Northern Ireland.
  (d) *Cuthona gymnota*, preserved length 9 mm, Helford Passage, Cornwall.
  (e) *Cuthona genovae*, preserved length 4 mm, Lough Ine, Ireland.
  (f) *Cuthona nana*, length 10 mm, Scapa Flow, Orkney.
  (g) *Cuthona pustulata*, length 18 mm, Skomer Island, S. Wales.
  (h) *Cuthona caerulea*, preserved length 11 mm, Lundy Island.
  (i) *Cuthona viridis*, preserved length 7 mm, Lundy Island.
  (j) *Cuthona foliata*, preserved length 4 mm, Scapa Flow, Orkney.
  (k) *Cuthona nana*, length 10 mm, Helford Passage, Cornwall.
  (l) *Tenellia adspersa*, length 4 mm, Clevedon, Somerset.
  (m) *Tergipes tergipes*, length 3 mm, Helford Passage, Cornwall.

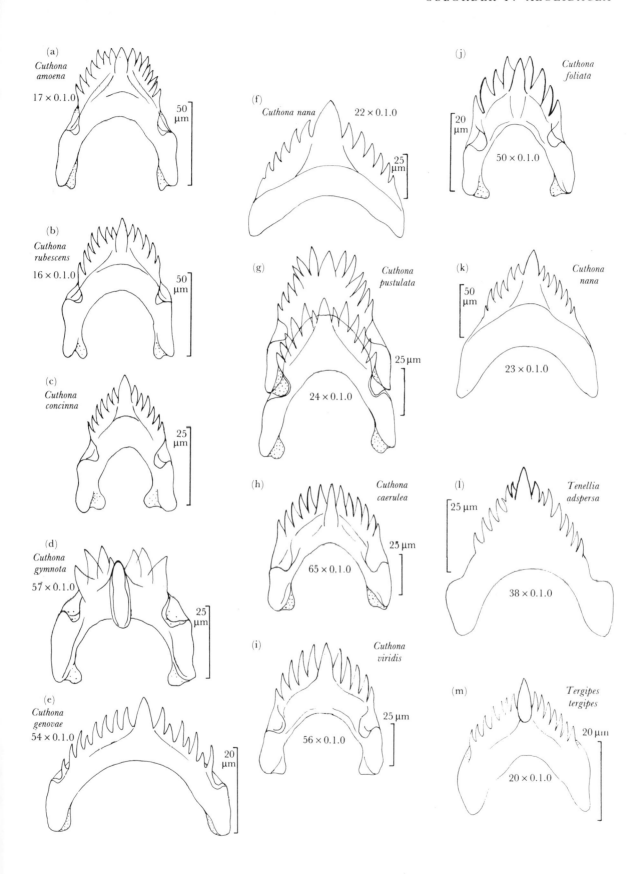

(a) *Cuthona amoena* 17 × 0.1.0 50 µm

(b) *Cuthona rubescens* 16 × 0.1.0 50 µm

(c) *Cuthona concinna* 25 µm

(d) *Cuthona gymnota* 57 × 0.1.0 25 µm

(e) *Cuthona genovae* 54 × 0.1.0 20 µm

(f) *Cuthona nana* 22 × 0.1.0 25 µm

(g) *Cuthona pustulata* 25 µm 24 × 0.1.0

(h) *Cuthona caerulea* 25 µm 65 × 0.1.0

(i) *Cuthona viridis* 25 µm 56 × 0.1.0

(j) *Cuthona foliata* 20 µm 50 × 0.1.0

(k) *Cuthona nana* 50 µm 23 × 0.1.0

(l) *Tenellia adspersa* 25 µm 38 × 0.1.0

(m) *Tergipes tergipes* 20 µm 20 × 0.1.0

75. **Cuthona caerulea** (Montagu, 1804)
   Doris caerulea Montagu, 1804
   Montagua caerulea; Fleming, 1822
   Eolis caerulea; Alder & Hancock, 1845a
   Eolis glotensis Alder & Hancock, 1846
   Eolidia bassi Vérany, 1846
   Eolis deaurata Dalyell, 1853
   Eolis glottensis Alder & Hancock, 1854b
   Cratena caerulea; Bergh, 1864
   Cavolina caerulea; Alder, 1869
   Cavolina glottensis; Alder, 1869
   Amphorina caerulea; Trinchese, 1879
   Trinchesia caerulea; Ihering, 1879
   Aeolis molios Herdman, 1881
   Amphorina molios; Eliot, 1906
   Cratena glotensis; Winkworth, 1932
   Trinchesia glotensis; Winkworth, 1951
   Trinchesia glottensis; Pruvot-Fol, 1954

APPEARANCE IN LIFE (Plate 30). The delicate translucent body may reach a length of 26 mm in extended length in a live animal. The oral tentacles and the slightly longer rhinophores are suffused with pale lemon-yellow pigment. Opaque white pigment often forms an indistinct streak running from the head to the posterior tip of the foot (dividing to pass on either side of the pericardial swelling). The foot is narrower than the body but the propodium is drawn out into rounded lateral projections, especially in larger individuals. The hindgut is often visible as a bright red tube leading to the anus situated in front of the innermost ceras of the first post-cardiac row. Variation in the colour of the hepatic lobe within each ceras and in the pigmentation of the ceratal epidermis result in a wide morphological range. The hepatic lobe may be almost black, various shades of blue or brilliant green. The epidermal pigment is fully expressed as three bands of colour. A subterminal red, orange or yellow band fading towards the colourless tip is always present. Usually, a blue, turquoise or green band encircles the ceras below this, but in some individuals this second band may be absent, allowing the hepatic coloration to show through. A third band below fades towards the base of the ceras. It is often less vivid than the most apical band but always the same colour.

The small cnidosac is opaque white and several other opaque bodies may be discerned, clustered in an irregular ring below the cnidosac. Histological details of the formation and variability of these ceratal colours have been studied by Bürgin-Wyss (1961). The cerata may be bulbous, especially in juveniles, or cylindrical in shape; they always taper to an acute point. Specimens up to 15 mm in length have up to 8 rows of cerata; up to 6 cerata are grouped in each row. Rare larger specimens may possess up to 12 rows of cerata on either side of the body, with a maximum of 8 cerata per row.

ANATOMY (Fig. 27h). The radular ribbon widens appreciably with age. An 11 mm preserved specimen from Lundy I. had a radula which tapered from 75 down to 35 μm. The adult formula varies from 45 to 96 × 0.1.0 and the median cusp is of an equivalent length to the 5–7 lateral denticles on each side. The flimsy jaws bear a single row of denticles along the masticatory margin. The penis is armed with a straight chitinous tube, and the vas deferens exhibits a distinct prostatic region. The genital system was illustrated by Schmekel (1970).

HABITS. Known diet hydroids are *Sertularella polyzonias* (L.), *S. gayi* (Lamouroux), *Halecium halecinum* (L.) and *Hydrallmania falcata* (L.). Miller (1961) found *C. caerulea* to have a strong preference for *Sertularella*, both in laboratory tests and in the field. In the vicinity of Marseilles, it has been found feeding on *Eudendrium racemosum* (Vicente, 1963).

Spawning has been observed between April and September in British waters. The spawn mass contains less than a thousand eggs (Miller, 1958), laid in a small ribbon curling anticlockwise.

DISTRIBUTION (Map 11). This species is reliably recorded from the western Mediterranean (Pruvot-Fol, 1954), the Atlantic coasts of France and Portugal, numerous localities all around the British Isles, and rarely from Sweden and southern Norway. Marcus (1955) extended the range to include São Paulo, Brazil, and Clark (personal communication, 1977) has recently found specimens around the shores of southern Florida, U.S.A. There are few intertidal records; most records have come from the sublittoral, down to 270 m.

DISCUSSION. Collections made available to us by divers have in recent years provided sufficient intermediates to satisfy us that, despite the great variability shown, all the extremes are reducible to a single species. The extremes are exemplified by the animals illustrated by Trinchese (1879), Eliot (1910) and Bürgin-Wyss (1961). Many intermediates between these (from both the Mediterranean and the Biscayan fauna), as well as other specimens which approached the type of *Eolis glotensis* Alder & Hancock, 1846, have been collected and studied during the preparation of this account.

76. ***Cuthona concinna*** (Alder & Hancock, 1843)
    Eolis concinna Alder & Hancock, 1843
    Cuthona concinna; Alder, 1879
    Cratenopsis concinna; Lemche, 1936
    Trinchesia concinna; Pruvot-Fol, 1954

APPEARANCE IN LIFE (Plate 31). This drab and inconspicuous tergipedid rarely exceeds 12 mm in length in British waters. The body is translucent white, but often has a yellowish hue, especially posterior to the pericardium. The oral tentacles and rhinophores display white tips but are otherwise almost colourless. The cerata are clavate, with dull brown hepatic lobes, and are set in up to 10 vertical rows, with up to 5 cerata per row. Darker granules can often be discerned within the lobes. Small white dots concentrated subapically in the ceratal epidermis are probably glandular. Some white pigment may also be present around the small cnidosac and may even form an indistinct streak running down the front of the ceras. The foot is of a similar width to the body for most of its length, but expands anteriorly to form distinct propodial lobes.

ANATOMY (Fig. 27c). The radula tapers only slightly; the ribbon of a 4 mm preserved specimen from Strangford Lough, Ireland, widened from 29 to 38 µm. In specimens over 5 mm in length the radula contains 25–45 teeth, each with 4 or 5 denticles on either side of the non-protruding median cusp. The jaws are denticulate.

Although the reproductive organs conform in most important respects to the tergipedid pattern, the penis is unarmed.

HABITS. Diet species recorded from the north-eastern Atlantic include *Laomedea gelatinosa* (Pallas), *Obelia longissima* (Pallas), *Sertularia cupressina* (L.) and *S. argentea* (L.). According to Meyer (1971), *Dynamena pumila* (L.) may be consumed around Canada and the U.S.A. Spawning occurs in May (Swennen, 1961b).

DISTRIBUTION (Map 12). This amphiatlantic boreo-arctic species has a range from the Normandy coast of France to Finmarken, Norway (Odhner, 1907), including localities all around Britain, and along the Netherlands coast (Swennen, 1961). American localities include the Atlantic coast of New England (Franz, 1970a) and Vancouver Island, Canada (O'Donoghue, 1922).

77. ***Cuthona foliata*** (Forbes & Goodsir, 1839)
    Eolis foliata Forbes & Goodsir, 1839
    Eolis olivacea Alder & Hancock, 1842
    Eolis conspersa Dalyell, 1853
    Cratena olivacea; Bergh, 1864
    Cuthona olivacea; M. Sars, 1870
    Cavolina olivacea; Beaumont, 1900

Amphorina olivacea; Eliot, 1906
Cratena foliata; Lemche, 1929
Cuthona foliata; Lemche, 1938
Trinchesia foliata; Winkworth, 1951

APPEARANCE IN LIFE (Plate 30). This is a distinctive species, reaching a maximal length of only 11 mm. A pale orange band may encircle each oral tentacle but equivalent bands are more distinct on the rhinophores. A characteristic crescent-shaped orange, red or brown patch is situated behind each rhinophore, curving outwards to the bases of the most anterior cerata. Similar patches immediately anterior to the rhinophores are less obvious. Another patch of pigment usually colours the dorsum behind the prominent pericardial swelling. Small white glistening dots are scattered liberally over the body and the cerata. The cnidosacs are large and fairly conspicuous. The hepatic lobe within each ceras may be olive green, dark brown or yellowish brown. In adults, seven to eight simple rows of cerata are set symmetrically on either side of the body. The longest row (from the first post-cardiac duct) rarely has more than 5 cerata. The body and head tend to be relatively broader than in other British tergipedids. The foot is slightly expanded anteriorly and is wider than the body for most of its length, tapering to a blunt tip posteriorly.

ANATOMY (Fig. 27j). Between 51 and 65 teeth have been seen in preparations of the gradually tapering uniseriate radula. A specimen from the Orkneys, 4 mm in preserved length, had a ribbon width tapering from 29 to 20 μm. Each tooth bears 4–7 denticles on either side of the non-protruding median cusp. The jaw plates are flimsy, with a short masticatory edge bearing one row of about 13 strong denticles.

The penis is armed with a hooked chitinous stylet and there is a short prostatic vas deferens. The female aperture lies beneath the cerata of the right hepatic duct, rather than anterior to them as demonstrated for the closely related *C. genovae* by Bouchet (1976).

HABITS. Our collections have been made intertidally, and in shallow sublittoral localities, and the specimens were feeding upon a variety of calyptoblastic hydroids including *Abietinaria abietina* (L.), *Sertularella polyzonias* (L.) and *S. gayi* (Lamouroux). Miller (1961) also recorded predation on *Tubularia*, *Obelia geniculata* (L.) and *Dynamena pumila* (L.). Spawning has been reported around Britain from May to August, and the egg mass can be described as a broad, semicircular coil.

DISTRIBUTION (Map 11). *Cuthona foliata* reaches its northern limits at the Faeroes and southern Norway (Lemche, 1929); reliable records exist also for the French Atlantic coast. Because of the possibility of confusion with records of *C. genovae*, it is not possible to be sure of the veracity of some Mediterranean records of *C. foliata*. Records exist for shallow localities all around the British Isles (Seaward, 1982).

DISCUSSION. Bouchet (1976) has established the separateness of *C. foliata* and *C. genovae*. He considered all Mediterranean records of *C. foliata* (Pruvot-Fol, 1951; Haefelfinger, 1960a; Swennen, 1961a; Vicente, 1967; Ros, 1975; Schmekel, 1968) to relate to *C. genovae*, while *C. foliata* s.s. was restricted to the open Atlantic coasts of Europe. A record from Morocco (Pruvot-Fol, 1953) may be a third, closely related species.

78. ***Cuthona genovae*** (O'Donoghue, 1926)
Amphorina alberti Trinchese, 1879, not Quatrefages, 1844
Cratena genovae O'Donoghue, 1926
Trinchesia genovae; Bouchet, 1976

APPEARANCE IN LIFE (Plate 9). This species rarely exceeds 6 mm in length. The general body shape is similar to that of *C. foliata* but there are slight differences in the anatomy and in significant pigmentation characteristics. The rhinophores are primarily a yellow shade over their upper halves, but banded with orange pigment sub-apically, having a less distinct orange band below. Streaks and patches of orange-red pigment are present on the head and in front of the pericardial hump, in a similar fashion to *C. foliata*, but in addition an orange-red stripe traverses the head between the oral tentacles, while a yellow streak runs longitudinally from the front of the head between the rhinophores, towards the pericardium. Vayssière (1888) and Bouchet (1976) noted an iridescent blue sheen over parts of the oral tentacles, but this was not obvious in the 3 mm specimens collected by Mr B. E. Picton, which at present constitute the only British records. The cerata are as numerous as in *C. foliata* of equivalent size. The digestive gland lobules are often salmon-pink in colour, with a distinctly granular apppearance.

ANATOMY (Fig. 27e). The tapering radula is small but is usually a ribbon of more than 46 teeth. Each tooth has a slightly protruding median cusp, flanked on either side by 6–9 denticles. Bouchet (1976) described the cutting edges of the jaws as relatively shorter than the corresponding edges of *C. foliata*. Schmekel & Portmann's (1982) 9 mm specimen from the Mediterranean Sea had the formula $57 \times 0.1.0$.

The female genital aperture is situated laterally, below the most anterior row of cerata on the right side, thus it is somewhat nearer the head than in *C. foliata*.

HABITS. More information is urgently needed about *C. genovae* which may in the past have been confused with *C. foliata*.

DISTRIBUTION. Several collections from the western Mediterranean have been assigned to this species (Bouchet, 1976); recent finds in Lough Ine (western Ireland) have extended the known range to the British Isles (Wilson & Picton, 1983).

79. **Cuthona gymnota** (Couthouy, 1838)
   Eolis gymnota Couthouy, 1838
   Eolis aurantia Alder & Hancock, 1842
   Eolis bellula Lovén, 1846
   Eolis aurantiaca Alder & Hancock, 1851
   Eolis sanguifer Dalyell, 1853
   Cavolina aurantiaca; Alder & Hancock, 1855
   Cratena gymnota; Gould, 1862
   Cuthona aurantiaca; Bergh, 1864
   Montagua gouldi Verrill, 1873
   Amphorina aurantiaca; Eliot, 1906
   Cratena aurantiaca; Iredale & O'Donoghue, 1923
   Cratena aurantia; Winckworth, 1932
   Cuthona aurantia; Lemche, 1938
   Catriona aurantia; Winckworth, 1941
   Trinchesia aurantia; Winckworth, 1951
   Trinchesia aurantiaca; Pruvot-Fol, 1954
   Catriona gymnota; Williams & Gosliner, 1979; Schmekel & Portmann, 1982

APPEARANCE IN LIFE (Plate 31). The maximal recorded body-length is 22 mm; it may be translucent and colourless, or lightly tinted by an orange or rosy hue. The rhinophores are consistently deeper orange or brown, darkening towards the tips. The cerata are arranged in as many as 12 rows along the sides of the body, with up to 7 cerata in the longest half-rows, near the pericardium. Each ceras contains a pink-orange to red-brown lobule of the digestive gland; epidermal pigmentation is confined to a subapical band of powdery dull white, yellow or orange. Some examples with white pigment on the cerata led Loman (1893) to propose the subspecific name *pallida*, but this seems inappropriate because we have found a continuous range of ceratal coloration. The foot is rounded anteriorly; the metapodial tip is short and rounded.

ANATOMY (Fig. 27d). The radular teeth have a distinctive recessed median cusp, flanked by 3 or 4 (rarely only 2) denticles on either side. The radular ribbon tapers markedly and consists of up to 87 teeth. In a 9 mm preserved specimen the width ranged from 5 to 80 μm. The pre-radular tooth is not always retained, but may be present even in large individuals, indicating a potential for the retention of all the radular elements throughout life. The cutting edge of the jaw is composed of fine bristles. The penis is armed with a straight, tubular spine; there is a small penial sac and the vas deferens has a prostatic section.

HABITS. The adults with their egg masses are often found near the bases of clumps of *Tubularia indivisa* L. It is sometimes found, especially in more exposed localities, on *T. larynx* Ellis & Solander, and from the American Atlantic coast on *T. crocea* (Agassiz). Certain other hydroids are also preyed upon, including *Bougainvillia ramosa* (Beneden), *Sarsia eximia* (Allman), *Garveia nutans* Wright and *Obelia longissima* (Pallas). The kidney-shaped spawn mass contains between 200 and 1000 eggs, 110–122 µm in diameter (Miller, 1958). Planktotrophic larvae hatch after 7–10 days at 16 °C (Schmekel & Portmann, 1982). Breeding has been noted in most months of the year (Miller, 1958; Swennen, 1961a). It passes through several generations each year (Miller, 1962).

DISTRIBUTION (Map 12). This species has a boreal amphiatlantic distribution; it has been recorded as far south as Arcachon (Bouchet & Tardy, 1976) and the western Mediterranean (Schmekel & Portmann, 1982). It is common in shallow waters all around the British Isles (Seaward, 1982), from Norway as far north as Finmarken (Odhner, 1939), and from Massachusetts and the New Jersey coast of the U.S.A. (Couthouy, 1838; Clark, 1975). A record from the American N.E. Pacific (Hurst, 1967) requires confirmation.

80. **Cuthona pustulata** (Alder & Hancock, 1854)
    Eolis pustulata Alder & Hancock, 1854a
    Cuthona pustulata; Alder, 1869
    Cratena longicauda Heincke, 1897
    Cratena pustulata; Eliot, 1906
    Cratenopsis pustulata; Lemche, 1936
    Trinchesia pustulata; Pruvot-Fol, 1954

APPEARANCE IN LIFE (Plate 29). The distinctness of this species, which had often (e.g. Thompson & Brown, 1976) been placed in the synonymy of *C. nana* (Alder & Hancock, 1842), has now been proved by the examination of 13 specimens measuring between 3 and 18 mm collected around Skomer I., Pembrokeshire, and of numerous juveniles up to 5 mm in length near Lundy I. A further juvenile, 5 mm in length, was collected off Mull, W. Scotland. At present, we are uncertain whether Roginsky's (1962) specimens from the White Sea, U.S.S.R., correspond with our *C. pustulata*.

The body is slender and translucent white, and the opaque white ovotestis can clearly be seen within the body cavity; the brown mandibles can also be discerned. There are white patches on the rhinophores and oral tentacles, sometimes concentrated distally, and there are scattered white specks on the cerata and around the bases of these appendages. The ceratal contents are light brown, pink or yellow.

An 18 mm specimen had 12 simple rows of cerata on either side of the body, including 6 precardiac rows. There was a total of 126 cerata with up to 8 in a single row. The cerata tend to be long and club-shaped, but may have swollen ends and very prominent epidermal glands obscuring the small cnidosacs. Often these glands are invisible to the unaided eye, unless the animal sickens. Alder (in Jeffreys, 1867) remarks that the cerata are capable of great distension, and the animal has the "peculiar power of bending them at right angles". Heincke's (1897) description of *Cratena longicauda* corresponds well with the rare examples of *Cuthona pustulata* which exhibit a greatly elongated metapodium.

ANATOMY (Fig. 27g). The tapering, uniseriate radulae of specimens between 8 and 18 mm in length had between 15 and 24 teeth. The central cusp appears slightly shorter than the 4–7 (usually 6) denticles on either side. The jaw plates are flimsy, but have a finely denticulated cutting edge. Contrary to Odhner's (1939) opinion, we have found a conical penial stylet, and the vas deferens is divided into a muscular and a prostatic zone. There is a single peripheral bursa copulatrix (Brown, 1980), in common with all British tergipedids except *Cuthona nana*.

HABITS. Alder & Hancock's specimens came from deep water "zoophytes" brought in by Cullercoats fishing vessels. Our specimens all came from *Halecium muricatum* growing on rock substrates between 10 and 33 m depth. Roginsky (1962) studied the development of a species supposed to be *C. pustulata* but said that "the papillae are greyish-blue with a rather steely tinge". Other inconsistencies include "brick-red eggs" and a diet of *Clava* and *Coryne*. Roginsky (1965) later became uncertain of the identification and added a footnote that her animals possibly belonged to a new species.

We have observed spawning from May until August; the ova measure approximately 180 µm in diameter. They are white and attached to the *Halecium* prey in small coils containing up to approximately 700 eggs. Hatching has not been witnessed, but the finding of clusters of as many as 30 juveniles (2–3 mm) on tufts of *Halecium* might suggest non-pelagic development.

DISTRIBUTION (Map 11). Because of doubts surrounding the identification of Norwegian material by Odhner (1939) and Russian records by Roginsky (1962), it is uncertain how far this species penetrates Arctic waters. A record from Iceland collected by Dr J. Sigurdson has, however, been verified by the authors, and Dr H. Lemche certainly collected this species in the Kattegat. Bouchet & Tardy (1976) record *C. pustulata* from as far south as Brittany. Specimens apparently resembling this species have been collected along the Pacific coast of north America, but we have not been able to confirm this.

British records have been few, but come from well-dispersed localities around the British Isles. Several references to this species in the literature are of dubious validity (e.g. Hamond, 1972). Some confusion seems to have arisen because some of the hand-coloured copies of the Ray Society monograph containing the type description depict pink cerata while others (and the original plate in the Hancock Museum, Newcastle upon Tyne) depict yellow cerata. The text describes them as "yellowish orange".

81. ***Cuthona rubescens*** Picton & Brown, 1978
   Eolis amoena Alder & Hancock, 1855 (part)
   Cratena amoena; Walton, 1908

APPEARANCE IN LIFE (Plate 31). This novel species, discovered recently, is based upon 28 specimens collected sublittorally near the coast of Devon, Cornwall, Pembrokeshire, Donegal, the Orkneys and Mull. One specimen was collected intertidally by Dr C. D. Todd at Robin Hood's Bay, Yorkshire. The maximal recorded length was 12 mm. The overall shape is like that of *Cuthona amoena*, but there are consistent differences in pigmentation.

Opaque white ovotestis lobules are visible through the translucent skin, as are the eyes and the dark brown mandibles. Small groups of white pigment specks are scattered over the dorsum, mainly in front of the ceratal bases. The smooth rhinophores are streaked on the rear face with opaque white surface pigment, interrupted by a crimson band. The blunt oral tentacles have a similar white streak but usually lack any counterpart of the crimson zone. Pigment on the cerata is restricted to the dorsal surfaces; it consists of a basal crimson patch and scattered white blotches, often concentrated over the cnidosac. The ceratal contents can be various shades of brown or olive-green; the latter is the colour adopted after a spell of feeding.

The cerata are cylindrical, but narrow at their bases and are rounded at the tip. They are set in up to 11 vertical rows with a maximum of 6 cerata in the two longest rows, on either side of the cardiac swelling. The foot is expanded to form rounded propodial lobes, more pronounced than those of *C. amoena*.

ANATOMY (Fig. 27b). The radula tapers only moderately; the ribbon of an 8 mm specimen ranged in width from 40 to 66 µm. In a 12 mm specimen, the radula consisted of 17 teeth; other specimens had the following radular formulae: length 8 mm, $16 \times 0.1.0$; 10 mm, $18 \times 0.1.0$; 10 mm, $16 \times 0.1.0$; 12 mm, $17 \times 0.1.0$. Each tooth may exhibit up to 8 denticles on either side of the stout median cusp.

The vas deferens proved to have a short prostatic section, and the penis bears a hooked terminal spine.

HABITS. Picton & Brown (1978) list all the records of this novel species; all except one came from scuba divers exploring at depths between 10 and 30 m; the exception was Dr Todd's single find made at extreme low water on the Yorkshire seashore. Recent reports have invariably suggested the diet of *Halecium halecinum* (L.), but Walton's (1908) two captures from the North Sea were found on *Sertularia argentea* (L.). It is noteworthy that *Cuthona amoena* also feeds on *Halecium halecinum*.

Spawn was found from May to August; it is a thin, white, irregularly coiled thread resembling closely that of *C. amoena*.

DISTRIBUTION (Map 12). Although described by Picton & Brown (1978) as a new species, it does appear that specimens were first recorded long before (Alder & Hancock, 1845–55, one specimen from Fowey Harbour in Cornwall which they named *Eolis amoena* although "the bases of the papillae [were] reddish and blotched with opaque white down the front of each"). Again in 1908 Walton described two captures from the North Sea that also differed from the type of *E. amoena* in that "the oral tentacles were white without the brown band", "the red band on the rhinophores was broad" and "the foot was more bilobed and produced into rounded lobes at the sides". More recent localities include the southernmost and northernmost parts of the British Isles, as well as western Ireland, the Irish Sea and the North Sea. There are no records from outside the British Isles.

82. ***Cuthona viridis*** (Forbes, 1840)
    Montagua viridis Forbes, 1840
    Eolis northumbrica Alder & Hancock, 1844a
    Eolis arenicola Alder & Hancock, 1847
    Eolis viridis; Forbes & Hanley, 1848
    Cavolina viridis; Alder & Hancock, 1855
    Eolis stipata Alder & Hancock, 1855
    Aeolis olriki Mörch, 1857
    Cratena viridis; Bergh, 1864
    Cratena olriki; Bergh, 1864; Bardärson, 1919
    Cavolina northumbrica; Alder, 1869
    Cuthona stipata; Alder, 1869
    Amphorina viridis; Eliot, 1910
    Diaphoreolis northumbrica; Iredale & O'Donoghue, 1923
    Cratena stipata; Winckworth, 1932
    Cuthona olriki; Lemche, 1938
    Cuthona viridis; Lemche, 1938
    Trinchesia viridis; Winckworth, 1951
    Trinchesia stipata; Winckworth, 1951
    Trinchesia caerulea Miller, 1958; Lemche, 1964 (part); not ***Doris caerulea*** Montagu, 1804

APPEARANCE IN LIFE (Plate 30). Specimens from British waters rarely reach 20 mm in length, although larger animals approaching 30 mm have been recorded from Arctic waters (as *C. olriki*). The semi-translucent body is either white or pale yellow. Opaque white pigment is concentrated distally on the blunt oral tentacles and the slightly more elongated rhinophores. The cerata each contain a dark green digestive gland lobule interspersed with black specks. The white or pale yellow cnidosacs are swollen, almost completely filling the tip of each ceras. These opaque pale cnidosacs are an important diagnostic character. White, occasionally pale yellow, surface pigment begins to make its appearance when the juveniles reach 2–3 mm in length; at first it is distributed unevenly on the frontal aspects of the cerata. In larger specimens it concentrates into an indistinct streak below a sub-apical ring. The cerata are generally elongated and cylindrical, but are slightly more bulbous in juveniles less than 4 mm in length. They are arranged in up to 9 rows (of which 4 are precardiac) with as many as seven cerata in the longest rows. The rows of cerata are set almost vertically on slightly elevated ridges (resembling *C. caerulea* in this respect).

ANATOMY (Fig. 27i). The uniseriate radula usually consists of between 35 and 54 teeth with five or six denticles on either side of the median cusp. The ribbon width of a 7 mm specimen ranged from 35 to 75 μm. The cutting margins of the jaw plates are always denticulated, but to a varying degree.

The penis is armed with a conical spine, while there is a prostatic vas deferens and a penial bulb.

HABITS. In addition to the calyptoblastic hydroids *Abietinaria abietina* (L.) and *Sertularella* spp. recorded by Brown (1980) as diet species, *Cuthona viridis* has been seen to attack also *Nemertesia*, *Sertularia* and, in the intertidal zone, *Clytia johnstoni* (Alder).

Spawning has been observed from February to August around British shores.

DISTRIBUTION (Map 11). This is a boreo-arctic species, reaching southward only as far as Roscoff on the French coast. It is well known from many localities around the British Isles, to 100 m, and penetrates the Arctic Ocean to Bear Island, as well as the seas around East and West Greenland and Iceland. (Many of the Arctic records were described under the name *C. olriki*.)

Similar, almost indistinguishable, species have been described from the west coast of the United States (*Cuthona albocrusta* (MacFarland, 1966)), Japan (*Cuthona signifera* (Baba, 1961)) and New Zealand (*Cuthona scintillans* Miller, 1977).

DISCUSSION. Both Miller (1958) and Lemche (1964) temporarily considered *C. viridis* to be synonymous with the variable *C. caerulea*. Although they are undeniably closely related, we have found them to be consistently distinct, even when only 1·5 mm in length. The large, opaque cnidosacs are an immediate distinction, while the bands of colour displayed by *C. caerulea* are never present. Both Lemche (personal communication) and Miller (1977) have since accepted the validity of *C. viridis*. No animal corresponding to the description of *Eolis northumbrica* has ever been refound in the northern Atlantic, and this is presumed to have been a specimen of *C. viridis* with damaged rhinophores. Photographs and preserved specimens of *Cuthona olriki* from Iceland were kindly placed at our disposal by Dr H. Lemche, but failed to show any significant deviation from British specimens of *C. viridis*. *Eolis stipata* was considered by Lemche (1941) to be a synonym of *C. olriki*.

Tergipes Cuvier, 1805

(type *Limax tergipes* Forskål, 1775; validated in opinion 773 of the I.C.Z.N., *Bull. zool. Nomencl.* **23**: 84 (1966))

Tergipedids with distinct oral tentacles, and sparse cerata so that only a single ceras originates from each hepatic branch.

83. *Tergipes tergipes* (Forskål, 1775)

Limax tergipes Forskål, 1775
Tergipes tergipes; Cuvier, 1805
Tergipes lacinulatus Blainville, 1824
Eolis despecta Johnston, 1828
Aeolis neglecta Lovén, 1846
Tergipes despectus Alder & Hancock, 1855

APPEARANCE IN LIFE (Plate 31). This minute aeolid reaches a maximal length of only 8 mm, and is immediately recognizable because each branch of the digestive gland leads to only a single ceras. A maximum of two cerata, usually only one, originate from the right hepatic lobe; the remainder of the cerata spring from the extensive left lobe. The body is translucent so that the digestive tract can be easily discerned, from the stomach to the ceratal contents and the opaque ovotestis. Streaks of red-brown pigment originate on the sides of the head and run posteriorly along each flank. A red-brown streak is usually to be seen at the base of each rhinophore, and descends to link with the streak on the flank. Similar red-brown pigment forms a subapical band covering the cnidosac of each ceras. The rhinophores are considerably longer than the oral tentacles, while the foot is narrower than the body, tapering to a long delicate tail beyond the last ceras.

Up to 8 cerata (rarely 10) may be present, the first two opposite, the others alternating. The undulating digestive gland shows greenish through the skin of the dorsum and the cerata.

Many authors have misleadingly illustrated this species in mirror-image, starting with Alder & Hancock (1845–55).

ANATOMY (Fig. 27m). The radula tapers greatly and rarely has more than 30 teeth, each with a slightly projecting median cusp and a variable (3–7, exceptionally 9–11) number of lateral denticles. A 3 mm preserved specimen had 20 teeth, increasing in width from 5 μm to 35 μm. The jaws have a serrated masticatory margin.

The penis is armed with a straight chitinous spine; other features of the reproductive organs were described by Schmekel (1970).

HABITS. This species is common on hydroids attached to kelp or on moored rafts, piers or ships, often in company with *Eubranchus exiguus*. The wide range of diet species includes *Obelia* and *Laomedea* species, together with *Gonothyraea loveni* (Allman), *Aglaophenia pluma* (L.), *Clava multicornis* (Forskål) and *Sarsia eximia* (Allman) (reviewed by Thompson, 1964).

The spawn mass is a small packet of white eggs, laid in March–August, attached to the hydroid stems. The spawn is more conspicuous to the human eye than are the adults. According to Giard (quoted by Garstang, 1889), the spawn masses resemble protectively the reproductive bodies of the hydroid. Development of the 80 μm ova is rapid (5 days at 20 °C), and young post-larvae are soon found in the field, sheltering in calyptoblast thecae (Tardy, 1964a; Swennen, 1961). The population density of *Tergipes* may be locally very high; Swennen (1959) counted 500/m² over 20 m² in March 1950 near Den Helder in the Netherlands. The generation time is short; Swennen (1961b) reared spawning adults in 5 weeks. *Tergipes* is euryhaline, and exceptionally occurs in localities where the salinity may be as low as 10‰ (Swennen, 1961).

DISTRIBUTION (Map 12). This species occurs all around the British Isles, to 20 m. It is widely distributed through the Atlantic Ocean from Brazil (Marcus, 1977) to Iceland and Norway (Lemche, 1938), the eastern American seaboard (Franz, 1970a) and the Mediterranean (Schmekel & Portmann, 1982).

Tenellia Costa, 1866
   (type *Tenellia mediterranea* Costa, 1866)
   Tergipedids with a domed anterior veil, produced laterally to form a lobe or a short tentacle on either side; having rows of cerata (not single cerata) along the upper sides of the body.

   (It is misleading to claim that *Tenellia* lacks oral tentacles; they are present in many specimens though not in all. They are not like the oral tentacles of other tergipedids, i.e. they are short and laterally or posteriorly directed, not extending towards the front as in, for instance, *Cuthona*. But "absence" of oral tentacles cannot be legitimately used to characterize the genus, as was done by Miller (1977), for example.)

84. **Tenellia adspersa** (Nordmann, 1845)
   Tergipes adspersa Nordmann, 1845
   Tergipes lacinulatus Schultze, 1849, not Blainville, 1824
   Eolis ventilabrum Dalyell, 1853
   Embletonia pallida Alder & Hancock, 1854a
   Tenellia mediterranea Costa, 1866
   Embletonia grayi Kent, 1869
   Eubranchus pallidus Jaeckel, 1952, not Alder & Hancock, 1842

Embletonia mediterranea; Vannucci & Hosoe, 1953
Tenellia pallida; Thompson & Brown, 1976
Schmekel & Portmann (1982) mistakenly cite as a synonym *Eolis pallida* Alder & Hancock, 1842, but this is a *Eubranchus* (Edmunds & Kress, 1969).

APPEARANCE IN LIFE (Plate 31). This inconspicuous species rarely exceeds 8 mm in length. The head is domed and spatulate, with lateral lobes which are sometimes produced into short, lateral or postero-lateral oral tentacles. The rhinophores are smooth and cylindrical. The general body colour varies from pale yellow to dull brown and there is a superficial black stippling over much of the body and the cerata. The tentacles are colourless but the cerata are pale yellow to orange. Small but distinct cnidosacs are detectable at the tips of the cerata. The cerata are arranged in up to 6 rows, each of up to 3 cerata on either side of the body (Coomans & Coninck, 1962). The anus is situated behind the second row of cerata on the right of the pericardium, while the genital openings are immediately ventral to the first ceratal row.

ANATOMY (Fig. 27l). The tapering radula consists of up to 45 teeth, each with a projecting median cusp and a variable number (up to 7) of denticles on either side.

The penis is armed with an apical stylet.

HABITS. All recorded elements of the diet have been listed by Roginsky (1970); they are mainly hydroids such as *Laomedea longissima* (Pallas), *L. loveni* Allman, *Cordylophora lacustris* Allman and the itinerant *Protohydra leuckarti* Greff (quoted from many authors, including Naville, 1926; Rasmussen, 1944; Jaeckel, 1952; Marcus & Marcus, 1955; Swennen, 1961b). Roginsky quotes a paper by Turpaeva (in the Russian language) which showed that Azov Sea specimens could flourish in a salinity range of 3–40‰. Acclimated specimens would live and lay their eggs in water of 50‰. It lives in shallow water, sometimes on the sea shore, often in harbours, estuaries and canals (such as the Canal de Caen, on *Cordylophora*, where Naville (1926) carried out his studies).

The spawn masses contain from 15 to 200 ova which may be white or pale pink in colour (Naville, 1926; Swennen, 1961b; Eyster, 1979). They develop into veliger stages which may then follow development-types 2 or 3; this is one of the few species of opisthobranchs capable of varying its mode of development (Roginsky, 1970), sometimes within a single population (Eyster, 1979). The parent nudibranch can produce either numerous batches of smaller eggs (approximately 70 μm in diameter) which hatched after 66–122 h in the form of veliger larvae, or less numerous batches of larger eggs (100 μm) that hatched after 140–190 hours as creeping stages which soon cast the embryonic shell and commenced benthic life directly. The temperature regime was 15–25 °C.

DISTRIBUTION (Map 12). There are very few British records of *Tenellia*, but it is probably common in areas like the Fleet in Dorset, and the Bristol Channel shores. It would be interesting to know if it will return to the upper Thames estuary; it was from the Victoria Docks, Rotherhithe, that the type of *grayi* Kent, 1869 (a synonym of *T. adspersa*) was first found, in company with *Daphnia* and many other freshwater crustaceans and rotifers.

The cosmopolitan global distribution between Japan (Baba & Hamatani, 1963), Brazil (Marcus & Marcus, 1955) eastern American and European shores (Eyster, 1979; Schmekel & Portmann, 1982) has probably been assisted by transportation in fouling communities attached to ships. The euryhaline and eurythermic characteristics of *T. adspersa* allow it to survive well in harbours and estuaries where it can feed on a wide variety of hydroids. The propensity for varying the reproductive mode may also enhance its capability for geographical spread.

EUBRANCHIDAE

Active aeolidaceans having an acleioproctic anal position; cerata swollen, sometimes with annular constrictions and rings of tubercles, and arranged in simple or branched rows. Propodial tentacles are absent, but smooth oral and rhinophoral (the latter usually much longer than the former) tentacles are conspicuous. The penis is unarmed, or armed with a stylet or thorns; having a bulb-like accessory gland near the base. A bursa copulatrix is present, opening into the female atrium.

The cutting edges of the jaws may be smooth or denticulated. The radula is triseriate; the lateral teeth are rectangular and each bears a single cusp. The larval shell is of type 1.

Eubranchus Forbes, 1838
   (type *Eubranchus tricolor* Forbes, 1838; validated in I.C.Z.N. Opinion number 774)
   = **Amphorina** Quatrefages, 1844
      (type *Amphorina alberti* Quatrefages, 1844)
   = **Galvina** Alder & Hancock, 1855
      (type *Eubranchus tricolor* Forbes, 1838)
   = **Capellinia** Trinchese, 1874
      (type *Capellinia doriae* Trinchese, 1874)
   = **Dunga** Eliot, 1902
      (type *Dunga nodulosa* Eliot, 1902)
   = **Egalvina** Odhner, 1929
      (type *Galvina viridula* Bergh, 1873)
   = **Eubranchopsis** Baba, 1949
      (type *Eubranchopsis virginalis* Baba, 1949)
   With the characters of the family.
The British species of *Eubranchus* have been reviewed by Edmunds & Kress (1969).

85. **Eubranchus cingulatus** (Alder & Hancock, 1847)
   Eolis hystrix Alder & Hancock, 1842) (suppressed in I.C.Z.N. Opinion 774 (1966))
   Eolis cingulata Alder & Hancock, 1847
   Galvina cingulata; Alder & Hancock, 1855

APPEARANCE IN LIFE (Plate 34). This slender species may reach 29 mm in length. The body is grey-white with blotches of superficial olive-green or brown. The cerata are arranged in up to 10 rows, each containing up to 13 (rarely 18) cerata in each half-row (Wilson & Picton, 1983). They are long, cylindrical, and only slightly inflated, having a white subterminal band and 2–3 interrupted olive-green or brown bands lower down. The cerata may pucker when contracted, but are not normally tuberculated (contrary to *E. doriae*). Character-istically the hepatic contents of the cerata are white or pale yellow, and the olive coloration on the dorsum is especially strongly developed between the bases of the cerata. The rhinophores are approximately double the length of the oral tentacles; both types exhibit a white tip and a conspicuous olive band (crimson according to Alder & Hancock).

ANATOMY (Fig. 28a). Some observations on anatomy and histology were published by Pelseneer (1894). Edmunds & Kress (1969) investigated the reproductive system and showed that a 30 µm long penial spine is present. It is difficult to be sure of the reliability of descriptions of the radulae because of the past confusion with *E. doriae*. We illustrate teeth from *E. cingulatus*, 23 mm in preserved length, from Salcombe (Devon); the formula is $54 \times 1.1.1$.

HABITS. Spawning has been reported in the months of September (Cullercoats: Alder & Hancock, 1845–55) and July (Plymouth: Marine Biological Association, 1957). This delicate species feeds upon the calyptoblast *Kirchenpaueria pinnata* (L.) (Hecht, 1895; Miller, 1961). We have found it infected with the copepod *Splanchnotrophus*.

DISTRIBUTION (Map 10). Records of this species are chaotic because of confusion with *E. doriae*. Schmekel & Portmann's (1982) report from the Bay of Naples plainly relates to *E. doriae* not to *cingulatus* as they claim. Edmunds & Kress's (1969) description of *cingulatus* from a number of British sources may be correct, but they introduce some uncertainty by illustrating this species with a line-drawing undoubtedly of a specimen of *E. doriae* (Text-fig. 1C, page 881 of their paper). We ourselves have fallen into the same error (Thompson & Brown, 1976, Fig. 89,

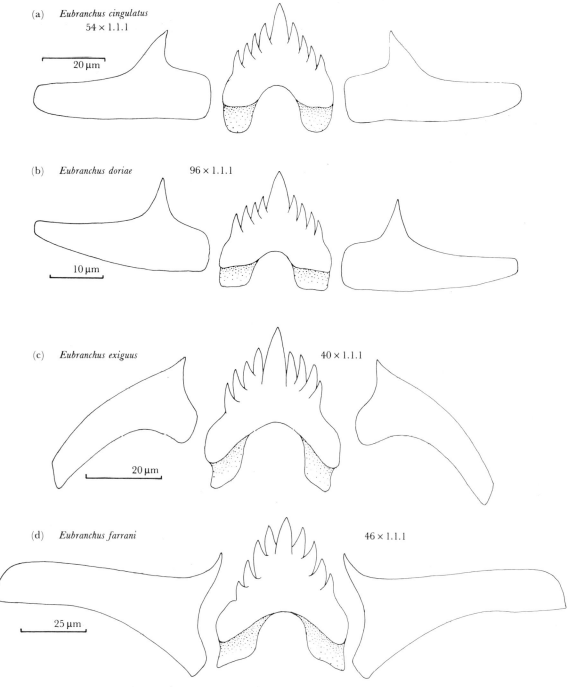

Fɪɢ. 28. Camera lucida drawings of aeolidacean radulae: *Eubranchus*.
(a) *Eubranchus cingulatus*, preserved length 23 mm, Salcombe, Devon.
(b) *Eubranchus doriae*, preserved length 4 mm, Lundy Island.
(c) *Eubranchus exiguus*, preserved length 4 mm, Kerrera, Scotland.
(d) *Eubranchus farrani*, preserved length 5·5 mm, Skomer Island, S. Wales.

page 167). Wilson & Picton (1983) promise a detailed re-investigation of published records of *E. cingulatus* and *E. doriae*.

86. *Eubranchus doriae* (Trinchese, 1874)
Capellinia doriae Trinchese, 1874
Tergipes doriae; Trinchese, 1879
Tergipes capellinii Trinchese, 1879
Capellinia capellinii; Vayssière, 1913.

APPEARANCE IN LIFE (Plate 34). This is a rare species in the British Isles, partly, no doubt, because it is so well camouflaged. The slender body may reach 6 mm in length (8 mm in the Bay of Naples, according to Schmekel & Portmann); it is white mottled with brown or olive, especially between the bases of the cerata. The cerata are mottled with brown, interrupted by 2–3 rings of pale nodular tubercles. The apex of each ceras exhibits a pale area overlain by a faint brown ring. The hepatic lobule within the ceras is white or pale yellow. There may be up to 7 transverse rows of cerata, each with 5–6 (maximally 8) cerata in each half-row. The rhinophoral and oral tentacles have white tips and a dark brown band.

ANATOMY (Fig. 28b). This is a much smaller species than *E. cingulatus*, with which it has often been confused, but the radular formulae obtained by Edmunds & Kress (1969) and ourselves are distinctive. Edmunds & Kress examined the radulae of two specimens from Banyuls-sur-Mer, on the Mediterranean coast of France, and published the following formulae: $61 \times 1.1.1$; $65 \times 1.1.1$. Our 5 mm specimen from Lundy I. had the formula $96 \times 1.1.1$. These formulae are very different from that described for *E. cingulatus* (body-length 23 mm in preservative, formula $54 \times 1.1.1$); moreover, it should be noted that the teeth of *E. doriae* are only half as large. The penial spine of *E. doriae* is only 10 μm long, compared with 30 μm in *E. cingulatus* (Edmunds & Kress, 1969).

HABITS. We have taken *E. doriae* on the hydroid *Plumularia setacea* (Ellis & Solander) at Pagham Harbour in Sussex; spawn was deposited in the month of November. Spawn described from the Bay of Naples contained white ova which measured 90 μm in diameter (Schmekel & Portmann, 1982). Edmunds & Kress (1969) believed that Tardy's account of the veliger development of what he called *Capellinia exigua* was actually based upon *Eubranchus doriae*; Tardy accepted this, but the confusion persists because the ova of Tardy's *Capellinia exigua* measured 120 μm in diameter, much larger than the Naples material. Some experiments on sexuality were described by Tardy & Dufrenne (1978), now using the name *Eubranchus doriae*.

DISTRIBUTION (Map 11). There have been very few authenticated records of *E. doriae* from the British Isles, principally from southern or south-western coasts. Wilson & Picton (1983) have published records for several Irish localities in Co. Cork and Co. Donegal; before that, this elusive species had been noted only for Pagham Harbour, Sussex, and for Lundy I. in the Bristol Channel. Elsewhere it is known from the Atlantic coast of France near La Rochelle and from the Bay of Naples (Tardy & Dufrenne, 1978; Schmekel & Portmann, 1982).

DISCUSSION. Trinchese's (1874) original material from the vicinity of Genoa (*Capellinia doriae*) was supplemented by his book on the aeolidaceans of the area (Trinchese, 1879, as *Tergipes doriae* and *Tergipes capellinii*). These descriptions and colour plates, together with Vayssière's (1913) material from Marseilles, give a clear indication of the distinctness of this species, with its irregular knobbly cerata. Despite this, several authors have confused *doriae* with *cingulatus*, which has banded, smoothly digitiform cerata. Edmunds & Kress (1969), Thompson & Brown (1976) and Schmekel & Portmann (1982) have all published illustrations of *E. doriae* incorrectly labelled *E. cingulatus*.

87. *Eubranchus exiguus* (Alder & Hancock, 1848a)
Eolis exigua Alder & Hancock, 1848a
Galvina exigua; Alder & Hancock, 1855
Aeolis exigua; Meyer & Möbius, 1865 (part)

APPEARANCE IN LIFE (Plate 34). This species does not usually exceed 10 mm in length, but exceptional individuals have been reported from the Isle of Man (15 mm: Miller, 1962) and from Den Helder in the Netherlands (18 mm: Swennen, 1961b). The body is greyish or yellowish white, with speckles and larger blotches of brown or olive-green; there are no extensive areas of white superficial pigment on the dorsum or the sides of the body. The inflated, urn-shaped, smooth cerata are sparse. Specimens we have examined possessed no more than 10 cerata in all, although Odhner (1907) examined material which exhibited up to 7 rows of cerata with 1–3 in each half-row. The shape of the cerata is very variable from minute to minute, but they are never knobbly. Their colour matches that of the body with brown or olive-green pigment forming 2–3 vague rings. Typically the tip of each ceras bears a ring of chalk-white which obscures the cnidosac within; there is usually a small band of brown apical to this. The hepatic lobes within the cerata are pale brown, deepening in colour towards the apex.

The oral and rhinophoral tentacles are pale, with 1–2 bands of brown and a subterminal zone of white. The rhinophores are nearly double the length of the oral tentacles.

The most notable distinguishing features of *E. exiguus* are the sparse, inflated, urn-shaped cerata and the distinct tendency towards a greenish tinge (sometimes referred to as olive or olive-brown) in the dark pigment of the body and its processes.

ANATOMY (Fig. 28c). There are scattered observations dealing with the anatomy and histology of *E. exiguus*. Hecht (1895) and Edmunds (1966) have contributed detailed reports on the structure of the cerata. Edmunds & Kress (1969) investigated the reproductive organs (they found the penis to be unarmed). Many authors have examined radulae (Colgan, 1914; Miller, 1958; Schmekel & Portmann, 1982). The formulae they derived were: length 1 mm: $33 \times 1.1.1$; 3 mm: $47 \times 1.1.1$ (Mediterranean specimen); 4·5 mm: $41 \times 1.1.1$; 6 mm: $52 \times 1.1.1$. On either side of the pronounced central cusp of the median tooth were 3–6 tiny denticles. The jaws were sometimes said to be smooth-edged, sometimes denticulate.

HABITS. This delicate aeolidacean is often found in company with the even more delicate *Tergipes tergipes*, feeding upon a variety of calyptoblast hydroids, including *Obelia geniculata* (L.), *O. dichotoma* (L.), *O. flabellata* (L.), *Laomedea flexuosa* Hincks, *Hydrallmania falcata* (L.) and *Kirchenpaueria pinnata* (L.) (Colgan, 1914; Cuénot, 1927a; Miller, 1961; Tardy, 1962). Edmunds (1966) reports feeding upon the gymnoblast *Coryne*.

Most European records have come from the lower shore or shallow sublittoral waters to 40 m; it may reach densities of up to 200/m² at certain seasons near Den Helder (Swennen, 1959). Miller (1958) believes that several generations may be passed through in the course of a year. Spawning occurs from March to July in the British Isles (McMillan, 1944; Marine Biological Association, 1957; Miller, 1961); there is a single report of autumnal spawning (October in Co. Galway, Farran, 1903). There have been a number of attempts to investigate the embryology, but they have been plagued by arguments about identification of the species used. The fullest account was given by Tardy (1962), including details of metamorphosis of the veliger larvae upon the adult prey, the hydroid *Kirchenpaueria*, but Edmunds & Kress (1969) give sound reasons for believing that Tardy's material was not *exiguus* but *doriae*. It is possible too that Hadfield's (1963) and Fischer's (1892) observations on larval and post-larval biology relate to a eubranchid other than *E. exiguus*. The kidney-shaped spawn mass characteristic of *E. exiguus* will distinguish this species from the spiral spawn of *E. cingulatus*; but the spawn of *E. doriae* has not been reliably described. This kind of confusion shows itself, for example, in published estimates of ovum diameter as divergent as 120 μm (Tardy, 1962) and 90 μm (Schmekel & Portmann, 1982), and of maximal numbers of ova per spawn mass varying from 20–70 (Hadfield, 1963) and 15–90 (Tardy, 1962) to 175 (Miller, 1958). We are even beginning to doubt the identification of the veliger larva described in detail by one of us (Thompson, 1959) under the name *E. exiguus*.

DISTRIBUTION (Map 11). This has been reported from scattered localities all around the British Isles, and elsewhere from the White Sea (Roginsky, 1962), from Massachusetts, Connecticut, Norway, the Baltic (down to 5–7‰, according to Swennen, 1961b), Atlantic and Mediterranean coasts of France, Italy, to 140 m (Miller, 1961; Franz, 1970a; Clark, 1975; Schmekel & Portmann, 1982).

88. **Eubranchus farrani** (Alder & Hancock, 1844a)
    Eolis farrani Alder & Hancock, 1844a
    Amphorina alberti Quatrefages, 1844
    Galvina farrani; Alder & Hancock, 1855
    Aeolis adelaidae Thompson, 1860
    Eolis andreapolis M'Intosh, 1865
    Eolis robertianae M'Intosh, 1865
    Galvina flava Trinchese, 1879
    Cavolina farrani; Garstang, 1889
    not **Amphorina alberti** Trinchese, 1879, Bergh, 1882, Vayssière, 1888

APPEARANCE IN LIFE (Plate 32). This species has in the past often been confused with *Eubranchus tricolor*, but *farrani* is quite distinct, being more slender, smaller, with fewer cerata of less compressed shape, and different colour patterns.

The maximal length is 20 mm (23 mm in the Bay of Naples, according to Schmekel & Portmann, 1982). The body is typically white or grey-white, with orange-tipped rhinophores and oral tentacles (sometimes with white apices), with scattered orange spots and blotches on the dorsum, and white inflated cerata having a conspicuous sub-terminal orange or yellow ring. This corresponds to the Form A of Edmunds & Kress (1969) and is the variety illustrated by Alder & Hancock (1845–55). Occasional variants show individual exaggeration of certain of the component markings. The entire body, cerata and all, may be of a beautiful golden hue. In others there may be brown patches on the body, and in one specimen (from Pembrokeshire) nearly the whole body was dark chocolate brown in colour (approaching the Form B of Edmunds & Kress). Other specimens may have orange blotches ringed with blue (Garstang, 1890; Farran, 1903), and the juveniles may exhibit no colour at all other than the pale brown or salmon pink hepatic contents of the cerata.

The cerata are arranged in up to 10 diagonal rows, of up to 5 cerata in each half-row. Their shape may be variable from minute to minute but it can always be seen that they are moderately inflated.

ANATOMY (Fig. 28d). Edmunds & Kress (1969) investigated the reproductive organs and reported the presence of a tapering penial spine, 30 µm in length. Some aspects of the histology of *E. farrani* were described by Henneguy (1925).

A number of descriptions of radulae have been published (Edmunds & Kress, 1969; Schmekel & Portmann, 1982), yielding formulae: length 6 mm, 43 × 1.1.1; 16 mm, 47 × 1.1.1; 23 mm, 56 × 1.1.1. There are 2–5 denticles on either side of the prominent median cusp of the central tooth. The jaws exhibited fine denticles along the cutting edges.

HABITS. Several authors agree with our own observation that *E. farrani* occurs around the British Isles in the shallow sublittoral, to 30 m, feeding upon calyptoblasts, including *Obelia geniculata* (L.), attached to kelps. Spawning occurs during the warmer months (it is not possible to be more precise because of the taxonomic uncertainty which at one time surrounded this species and some of its congeners). The diameter of the white, cream or pinkish ova is 90 µm (Colgan, 1914; Schmekel & Portmann, 1982) and hatching occurs after 6 days at 16 °C.

DISTRIBUTION (Map 10). This species has been recorded from localities all around the British Isles. Elsewhere it occurs from Norway and the Atlantic coast of France to the western Mediterranean (Odhner, 1907; Cornet & Marche-Marchad, 1951; Pruvot-Fol, 1954; Haefelfinger, 1960a; Schmekel & Portmann, 1982). A record from the Suez Canal (O'Donoghue, 1929) requires confirmation.

89. **Eubranchus pallidus** (Alder & Hancock, 1842)
    Eolis pallida Alder & Hancock, 1842

Eolis minuta Alder & Hancock, 1842
Eolis picta Alder & Hancock, 1847
Galvina pallida; Alder & Hancock, 1855
Galvina picta; Bergh, 1873
Eolis flavescens Friele & Hansen, 1876
Galvina flavescens; Bergh, 1877

APPEARANCE IN LIFE (Plate 33). Despite its name, this attractive species is not usually pale in colour like the variety which Alder & Hancock chose to illustrate in their type description. Although with a microscope the external colour pattern appears vivid, in the field *E. pallidus* is well camouflaged and may easily be missed. The maximal length may reach 23 mm (Miller, 1958) and the pattern is composed of brown, red, orange, gold and white markings on a translucent grey-white background. The dorsum and flanks are mottled with red or brown (*never* greenish) and there may also be some white and orange blotches. The inflated cerata are arranged in up to 10 oblique rows with up to 7 cerata in each half-row. Each ceras typically has a pale distal tip then a subterminal white area obscuring the cnidosac; over part of this white zone is a conspicuous golden ring. The remainder of the ceras exhibits superficial white specks and red or rich brown blotches. The sacculated hepatic lobule is pale to medium brown.

The rhinophoral tentacles are approximately double the length of the oral tentacles. Both pairs exhibit a white distal tip, orange-brown basal mottling and a conspicuous dark band in an intermediate zone. The frontal margin of the foot is somewhat broader than in other British eubranchids.

Eubranchus pallidus approaches *farrani* in some of its features but it is nonetheless distinct in body-shape and in the rich brown mottling of *pallidus*. This mottling lacks the greenish tinge often described for *cingulatus*, *exiguus* and *doriae*, and *pallidus* lacks any trace of dark rings on the cerata.

ANATOMY (Fig. 29a). The structure and histology of the reproductive organs have been described by Lloyd (1952) and by Edmunds & Kress (1969). A 90 µm penial spine is present. Some observations on the histology of the cerata were presented by Edmunds (1966). A number of radulae have been described by Miller (1958) and by Edmunds & Kress (1969): length 2·5 mm, 48 × 1.1.1; 6 mm, 48 × 1.1.1; 18 mm, 60 × 1.1.1; 23 mm, 56 × 1.1.1. The median cusp of the central tooth exhibited up to 5 denticulations on either side. The jaws are distinctly denticulate.

HABITS. This is an opportunistic species, like other eubranchids, and has a generation time of about 6 weeks (Orton, 1914); the normal life span is 4–8 months (Miller, 1958). The diet consists of both gymnoblastic and calyptoblastic hydroids (Walton, 1908; McMillan, 1944; Swennen, 1961b; Miller, 1961), for example *Obelia geniculata* (L.), *O. longissima* (Pallas), *Sertularia cupressina* (L.), *Hydrallmania falcata* (L.), *Tubularia indivisa* L., *T. larynx* Ellis & Solander, *Hydractinia echinata* (Fleming).

Spawn has been found in the months of May (McMillan, 1944) and July (Miller, 1958); it takes the form of a loose coiled ribbon of up to $1\frac{3}{4}$ turns, containing up to 700 white ova (Miller, 1958).

DISTRIBUTION (Map 10). This species had already evidenced a wide distribution within the British Isles by the mid-nineteenth century; Alder & Hancock (1845–55) recorded it from off the coasts of Northumberland, Torbay, Dublin and the Menai Strait. Many of the intervening gaps have since been filled (Seaward, 1982). Elsewhere there are records from Norway, the Netherlands, Atlantic coasts of France, the western Mediterranean and New England (Odhner, 1907; Swennen, 1961; Pruvot-Fol, 1954; Brown, 1979a), to 60 m. A report from the Suez Canal requires confirmation (O'Donoghue, 1929).

90. **Eubranchus tricolor** Forbes, 1838
Eolis violacea Alder & Hancock, 1844a
Eolis amethystina Alder & Hancock, 1845a

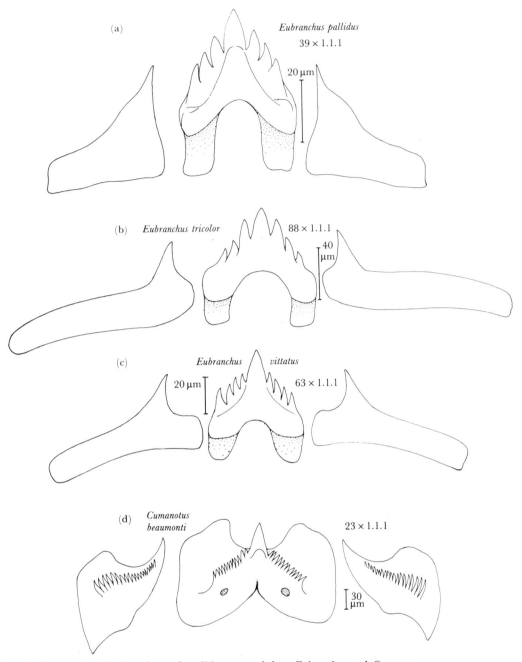

(a) *Eubranchus pallidus*
39 × 1.1.1
20 μm

(b) *Eubranchus tricolor* 88 × 1.1.1
40 μm

(c) *Eubranchus vittatus*
20 μm 63 × 1.1.1

(d) *Cumanotus beaumonti* 23 × 1.1.1
30 μm

Fig. 29. Camera lucida drawings of aeolidacean radulae: *Eubranchus* and *Cumanotus*.
(a) *Eubranchus pallidus*, preserved length 4 mm, Lundy Island.
(b) *Eubranchus tricolor*, preserved length 15 mm, Lundy Island.
(c) *Eubranchus vittatus*, preserved length 28 mm, Salcombe, Devon.
(d) *Cumanotus beaumonti*, preserved length 9 mm, coll. C. Eliot at Plymouth, Devon, in 1908.

Galvina tricolor; Alder & Hancock, 1855
Galvina viridula Bergh, 1873
Egalvina viridula; Odhner, 1929

APPEARANCE IN LIFE (Plate 32). The body-length may reach 45 mm (Miller, 1958), pale yellow
or greyish white in colour. Although Jeffreys (1867) attempted to convey the arrangement of
the cerata by writing of 13 or 14 transverse rows, more recent authors have stressed that this is

unhelpful because the hepatic ducts divide many times so that the 150 (maximally) cerata are closely packed together, without any simple pattern. So close is the packing that individual cerata are flattened antero-posteriorly; this gives them their unique flattened but inflated flask-shape. Early authors, such as Forbes (1838) and Herdman (1881), wrote of the animal's superficial resemblance to a hedgehog; Herdman went further and fancifully noted that when abruptly disturbed this nudibranch rolled up like a hedgehog.

The cerata are translucent white or cream, with a golden yellow subterminal ring, surrounded by opaque white pigment; deep within each ceras is the red-brown, smooth digestive gland lobule which is characteristically tinged with violet towards the tip. It is noteworthy that juveniles up to 13 mm in length may be pure white all over, except for a pale brown core to each ceras.

The rhinophoral and oral tentacles are pale yellow-brown with paler tips.

ANATOMY (Fig. 29b). Edmunds & Kress (1969) have described the reproductive organs; the penis is unarmed. The central teeth of the radula each exhibit a strong median cusp flanked on either side by 3–4 denticles. Details are available for numerous radular formulae: length 2 mm, $50 \times 1.1.1$; 5·5 mm, $62 \times 1.1.1$; 7 mm, $50 \times 1.1.1$; 15 mm, $83 \times 1.1.1$; 15 mm, $88 \times 1.1.1$; 30 mm, $79 \times 1.1.1$; 45 mm, $90 \times 1.1.1$ (Miller, 1958; Edmunds & Kress, 1969; personal observations). The jaws bear a single distinct row of denticles along each cutting edge.

HABITS. Miller (1961) listed a diet of both gymnoblast (*Tubularia indivisa* L.)- and calyptoblast hydroids, especially the rather coarse species, such as *Hydrallmania falcata* (L.), *Abietinaria abietina* (L.), *Nemertesia antennina* (L.) and *N. ramosa* (Lamouroux), but our experience, like that of Garstang (1890), is that *Eubranchus tricolor* is best found by searching the most delicate calyptoblasts such as *Obelia geniculata* (L.).

Spawn has been noted at Plymouth in May and November (Marine Biological Association, 1957); each spawn band may contain up to 650 pink ova (Miller, 1958). Roginsky (1962) noted field spawn in June in the White Sea.

DISTRIBUTION (Map 10). This species has been recorded all around the British Isles, although confusion exists in some of the older records because of the contemporary merging of *E. tricolor* with *E. farrani*. Elsewhere, it is reliably reported from Arctic waters (Roginsky, 1962; Lemche, 1941), and from the Atlantic coast of France (Cuénot, 1927a; Cornet & Marche-Marchad, 1951). It is not usually found on the shore, but occurs on hydroids in depths to 80 m.

91. **Eubranchus vittatus** (Alder & Hancock, 1842)
  Eolis vittata Alder & Hancock, 1842
  Galvina vittata; Bergh, 1873, Eliot, 1910
  Cratena vittata; Trinchese, 1881

APPEARANCE IN LIFE (Plate 34). This is one of the most rare British nudibranchs, reaching 29 mm in extended length but found on only three occasions. The body is pale brown in colour, with rust-coloured blotches. The cerata are also pale brown, and exhibit two or three rusty brown bands and a distal white or yellow zone. There are up to 7 transverse rows of cerata, with 4–7 in each half-row. They are elongated and slender in shape. Both oral and rhinophoral tentacles are brownish basally, with a rusty orange-brown band and a pale yellow tip.

ANATOMY (Fig. 29c). The radula of the largest known specimen, 29 mm long in life (28 mm in preservative) was examined. The formula was $63 \times 1.1.1$ (Alder & Hancock examined a radula which exhibited $67 \times 1.1.1$); each median tooth had a strong central cusp flanked on either side by 4 denticles. Alder & Hancock (1845–55) show the lateral teeth to be triangular in shape but this is based upon a misinterpretation; the true shape is shown in Fig. 29c.

HABITS. Farran (1903) recorded finding *E. vittatus* on the hydroid *Antennularia* ($=$ *Nemertesia*); nothing else is known except Alder & Hancock's (1845–55) observation that it occurs only in "deep" water (North Sea, off Cullercoats).

DISTRIBUTION (Map 10). This species was until recently known only from the North Sea and from Co. Galway, Ireland, but divers brought in a conspicuous specimen in 1980, from the sea off Salcombe, Devon. It confirmed that this is a valid and distinct species.

A record from Suez (O'Donoghue, 1929) was probably mistaken.

CALMIDAE

Slow-moving, semi-parasitic aeolidaceans, living in association with the eggs of marine animals, with numerous cerata, set in pedunculate clusters. Lobules of the digestive gland penetrate the lumina of the cerata, but cnidosacs are absent. Smooth oral, rhinophoral and propodial tentacles are present. The ovotestis takes the form of symmetrical lateral series of rosettes, visible through the dorsal skin in life. The genital openings are widely separated, male from female; the penis is unarmed.

The hindgut and anal opening are absent in the adult. Thin mandibles are present, each with a smooth cutting face. The radular teeth are mostly fused, so that the radula is a more or less rigid continuous band; the first-formed teeth are smooth and arched, resembling those of *Favorinus*. The larval shell is of type 2.

Calma Alder & Hancock, 1855
(type *Eolis glaucoides* Alder & Hancock, 1854a, validated in 1966 by Opinion 780 of the International Commission)
= *Forestia* Trinchese, 1881
(type *Forestia mirabilis* Trinchese, 1881)
With the characters of the family

92. ***Calma glaucoides*** (Alder & Hancock, 1854a)
Eolis glaucoides Alder & Hancock, 1854a
Eolis albicans Friele & Hansen, 1876
Forestia mirabilis Trinchese, 1881

APPEARANCE IN LIFE (Plate 36). This species can adopt a very flattened posture when among the eggs of fish, so that it is very difficult to see; the pale body-colour increases the cryptic effect.

The animal may reach 23 mm in maximal length, translucent white or pale cream in colour; the cerata, set in up to 12 pairs of dorso-lateral clusters (maximally 4 cerata in each cluster), each contain a digestive gland lobule of which the colour varies (according to diet) from pale yellow to brown. The shape of the cerata also varies so that each ceras may be swollen and pear-shaped after a meal, but can become slender and finger-like later. The ceratal clusters are held at a low angle laterally, and often appear to stem from common peduncles. The ceratal apices are rounded and speckled with white pigment; cnidosacs are absent. The genital apertures are separate, a condition found also in *Fiona* but rare in other aeolidaceans. The male opening is anterior to and below the right rhinophore, while the female opening lies between and below the 1st and 2nd ceratal clusters on the right side of the body. According to Evans (1922), Trinchese (1881) confused the nephroproct with the anus; there is no hindgut and no anus in the adult *Calma*.

In sexually mature individuals the pale yellow ovotestis lobules are clearly visible through the skin of the back; they look like serially repeated rosettes, lying symmetrically between the peduncles of the ceratal clusters. According to Evans (1922), the grey centres of the rosettes are male acini, with the yellowish female acini arranged around them. The serial arrangement of the components of the ovotestis is unusual in the nudibranchs.

The head bears smooth oral and rhinophoral tentacles which sometimes have a pinkish tinge (Miller, 1958); graceful, reflexed propodial tentacles are also present. The foot is broad and clings tenaciously to the food.

ANATOMY (Fig. 31a). The fullest account of the anatomy of *Calma* was that of Evans (1922), confirming and supplementing the investigations of Hecht (1895). Further details have been supplied by Henneguy (1925: histology) and Rowett (1946: buccal mass). The most distinctive feature of *Calma glaucoides* is its remarkable radula which consists of a continuous band of fused teeth (Fig. 31a); a 15 mm specimen possessed 197 such teeth in its radula. In shape, each tooth is smooth and arched. Evans (1922) detached some of the first-formed teeth from the ribbon and examined them; he concluded that they resembled the teeth of *Favorinus* (a favorinid which also favours fish eggs for its diet).

HABITS. This species feeds almost exclusively upon the eggs of littoral teleost fish, sometimes in considerable aggregations. Farran (1903), for example, found about 60 specimens on a batch of *Gobius niger* L. eggs on the coast of Co. Galway. The Plymouth Marine Fauna (Marine Biological Association, 1957) records a find of as many as 50 individuals among eggs of *Blennius ocellaris* L. in an empty *Buccinum* shell. Evans (1922) notes that *Calma* may attack the eggs of species of *Cottus*, *Lepadogaster*, *Liparis* and *Gobius*. In the Isle of Man Miller (1961) reports this nudibranch from the eggs of only *Lepadogaster lepadogaster* (Bonnaterre) and *L. gouanii* (probably synonymous with *L. lepadogaster*). Haefelfinger (1962b) alone claims that the diet of *Calma* may include also cephalopod eggs. The Plymouth Marine Fauna cites a report dated 1909 to the effect that this species feeds upon the hydroid *Hydractinia* growing on *Buccinum* shells, but confirmation is needed.

The spawn is a spiral ribbon, white or cream in colour containing ova 80 µm in diameter (Schmekel & Portmann, 1982), produced in the months of April (Marine Biological Association, 1957) and July (M.B.A. and Farran, 1903).

DISTRIBUTION (Map 12). British records are sparse and come from the west coasts (Seaward, 1982). Outside the U.K., there are reports from the Galician coast of Spain (Picton, pers. comm.), the western Mediterranean (Schmekel & Portmann, 1982) and Norway (Odhner, 1939). It is always found in shallow water, to 40 m.

FIONIDAE

Mobile, aggressive aeolidaceans, living in association with oceanic, drifting cirripedes or chondrophores, with a latero-dorsal anal position and numerous cerata which are scattered and form neither discrete clusters nor rows. The ducts of the digestive gland are symmetrical in the adult. Lobules of this system penetrate the lumina of the cerata, but cnidosacs are absent. Each ceras exhibits an undulating membrane down its mesial face. A latero-dorsal ridge runs around the body, corresponding to a parapodial rim, not a pallial or notal ridge such as is found in the doridacean and dendronotacean nudibranchs. Smooth oral and rhinophoral tentacles are present on the head, but propodial tentacles are absent. The genital apertures are widely separated, male from female; the penis is unarmed.

Elongated jaws are present. The radular teeth each have a prominent cusp and lateral denticles; radula uniseriate. The larval shell is of type 2.

Fiona Alder & Hancock, 1851 (in Forbes & Hanley, 1848–53)
    (type *Oithona nobilis* Alder & Hancock, 1851)
    = *Oithona* Alder & Hancock, 1851 (in Forbes & Hanley, 1848–53), not Baird, 1843
    (type *O. nobilis* Alder & Hancock, 1851 (in Forbes & Hanley, 1848–53)
    = *Hymenæolis* Costa, 1866
    (type *H. elegantissima* Costa, 1866)
    With the characters of the family.

93. *Fiona pinnata* (Eschscholtz, 1831)
    Limax marinus Forskål, 1775
    Eolidia pinnata Eschscholtz, 1831
    Eolidia longicauda Quoy & Gaimard, 1832
    Oithona nobilis Alder & Hancock, 1851 (in Forbes & Hanley, 1848–53)
    Fiona atlantica Bergh, 1857

Hymenæolis elegantissima Costa, 1866
Fiona marina; Casteel, 1904
Fiona marina; Eliot, 1910

APPEARANCE IN LIFE (Plate 36). This species is rare in the British Isles and the only published descriptions were made using dead or dying animals. Alder & Hancock's (1845–55) colour plate was based upon two specimens found near low water at Bar Point, Falmouth in 1850, but unfortunately these were poisoned inadvertently. We are forced to rely to a great extent on observations on *Fiona* made by Bayer (1963), Holleman (1972) and on Australian material examined during 1968 while one of us was resident in New South Wales.

The body-length may reach 60 mm (Odhner, 1907), but perhaps 50 mm would be more general (Alder & Hancock, 1845–55; Pruvot-Fol, 1954). Our Australian material did not exceed 18 mm in length. The body is pale flesh coloured, with white pigment flecks on the epidermis of the cerata; the tips of these organs are opaque bluish-white with a noticeable metallic glint. No cnidosacs are present within the cerata, although the lobules of the digestive gland penetrate their lumina in the usual aeolidacean fashion. The colour of these lobules varies with the diet. Holleman (1972) has investigated this colour change in some detail. When feeding upon the chondrophore *Velella*, the digestive gland becomes bright blue, but when given *Lepas anatifera* (L.) (another suitable item of the normal diet) the lobules soon change to a rich pink (pale brown, according to Eliot, 1910). The lobules of Alder & Hancock's Falmouth specimens were "rich dark brown"; their diet is open to conjecture.

The cerata emerge without apparent order all over the dorsum, which is an ovoid platform, projecting all around the body as a latero-dorsal ridge. It is probable that this is a pedal structure, not homologous with the notal ridge (or pallial skirt) of the dendronotacean and doridacean nudibranchs. The cerata are finger-like and very numerous in the adult; the most obviously diagnostic feature of *Fiona* is that the cerata each bear an undulating attached membrane on the mesial face. This is unique among the opisthobranch molluscs and its function is unknown. The undulating membranes, it should be emphasized, are immobile structures.

The rhinophoral and oral tentacles are pale, smooth and tapered. There are no propodial tentacles.

ANATOMY (Fig. 31b). The general anatomy has been investigated by Alder & Hancock (1845–55) and Pelseneer (1894). More specialized investigations were published by Henneguy (1925: histology) and Russell (1929: nervous anatomy). The principal point of interest is the apparent symmetry of the channels which connect the digestive gland lobules with the stomach; this must be secondary, because the veliger larva has, in common with other nudibranch larvae, a large left digestive gland and a vestigial right moiety (Haddon, 1882; Casteel, 1904; Holleman, 1972). Alder & Hancock (1845–55) describe the elongated jaws (graphically described as having the shape of a couple of valves of *Mytilus*) and the radula, with the formula 33–40 × 0.1.0. We have examined the radula of a specimen from San Pedro, California, 10 mm long in preservative. The formula was 39 × 0.1.0; each tooth exhibited 7 slender denticles on either side of the prominent median cusp (Fig. 31b). Schmekel & Portmann (1982) describe a radula from the Mediterranean, having the formula 44 × 0.1.0; the animal measured 38 mm in length.

HABITS. This is an opportunistic species settling upon floating objects, and attacking living components of the specialized, oceanic, drifting fauna (the pleuston). *Fiona* cannot swim after larval metamorphosis and it is incorrect to describe the adults as pelagic (the term used by Eliot, 1910). The veligers settle and metamorphose on goose barnacles and on chondrophores such as *Velella* and *Porpita*. Bayer (1963) and Holleman (1972) have described the feeding behaviour. The initial attack on a *Lepas anatifera* (L.) (*L. pectinata* Spengler is not acceptable) is made in the region of the soft stalk. After a time, the tergal and scutal plates relax and the nudibranchs are able to attack the viscera. *Fiona* will also attack *Velella* and *Porpita*, but always avoids the cnidarian tentacles and eats first the epithelium on the dorsal side of the float.

Nematocysts become prevalent in the faeces soon after the attack has begun, and, as remarked above, the digestive gland changes from brown or pink to bright blue in colour. The diet in British waters is unknown.

The spawn takes the form of a spiral of white ova. The embryonic period is 10–11 days at 20 °C (Schmekel & Portmann, 1982), for comparison with estimates at unknown temperatures of 5 days (Holleman, 1972) and 10 days (Mazzarelli, quoted by Pelseneer, 1935). Spawn may be attached to empty *Spirula* shells or to *Velella* sails (Tchang-Si, 1931); each capsule contains 1 or 2 eggs. Fecundity is remarkably high; in the Miami laboratory of Bayer (1963), 4 adults produced 89 spawn masses in the course of 3 weeks. Observations on the rapid post-larval growth rate have been detailed by Holleman (1972).

DISTRIBUTION (Map 12). Outside Britain, *F. pinnata* has been reported from the Mediterranean Sea, Florida, California, Japan, Australia and Madagascar (Bergh, 1880; O'Donoghue, 1926; Lance, 1961; Tchang-Si, 1931; Kropp, 1931; Schmekel & Portmann, 1982). Within the British Isles, there are sparse records, from the south and west of England (Seaward, 1982). Other than Alder & Hancock's first British records, this species is normally found at the surface of the ocean or attached to stranded flotsam, such as cuttlebones bearing *Lepas*, or empty *Spirula* shells.

PSEUDOVERMIDAE

Highly specialized aeolidaceans, degenerate and vermiform, living in the interstices of coarse sediments; reduced in size, usually less than 10 mm in length. No propodial, oral or rhinophoral tentacles are present, and the cerata are reduced to small fingers or buttons forming a paired dorso-lateral series. Each ceras contains a cnidosac, connected to the alimentary canal in the usual way.

The jaws may be smooth-edged or bear a single row of denticles. The median radular teeth each have a prominent cusp and lateral denticles; radula triseriate. The larval shell is unknown.

Pseudovermis Périaslavzeff, 1891
(type *Pseudovermis paradoxus* Périaslavzeff, 1891)
With the characters of the family.
The world's species of *Pseudovermis* have been reviewed by Salvini-Plawen & Sterrer (1968) and by Challis (1969).

94. *Pseudovermis boadeni* Salvini-Plawen & Sterrer, 1968
P. schultzi Boaden, 1961, not Marcus & Marcus, 1955

APPEARANCE IN LIFE (Plate 36). We have not examined this species alive; our description is based upon that of Boaden, who found 3 specimens in a gently sloping beach of coarse sand at Traeth Bychan, on the coast of Anglesey. In length these measured up to 3·5 mm. The body was elongated and vermiform, with a rounded, acorn-shaped head, adapted for burrowing within the substratum. The viscera can be discerned through the body-wall: a slightly asymmetrical series of dorso-lateral, ovoid cnidosacs (15 in all) lies beneath button-like excrescences of the back. These buttons correspond to the cerata of other aeolidaceans, but in *P. boadeni* they are not penetrated by branches of the digestive gland. Within each cnidosac, Boaden (1961) discovered nematocysts which resembled closely those of the prey, the cnidarian *Halammohydra*.

ANATOMY (Fig. 31c). A 3 mm specimen from Traeth Bychan was investigated, through the courtesy of Dr Boaden. The formula was 20 × 1.1.1. The lateral teeth each display a forked tip; the median teeth each had 6 fine denticulations on either side of a prominent central cusp. Stout jaws are present, each having a toothed cutting edge.

HABITS. Little is known about the mode of life of this species, except that it is believed to feed upon *Halammohydra vermiformis* Swedmark & Teissier.

DISTRIBUTION. The type locality was, as mentioned above, the coast of Anglesey (Boaden, 1961); the only subsequent report has been from the Bristol Channel, published by Seaward (1982).

DISCUSSION (Map 11). According to Poizat (1978), there are 5 other species of European *Pseudovermis*: *P. paradoxus* Kowalevsky, 1901 (Black Sea), *P. schultzi* Marcus & Marcus 1955 (northern Adriatic Sea; Bassin d'Arcachon), *P. axi* Marcus & Marcus, 1955 (Banyuls-sur-Mer), *P. setensis* Fize, 1961 (Golfe du Lion) and *P. papillifera* Kowalevsky, 1901 (Aegean and northern Adriatic Seas; Ile des Embiez). More observations on living animals will be needed before their validity can be attested.

EMBLETONIIDAE

Mobile aeolidaceans, dorso-ventrally compressed, with a mid-lateral anal position and swollen cerata arranged in a single series down each side of the body. The ceratal tips may be rounded (*Embletonia pulchra*) or bifid (*E. gracilis* Baba & Hamatani, 1963); cnidosacs are lacking, and are replaced by terminal pads of nematocysts (up to 4 pads on each ceras). Smooth rhinophores are present but the oral tentacles are represented only by a dilated, bilobed oral veil; propodial tentacles are absent.

Jaws are present and exhibit a toothed cutting edge. The uniseriate radula consists of arched teeth which have a prominent median cusp and numerous lateral denticles. Larval shell unknown.

Embletonia Alder & Hancock, 1851
(type *Pterochilus pulcher* Alder & Hancock, 1844, validated in 1966 by Opinion 782 of the International Commission)
With the characters of the family.

95. **Embletonia pulchra** Alder & Hancock, 1851
Eolidia minuta Forbes & Goodsir, 1839
Pterochilus pulcher Alder & Hancock, 1844
Embletonia faurei Labbé, 1923

APPEARANCE IN LIFE (Plate 36). This delicate flattened aeolidacean does not usually exceed 7 mm in length, although Labbé (1923) reported specimens from Croisic which attained 10 mm. The body is translucent cream or pink in colour, obscured by dense white epidermal stippling. The cerata, held at a low angle laterally, form a single series down each side of the dorsum consisting of up to 7 pairs in all (9 in the Croisic material). The first pairs are more or less opposite, but the larger cerata towards the rear tend to alternate. Each ceras is stout and elliptical, containing a pale yellow, brown, orange or red lobule of the digestive gland. No cnidosac is present, but each ceratal tip has up to four superficial pads which appear to contain nematocysts (Marcus & Marcus, 1958). The source of these organs is unknown. The genital openings are situated on the right side, between the first and second cerata; the anal papilla lies below the third ceras.

The rhinophores are smooth and are set widely apart on the head; propodial tentacles are lacking. The front of the head is bilobed, dilated to form a flattened oral veil; there are no oral tentacles.

ANATOMY (Fig. 31d). Miller (1958) described 2 specimens from the Isle of Man. Both had radular teeth which exhibited a prominent median cusp and up to 4 lateral denticles on either side; the formula was $57 \times 0.1.0$ (body-length 2·5 mm) or $65 \times 0.1.0$ (4·0 mm). Poizat (1978) noted specimens with up to 70 teeth. The cutting edges of the jaws each had a row of denticulations (approximately 8, according to Loyning (1927)). The jaws, radula and various aspects of the histology of the cerata were investigated by Marcus & Marcus (1958).

HABITS. Although Labbé's Croisic material came from fish eggs (which the nudibranchs were consuming) on an intertidal jetty, most finds have come from intertidal or shallow sublittoral coarse deposits. Poizat (1978) believes that the diet normally consists of detritus and bacteria, but other authors believe that *Embletonia* takes hydroids (Miller, 1961: *Hydrallmania falcata* (L.) and *Nemertesia antennina* (L.); Schmekel & Portmann, 1982: *Tubularia*).

Spawn has been described from the Isle of Man in the month of June (Miller, 1958); it is deposited in a spiral of up to 210 ova, each of which measures 67–73 μm.

DISTRIBUTION (Map 12). This species occurs all around the British Isles in very shallow waters (Jeffreys, 1867; Boaden, 1963; Poizat, 1978; Seaward, 1982). Elsewhere it has been reported from Norway and Heligoland to Capo di Sorrento, near Naples, and at Rovinj in the northern Adriatic Sea (Odhner, 1914; Loyning, 1927; Odhner, 1939; Starmühlner, 1955; Schmekel & Portmann, 1982).

CUMANOTIDAE

Mobile aeolidaceans, capable of swimming by vigorous movements of the elongated, cylindrical cerata. Lobules of the digestive gland penetrate the cerata and cnidosacs are present. The cerata are arranged in up to 13 pairs of simple transverse rows, some of which are in front of the rhinophores. Smooth rhinophores are present; propodial tentacles are present in the largest individuals; a dilated oral veil is present which in adults may bear lateral tentacular processes.

The anal opening is lateral, on the right side. The genital openings have two unique features: the grooved, inrolled penis, and the circular clasping organs adjacent to the female opening.

Strong, elongated jaws are present. The triseriate radula consists of teeth which possess prominent cusps flanked by fine denticles. The larval shell is of type 1.

Cumanotus Odhner, 1907
(type *C. laticeps* Odhner, 1907)
With the characters of the family.

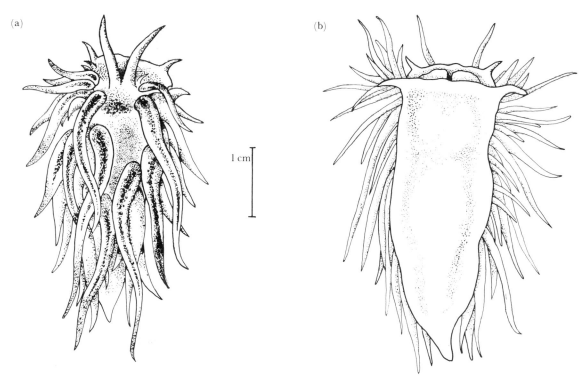

(a)

(b)

1 cm

FIG. 30. *Cumanotus beaumonti* (after Eliot, 1910): (a) dorsal view, (b) ventral view.

96. *Cumanotus beaumonti* (Eliot, 1906)
Coryphella beaumonti Eliot, 1906
Cumanotus laticeps Odhner, 1907

APPEARANCE IN LIFE (Fig. 30). This is one of the rarest of the British opisthobranchs, and has not been recorded in these islands since the first decade of the century. Our knowledge is based upon descriptions of Plymouth specimens supplied to Eliot, and Norwegian preserved material examined by Odhner. Despite its rarity, however, this species is so distinctive that there can be no doubt at all of its validity.

The largest British specimen known measured 20 mm in preservative (Eliot, 1906; 1910). In life, Eliot's correspondent Beaumont stated that the body-colour was red, tending to orange on the rhinophores, but elsewhere more rosy. The digestive gland lobules were rosy purple, sometimes with a greenish purple tinge at or near the base. The cerata are elongated (up to 15 mm long in the 20 mm specimen) and are set upon up to 8 narrow ridges (not horse-shoes as in many aeolidacean genera) which are obliquely transverse on the sides of the back. The first 2 or 3 ceratal rows are anterior to the bases of the smooth rhinophores; the longest cerata are situated towards the rear. There may be up to 6 (according to Eliot) or 9 (according to Odhner) cerata per half-row.

The head is nearly as broad as the propodium and bears small but distinct oral tentacles. The propodial tentacles are extremely conspicuous and curved gracefully towards the rear. The rhinophores are close together basally, and are usually held dorsally so that they are reflected towards the rear.

The anal papilla lies in front of the 7th ceratal row, on the right side. The penis is long, curved and inrolled; it lacks chitinous armature but Eliot (1906) commented that basally it exhibited numerous soft yellow cones. Near the female opening there are two small rosettes of chitinous hooks which function as a clasping organ during copulation (Eliot, 1910).

ANATOMY (Fig. 29d). The jaws have 1–3 rows of denticles along each cutting edge. The radula of a 9 mm preserved specimen from Eliot's collection at Plymouth in 1908 was examined in polyvinyl lactophenol. The formula was $23 \times 1.1.1$; each lateral tooth displayed a stout spine and up to 20 subterminal denticles; each median tooth had a prominent median cusp and up to 12 denticles on either side.

HABITS. Little is known about this species, but the problematic *Cumanotus cuenoti* Pruvot-Fol, 1948 is believed to feed upon *Tubularia* at Arcachon (Cuénot, 1927) and certainly takes *Ectopleura dumortieri* (Beneden) (Tardy & Gantès, 1980). The British species has been taken from depths of 1–2 m (Eliot, 1910), and the Norwegian records were from 2 to 3 m (Odhner, 1907). Swimming has not been described, although well known in Pacific Ocean specimens perhaps belonging to the present species, and authoritatively documented for *C. cuenoti*.

Spawning occurs in May and the mass takes the form of a helicoidal white jelly attached to the substratum by a delicate but tough stalk.

DISTRIBUTION (Map 11). The only certain records are from Norway (Odhner, 1907) and the Plymouth area (Eliot, 1906; 1910; Kress, 1964, personal communication).

DISCUSSION. *Cumanotus cuenoti* Pruvot-Fol, 1948 (recently re-described from the Arcachon area of France by Tardy & Gantès, 1980) appears to be distinct. It is of small size when sexually mature (evidenced by the production of spawn) and does not exceed 10 mm in length; it lacks oral and propodial tentacles; and the number of denticles on each of the teeth is small (up to 9 denticles on each lateral tooth and up to 9 on either side of the median cusp of each central tooth).

The situation of the species recorded by Hurst (1967) under the name *C. beaumonti* is disputable. We re-examined specimens which we collected from Hurst's locality, San Juan Island, on the American N.W. Pacific coast and came to the conclusion that they and the British material were conspecific as Hurst had implied. On that understanding, we published a drawing of swimming in such Puget Sound animals (Thompson & Brown, 1976), based on a colour photograph. But we recently examined a radula from a 20 mm preserved specimen taken near San Juan Island in July 1969, and noted important differences from British material, such that we now believe that the two forms are separate and distinct. The radula of the American specimen had the formula $26 \times 1.1.1$; each median tooth had a slender central cusp flanked by approximately 26 pointed denticles, many of which were bifid at the tip. The lateral teeth had a very short cusp, flanked by about 28 denticles. The principal differences from Plymouth radulae were the short cusp of the American laterals, and the absence in American material of the two oval markings on the median tooth illustrated for British material in Fig. 29d. There were differences too in the number of denticles on the median teeth, greater in the American specimen.

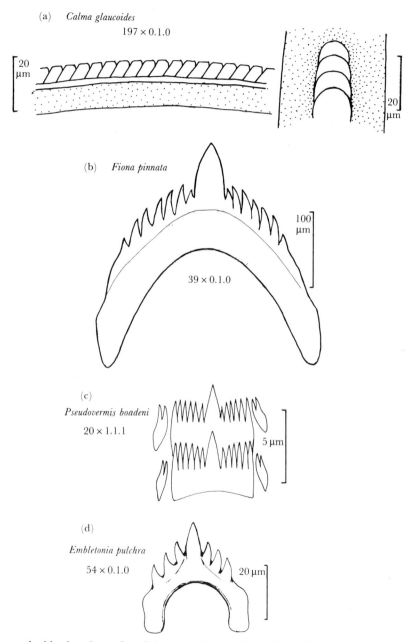

FIG. 31. Camera lucida drawings of aeolidacean radulae: *Calma, Fiona, Pseudovermis* and *Embletonia*.
(a) *Calma glaucoides*, length 15 mm, viewed from the side and from above the radula, St Anthony, S. Cornwall.
(b) *Fiona pinnata*, preserved length 10 mm, San Pedro, California.
(c) *Pseudovermis boadeni*, preserved length 1 mm, Traeth Bychan, N. Wales.
(d) *Embletonia pulchra*, preserved length 1·5 mm, Weymouth, Dorset.

Accordingly, it will be necessary to propose a specific name for the American Pacific *Cumanotus*; we suggest *fernaldi*, after the then Director of the Friday Harbor Laboratories on San Juan Island.

FACELINIDAE

Mobile, aggressive aeolidaceans, having a cleioproctic anal position (within the foremost ceratal group derived on the right side from the left digestive gland) and regular rows of cerata in both pre- and post-cardiac groups; the ducts of the digestive gland are irregularly branched. Propodial tentacles and long, agile oral tentacles are present; the rhinophores are lamellate or tuberculate, occasionally smooth. The penis is usually simple but occasionally armed with spines or hooks; there is usually one receptaculum, occasionally two.

The jaws may be smooth edged or bear a single row of denticles. The radular teeth each have a prominent cusp and lateral denticles; radula uniseriate. The larval shell is of type 1.

Facelina Alder & Hancock, 1855

(type *Eolis coronata* Forbes & Goodsir, 1839; validated in I.C.Z.N. Opinion number 775)

The head bears long, mobile oral tentacles, and rhinophores which are lamellate, rarely tuberculate or smooth. The cerata are grouped into distinct clusters, each consisting of numerous rows.

The right liver and all the subdivisions of the left liver are much branched. The nephroproct lies in the inter-hepatic space, closely anterior to the foremost branch of the second ceratal group.

The penis usually bears elevated chitinous hooked spines, rarely absent; an accessory penial sheath gland is often present.

Because so much confusion has surrounded the British species, we present a tabulation of the diagnostic features, to supplement the full species-descriptions which follow.

DIAGNOSTIC FEATURES OF THE BRITISH SPECIES OF *FACELINA*

	Facelina annulicornis	*Facelina bostoniensis*	*Facelina coronata*	*Facelina dubia*
Pigmentation of the body	No blue iridescence White spots generally distributed on dorsal surfaces.	Rare blue iridescence Very few irregular spots of white on the body; nearly always a patch of white between the rhinophores.	Distinct blue iridescence. Very few irregular spots of white on the body, never forming a patch between the rhinophores.	No blue iridescence. Very few or no white spots on the body, sometimes forming a patch between the rhinophores.
Pigmentation of the cerata	White spotted; distal tips white.	Shield-shaped white patch subapically; without a continuation of spots down the anterior surface of the ceras.	Amorphous white patch subapically; with irregular spots often continuing down the anterior surface.	Dull powdery white pigment in patches below opaque white tips.
Cerata/body proportions	Cerata rarely longer than $\frac{1}{4}$ of the body length in adults. Clusters of cerata are distinct and the animal has a slender appearance.	Cerata often $\frac{1}{2}$ to $\frac{1}{3}$ of the body length. Clusters are often obscured by the long anterior cerata, giving a broad, squat appearance.	Cerata rarely longer than $\frac{1}{4}$ body length in adults. Clusters are distinct and the animal appears slender.	Cerata often $\frac{1}{2}$ to $\frac{1}{3}$ of body length. Clusters often hidden by long anterior cerata and the animal has a broad appearance.

97. *Facelina annulicornis* (Chamisso & Eysenhardt, 1821)
Eolidia annulicornis Chamisso & Eysenhardt, 1821
Eolis punctata Alder & Hancock, 1845a

APPEARANCE IN LIFE (Plate 38). The translucent body may reach 40 mm in extended length alive. The anterior portion often displays a rosy pink blush, most intense around the mouth. Distinctive glistening white spots are scattered liberally over the body, the cerata and the upper surfaces of the foot. The distal third of each of the long, mobile oral tentacles is white. The rhinophores have white tips but are brown-red below, fading basally. Specimens between 30 and 40 mm in length have between 18 and 28 oblique lamellae on each rhinophore, although only about half of these form complete circles. The cerata are set in up to 8 clusters with the pre-cardiac clusters consisting of about 7 distinct rows in large individuals. The hepatic ceratal core is usually a shade of yellow, orange or brown, but rare specimens exhibit green contents. A red nephric duct leading to the nephroproct in the inter-hepatic space is often visible, but the cleioproctic anus is partially hidden behind the first ceratal row in the second cluster. The gonopore is situated below the last four rows in the right pre-cardiac cerata cluster.

The propodium is expanded into tentacles, separated slightly from the pedal sole by a post-tentacular constriction. The anterior margin of the foot and the propodial tentacles are grooved while the upper labial margin is medially notched.

ANATOMY (Fig. 32a & b). The short radular ribbon consists of up to 19 teeth, each with a stout central cusp, flanked on either side by 5–9 stout denticles. Manipulation of the prey is facilitated by strong jaws with a single row of serrations along the masticatory margin. The reproductive organs have been described by Schmekel (1970). The spatulate penis is unarmed but a large gland opens into the penial sheath. The vas deferens is prostatic for nearly its whole length. A single receptaculum connects to the oviduct.

HABITS. Specimens have frequently been found sublittorally between 20 and 38 m, creeping over the sea bed unattached to any obvious food source. Otherwise they have been taken from a variety of gymnoblastic and calyptoblastic hydroids, but no clear dietary preferences have been established. In the laboratory they will attack any hydroid presented to them as well as nudibranch eggs and other aeolid adults. The spawn mass is an irregularly coiled thread with up to 1500 eggs. The ova measure 80 µm in diameter (Schmekel & Portmann, 1982). Around Britain, spawning is known to occur between May and September. Planktotrophic veligers are produced with larval shells of type 1 (Thompson, 1961a).

DISTRIBUTION (Map 9). Favourable localities around Britain have included Skomer Island, and Lundy Island in the Bristol Channel. Almost all British records are from south west coasts, although rare collections have been made as far north as western Scotland. Elsewhere, this species is reliably recorded from Naples (Schmekel, 1968), Villefranche (Haefelfinger, 1960a) and the Atlantic coasts of France (Bouchet & Tardy, 1976).

DISCUSSION. There has been disagreement about both the taxonomy and the nomenclature of this species. Until 1951, when Winckworth's additions and corrections to his 1932 list of British Mollusca were published, this species was almost unanimously referred to by Alder & Hancock's (1845) name *punctata* (type locality Torbay, Devon). Winckworth (1951) substituted the name *annulicornis* Chamisso & Eysenhart, 1821 (*sic.*), introducing a mis-spelling of Eysenhardt's name that has been repeated many times (e.g. in the Concordance of the Conchological Society of Great Britain and Ireland, and the Linnean Society's opisthobranch synopsis (Thompson & Brown, 1976)). The Concordance also followed Pruvot-Fol's (1954) error in quoting the date of publication of the Chamisso & Eysenhardt paper as 1810. Another error was introduced by Russell's (1971) *Index Nudibranchia*; that author misunderstood the latinization of Eysenhardt's given name from Wilhelm to Guilelmus and consequently cites the authors of *annulicornis* as Chamisso & Guilelmus, 1821; this mistake was copied by Schmekel & Portmann (1982).

We employ the name *Facelina annulicornis* because we agree with Winckworth (1951) that the type-description left little doubt about its identity. This was in Latin and has been quoted in full by Pruvot-Fol (1954). Most of this description dealt with general features, but the presence of about 8 lamellae on each rhinophore and the brownish body-colour spotted with white are in our opinion sufficient for us to believe that Chamisso & Eysenhardt's name is valid. We have not seen their colour representation of their species, but Pruvot-Fol (1954), who apparently had examined it, considered that it was conspecific with Alder & Hancock's *punctata*. Consequently, *punctata* must fall into the synonymy of *annulicornis*.

98. ***Facelina bostoniensis*** (Couthouy, 1838)
    ?***Doris auriculata*** Müller, 1776 (part)
    Eolis bostoniensis Couthouy, 1838
    Eolis rufibranchialis Thompson, 1843
    Eolis drummondi Thompson, 1843
    Eolis curta Alder & Hancock, 1843
    Eolis tenuibranchialis Alder & Hancock, 1845a
    Eolis janii Vérany, 1846
    Eolis panizzae Vérany, 1846
    Eolis gigas Costa, 1866
    not ***Eolis rufibranchialis*** Johnston, 1832

APPEARANCE IN LIFE (Plate 37). This is the largest British facelinid, reaching a length of 55 mm. The mouth region is often rosy pink in colour, but the rest of the body is generally translucent white. The long, mobile oral tentacles can equal half the total length of the animal and they are relatively longer in juveniles than in mature adults. The distal third of each oral tentacle and the tips of the rhinophores are pigmented white. A few patches of white are scattered over the head and pericardium, with a larger patch between the rhinophores. The oesophagus is usually red and is clearly visible through the skin. Only rare individuals have any trace of blue iridescence over the head, whereas this is usually present in the related *Facelina coronata*. As many as 30 rhinophore lamellae have been counted in large individuals, and they tend to be of uniform width. The foot is wider than the body and the whole aspect of the animal is broad compared with *F. coronata*. The cerata from the first cluster can be almost as long as the body; although the ceratal arrangement is similar to that in *F. coronata*, the inter-hepatic space is usually obscured in *F. bostoniensis* by the relatively longer cerata. The hepatic lobes vary in colour from light brown to dark chocolate brown. Sub-apical white pigment is arranged in a shield-shaped patch, neater in appearance than the fragmented spots of *F. coronata*.

ANATOMY (Fig. 32e & f). The radula consists of up to 20 teeth and is similar to other British facelinids. Trinchese (1881) noticed a distinction between *F. bostoniensis* and *F. coronata*, concerning the masticatory margin of the well-developed jaws, such that *F. coronata* rarely had more than 20 (usually about 15) serrations, whereas *F. bostoniensis* often has more than 30.

A series of simple, hooked spines, set on thickened basal pads provides armature for the large penis.

HABITS. The spawn mass is a grecian key spiral of several turns; it may contain up to 8000 eggs. The normal sublittoral diet is *Tubularia larynx* Ellis & Solander or *T. indivisa* L. in British waters, although occasional intertidal specimens are taken on *Clava multicornis* (Forskål). If hungry, virtually any hydroid will be attacked.

DISTRIBUTION (Map 9). It has been recently established that this species is amphi-atlantic (Brown, 1981a), and that Couthouy's description from Massachusetts provides the senior synonym for *Eolis curta* Alder & Hancock, 1843, the name used by European authors. *Facelina bostoniensis* is the only facelinid occurring on the north east coast of America; it has an extended range in Europe, from shores and shallow waters from southern Norway, all around the British Isles, Atlantic France, Iberian Peninsula and the western Mediterranean Sea (Pruvot-Fol, 1954).

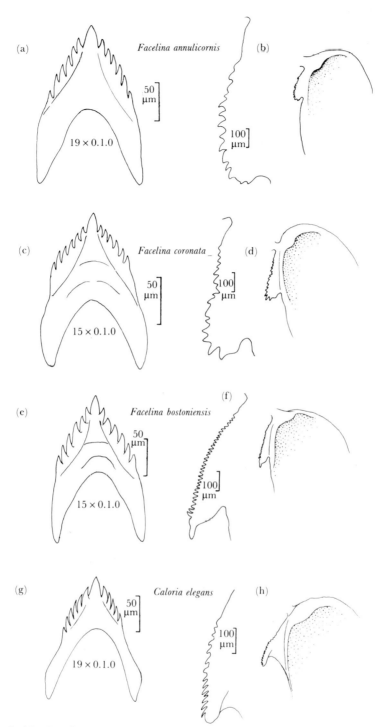

FIG. 32. Camera lucida drawings of aeolidacean jaws and radulae: *Facelina* and *Caloria*.
(a) *Facelina annulicornis*, preserved length 34 mm, Lundy Island, radula tooth.
(b) cutting edge of a jaw plate of the same, at two different magnifications.
(c) *Facelina coronata*, preserved length 28 mm, Lundy Island, radula tooth.
(d) cutting edge of a jaw plate of the same, at two different magnifications.
(e) *Facelina bostoniensis*, preserved length 32 mm, Lundy Island, radula tooth.
(f) cutting edge of a jaw plate of the same, at two different magnifications.
(g) *Caloria elegans*, preserved length 10 mm, Ischia, Bay of Naples, Italy, radula tooth.
(h) cutting edge of a jaw plate of the same, at two different magnifications.

99. *Facelina coronata* (Forbes & Goodsir, 1839)
  ? *Doris auriculata* Müller, 1776 (part)
  ? *Doris longicornis* Montagu, 1808
  ? *Eolida plumosa* Fleming, 1828
  Eolida coronata Forbes & Goodsir, 1839
  Facelina coronata; Alder & Hancock, 1855
  not *Eolis peregrina* Gmelin, 1791

APPEARANCE IN LIFE (Plate 37; vol. I, Plate 7). This colourful aeolidacean may reach a length of 38 mm in life. Much of the taxonomic confusion surrounding it arises from the variability of body coloration. The red oesophagus always shows through the skin behind the rhinophores; the oral region is often brilliant red but can be pale. The ground colour of the body surface is translucent white, exhibiting the opaque white ovotestis which fills the posterior two-thirds of the visceral cavity, behind the pericardium. There is a characteristic iridescent blue sheen in the head region, especially clearly seen at the outer bases of the oral tentacles. The lamellate rhinophores are light to dark brown, measuring about half the length of the oral tentacles. The lamellae tend to alternate between complete and incomplete hoops, as was noticed by Alder & Hancock (1845–55), but this character is not peculiar to *F. coronata*. The oral tentacles bear distal streaks of superficial white pigment; similar chalk-marks can be seen on the anterior surfaces of the rhinophores.

The cerata often exhibit the same iridescent blue sheen noted on the head, with superficial chalk-white pigment extending irregularly down the frontal aspect of each ceras. The hepatic lobes are broad and light brown in colour, or more narrow and dark brown tending to blood-red. The whole aspect of the body is more slender than that of *F. bostoniensis*; this difference is valid at all stages of growth.

ANATOMY (Plate 35b; Fig. 32c & d). The radula usually consists of 15–18 teeth; it cannot as yet be reliably distinguished from other *Facelina* species. The jaws have a single row of denticles (approximately 15 in number) along each cutting edge.

The penis is a large, spatulate organ, armed with a single row of 16–20 hooked spines along the distal edge. When everted, the penis exhibits a crenulated dorsal surface, contrasting with the smooth ventral side. There is a large gland opening into the penial sheath, and a single receptaculum connecting to the oviduct. The arrangement of the organs of the anterior genital mass is similar to that of *F. bostoniensis*, figured by Schmekel (1970, as *F. drummondi*).

HABITS. The largest collections have come from clumps of *Tubularia indivisa* L., growing in water deeper than 15 m; in shallower water some specimens have been taken from *Obelia* spp. and *Laomedea flexuosa* Hincks attached to *Laminaria* fronds. Maximal egg-counts have been 13,000 per spawn mass (Thompson, 1967) or 27,400 (Miller, 1958), but spawn found in the field seldom contains more than 2500 eggs per mass. Each spawn mass is typically a grecian key spiral of up to 6 whorls. The planktotrophic larvae have a veliger shell of type 1 (Thompson, 1961a) and have been described in some detail by Tardy (1970). The spawning season extends from May to October.

DISTRIBUTION (Map 9). This boreo-lusitanian species extends from Norway to the western Mediterranean (Pruvot-Fol, 1954); it is well known from localities all around the British Isles (Seaward, 1982). Picton (personal communication) recently reported it from Santander and Galicia in Spain.

DISCUSSION. Forbes & Goodsir (1839) described *Eolida coronata* from the Shetland Islands 5 years before Thompson (1843) described *Eolis drummondi* from Northern Ireland. Alder & Hancock (1845–55) redescribed both as separate species of *Eolis*, but in an appendix introduced the genus *Facelina* Alder & Hancock, 1855, with *coronata* as the type species. They also recognized that *E. drummondi* was a junior synonym of their earlier *E. curta* Alder & Hancock, 1843.

Facelina coronata and *F. curta* were thenceforth considered correctly to be distinct species, although various junior synonyms continued to be used, until Odhner (1939) mistakenly decided to merge them, re-introducing a disused name, *F. auriculata* (Müller, 1776). This name had previously been neglected for the valid reason that the type description contained insufficient detail to allow identification with any known species, although it was obviously a facelinid. Müller's species may embrace either *coronata* or *curta*, perhaps both, but we can never be sure, because his type material has not survived and because both *coronata* and *curta* occur in the vicinity of the type locality, Stad in Norway.

Similarly, the type descriptions of *Doris longicornis* Montagu, 1808 and *Eolida plumosa* Fleming, 1828 may have been based upon material of *coronata* or *curta* or both. Material collected by Grant provided the basis for the first adequate description of what was later described as *Eolida coronata* by Forbes & Goodsir (1939), but Grant's suggested name, *Eolis peregrina*, was invalid because it was a junior homonym of Gmelin's (1771) species (later transferred to the genus *Cratena*).

Forbes & Goodsir (1839) were thus the first to apply a valid specific name, *coronata*, to animals of this well-defined taxon, similar to, but distinct from, the closely related amphiatlantic *F. curta*. As explained earlier in this account of the British facelinids, *F. curta* must be merged into the synonymy of the senior *bostoniensis* Couthouy, 1838.

100. *Facelina dubia* Pruvot-Fol, 1948

APPEARANCE IN LIFE (Plate 39). Mature animals measure between 17 and 38 mm in length. The head bears mobile oral tentacles which may be half the length of the body. The rhinophores are shorter, about a quarter of the body length, smooth in outline, tapering to a blunt tip.

In external features, *F. dubia* is reminiscent of *F. bostoniensis*, although this latter species has distinctive lamellate rhinophores which are immediately diagnostic. The colour pattern changes slightly with age. Juvenile *F. dubia* have a translucent body with an orange-brown tint over the head and tentacles, fading behind the pericardial region. Adults over 30 mm are more deeply pigmented, with a salmon-pink colour. The orange-red oesophagus is clearly visible through the skin immediately behind the rhinophores. Opaque white specks are dotted over the head and tentacles, most concentrated on the frontal surfaces of the rhinophores and towards all the tentacular tips. A larger patch of white is often located between the rhinophore bases. There are indistinct patches of superficial white pigment over the cerata; these organs are capped with the same pigment but in the adults this has a yellow tint. The hepatic lobes within the cerata are dark grey, brown or black.

The cerata are very long in life, masking the details of their clustered bases. Investigation of preserved specimens with the cerata removed reveals that there are 7 or 8 groups arranged in oblique rows on raised pads. In an individual which in life measured 34 mm, there were 31 rows of cerata on each flank.

The body of *F. dubia* is broader than other British facelinids, and the metapodium is very short. Elongated, mobile propodial tentacles are present. The genital apertures lie beneath the first ceratal group on the right side; the anus is situated in front of the last row of cerata in the second ceratal cluster, while the nephroproct is in the inter-hepatic space.

ANATOMY (Fig. 33a & b). The uniseriate radula of a 14 mm specimen consisted of 15 teeth; the strong jaws have a single row of denticles along each cutting edge. The penis is distinctive, having a wrinkled stem and a bulbous head composed of two flaps, the ends of which are bent back against the dorsal face of the organ. The inner faces of these flaps bear a row of chitinous hooks (Schonenberger & Schonenberger, 1969). The penial sheath gland found in other British facelinids is absent in *F. dubia*, thus approaching the genera *Facelinella* Baba, 1949 and *Caloria* Trinchese, 1888.

HABITS. Specimens from Lough Ine were feeding and spawning upon *Tubularia larynx* Ellis & Solander in August and September 1980, attached to mooring ropes, to 20 m. The spiral spawn band contains 100 μm ova which develop to hatching in 6–8 days at 16 °C (Schmekel & Portmann, 1982).

DISTRIBUTION (Map 9). The only British records come from Lough Ine, S.W. Ireland (Picton & Brown, 1981). The type was described by Pruvot-Fol (1948) from preserved material collected

151

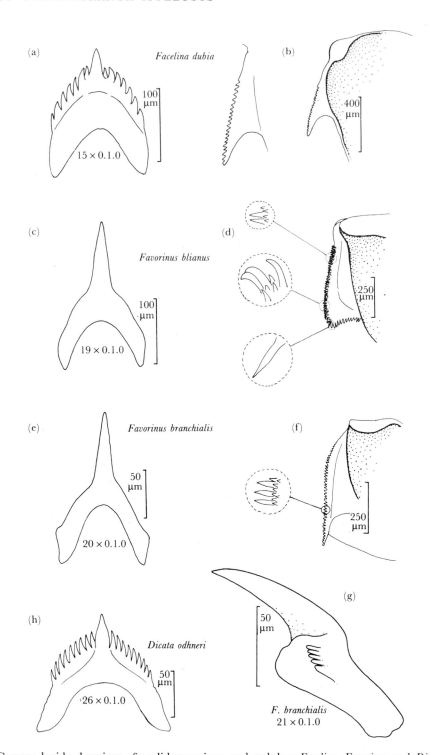

FIG. 33. Camera lucida drawings of aeolidacean jaws and radulae: *Facelina*, *Favorinus* and *Dicata*.
  (a) *Facelina dubia*, preserved length 14 mm, Lough Ine, Ireland, radula tooth.
  (b) cutting edge of a jaw plate of the same, at two different magnifications.
  (c) *Favorinus blianus*, preserved length 13 mm, Lundy Island, radula tooth.
  (d) cutting edge of a jaw plate of the same, at different magnifications.
  (e) *Favorinus branchialis*, preserved length 9 mm, Loch Fyne, Scotland.
  (f) cutting edge of a jaw plate of the same, at two different magnifications.
  (g) side view of a denticulate tooth from a juvenile, preserved length 4 mm, Port Erin Bay.
  (h) *Dicata odhneri*, preserved length 9 mm, Lough Ine, Ireland, radula tooth.

near Arcachon by Cuénot. This description has been augmented by Schonenberger & Schonenberger (1969) at Naples. The British material agrees with the Neapolitan in every particular. Notes left by Cuénot indicate that some individuals may have a green tinge; we cannot confirm this.

Caloria Trinchese, 1888
(type *Caloria maculata* Trinchese, 1888)
Facelinids resembling *Facelina* except that (a) the rhinophores are papillate in the adult state; (b) black "glands" are present in the cerata, below the cnidosacs; (c) the rear margin of each jaw is deeply indented; (d) the penial sheath gland is absent; and (e) the cerata are arranged in simple rows, not arches or horse-shoes (Schmekel & Portmann, 1982).

101. **Caloria elegans** (Alder & Hancock, 1845a)
  Eolis elegans Alder & Hancock, 1845a
  Caloria maculata Trinchese, 1888
  Acanthopsole quatrefagesi Vayssière, 1888
  Facelina elegans; Iredale & O'Donoghue, 1923

APPEARANCE IN LIFE (Plate 38). This slender facelinid reaches a length of 34 mm in British waters, in contrast to Mediterranean specimens, which have only rarely reached 20 mm. The body is pellucid white, terminating at the rear in the slender metapodium, which bears an opaque white dorsal longitudinal ridge. The propodial tentacles are long and gracefully recurved, with an opaque white transverse streak; a similar streak covers the frontal surface of each oral tentacle and the front of the head. The fawn rhinophores measure just over half the length of the oral tentacles; the rear surfaces of the former are in adult specimens covered with small, irregularly distributed papillae. These papillae appear to be absent in most of the Mediterranean records (Haefelfinger, 1960b; Schmekel & Portmann, 1982), perhaps because they exhibited a juvenile condition. Trinchese's 20 mm specimen showed an intermediate condition, "presentano delle rughe irregolari" (which Picton, 1979 translates as "with irregular wrinkles"). Such a radical difference between the rhinophores of juveniles and of adults would not be without precedent in aeolidaceans (Miller, 1971).

The cerata are arranged in 8 or 9 groups in the adult; the first 5 groups are clearly separated, but they appear to merge posteriorly. Each group consists of numerous rows of elongated, slender cerata, containing orange pink lobes of the digestive gland, dark at the base and black distally. Above the black zone, sometimes said to be a gland (Picton, 1979), is an elongated, sometimes curved cnidosac. The tips of the cerata are normally curved but can be straightened on alarm. There is a subterminal superficial opaque white ring around each ceras, partially obscuring the cnidosac. Mediterranean specimens differ in that the digestive gland lobes may be brown, orange or red in colour (Schmekel & Portmann, 1982).

ANATOMY (Fig. 32g & h). The radula of a 10 mm specimen from Naples had the formula 20 × 0.1.0, (Schmekel & Portmann, 1982), whereas the 34 mm Lundy individual had 24 teeth. There are 5 or 6 lateral denticles on either side of the prominent median denticle. The jaws have a characteristically deeply indented rear margin (illustrated by Picton (1979), and up to 22 denticles on each cutting face.

The reproductive organs have been investigated by Schmekel (1970) and by Picton (1979); a ring of small spines or papillae can be discerned surrounding the internal opening of the vas deferens into the lumen of the penis.

HABITS. Little is known concerning the biology of *C. elegans*. Spawn was found in July at Lundy I. (Picton, 1979); it consisted of a narrow spiral containing a thin strand of white eggs. Adults were found to a depth of 30 m.

DISTRIBUTION (Map 9). This is a Lusitanian species, with records from the Bay of Naples (Schmekel & Portmann, 1982) and Villefranche (Vayssière, 1888; Haefelfinger, 1960b) and from Berry Head, Torbay (Alder & Hancock, 1845a) and Lundy I. (Picton, 1979). Fisher's (1936) report from Co. Antrim is dismissed as a misidentification by Picton (1979).

FAVORINIDAE

Mobile, aggressive aeolidaceans, having a cleioproctic anal position (within the foremost ceratal arch derived on the right side from the left digestive gland). All the branches of the digestive gland take the form of arches (often called horse-shoes), sometimes reduced to simple rows at the rear. Propodial tentacles and long, agile oral tentacles are present; the rhinophores are smooth, sometimes with one or more bulbous swellings. The penis is short, conical and unarmed. Two receptacula are present.

The jaws may be smooth-edged or bear denticles on the cutting edge. The teeth of the uniseriate radula have a slender median cusp and faint lateral denticles, sometimes lacking. The larval shell is of type 1.

Favorinus Gray, 1850

(type *Eolis alba* Alder & Hancock, 1844a; validated in I.C.Z.N. Opinion number 783).

Favorinids having one or more bulbous swellings on the rhinophores. The radular teeth exhibit weak lateral denticles which are sometimes altogether absent. The cutting edge of the jaws is denticulate.

All the Atlantic species of *Favorinus* are illustrated by Ortea (1982).

102. *Favorinus blianus* Lemche & Thompson, 1974

APPEARANCE IN LIFE (Plate 40). This species can reach a length of 30 mm. The moderately broad body is pale straw-yellow in colour with a characteristic pattern of dark and light superficial pigment. The yellowish rhinophores each exhibit a dark brown strip running from tip to base on the rear face (absent in juveniles and in rare adults). Opaque white pigment streaks the dorsal sides of the oral tentacles, and of the propodial tentacles and the metapodium, as well as covering the distal tips of the cerata. There is a tendency for white pigment to form a median longitudinal line between the united bases of the rhinophores, but this line does not widen into a diamond-shape, contrary to the pattern of *Favorinus branchialis*.

The cerata are stout and tend to be reflected mesially so as to conceal the dorsum. They are arranged in clusters, each of which takes its origin from a horse-shoe shaped digestive gland lobe. The first clusters each contain up to 12 cerata and originate immediately posterior to the rhinophoral bases.

The tentacles of the head are elongated and conspicuous. The propodial tentacles are stout and curved towards the rear. The oral tentacles are by far the longest, dwarfing the rhinophores. But the rhinophores are by far the most conspicuous features of the body by virtue of three massive shelf-like swellings or bulbs which widen on the lower half of each tentacle.

ANATOMY (Fig. 33c & d). Several radulae were examined. A 9 mm specimen from Denmark had the formula $15 \times 0.1.0$; a 13 mm specimen from Lundy I. showed $19 \times 0.1.0$, while a 30 mm "giant" from Pembrokeshire had the formula $26 \times 0.1.0$. All the teeth exhibited an elongated median cusp flanked by smooth shoulders; the overall length of the teeth is considerably greater than that of teeth from comparable specimens of *F. branchialis*. We have not examined the teeth of juveniles of *F. blianus*; it would be interesting to see whether a dental metamorphosis occurs, comparable with that described for *F. branchialis*.

Thin jaws are present, with denticulate cutting edges (Fig. 33d). We have established that cnidosacs are present in the ceratal tips. Nothing is known about the reproductive organs.

HABITS. This species is known to consume cnidarians such as *Tubularia* (Lemche & Thompson, 1974). The spawn of other nudibranchs (e.g. *Doto coronata* (Gmelin)) is also eaten and fully formed embryos of the prey may be seen within the digestive gland lobes of *F. blianus*. There is an urgent need for more information about dietary preferences in this species. We also need information about reproduction; the spawn is unknown.

DISTRIBUTION (Map 9). Around Britain there are records for several localities on the west coasts (Seaward, 1982), from Lundy I., from Pembrokeshire, N. Devon, Cornwall and from western Ireland (Killary Bay) down to 35 m, as well as from Norway (Bergen) and Denmark (Kattegat) (Lemche & Thompson, 1974). We think it probable that it occurs on the Atlantic coast of France; it has recently been reported from the Galician coast of Spain (Picton, personal communication).

103. **Favorinus branchialis** (Rathke, 1806)
   Doris branchialis Rathke, 1806
   Eolis alba Alder & Hancock, 1844a
   Eolis carnea Alder & Hancock, 1854a
   Favorinus versicolor Costa, 1866
   Matharena oxyacantha Bergh, 1871b
   Favorinus albus; Trinchese, 1881
   Favorinus albidus Iredale & O'Donoghue, 1923

APPEARANCE IN LIFE (Plate 40). The maximal length of this delicate, slender aeolidacean is 25 mm. The body is white, frosted superficially by opaque white surface pigment. The cerata are usually pale, with the yellow or pale brown digestive gland lobule showing through the white epidermis, which is interrupted subapically; this distal ring often has an olive-green or brown tinge. There are up to 6 clusters of cerata, each of 6–7 cerata. The first 2 clusters are in horseshoes, but further back each cluster arises from a single row. The anus lies partially concealed between the bases of cerata in the second cluster on the right side; the genital apertures lie ventral to the first ceratal group.

Conspicuous, pale, gracefully curved propodial and oral tentacles are present. A notable feature of the rhinophores is the presence of a subterminal bulb (in uncommon cases there may be two such swellings on each rhinophore). The lower part of each rhinophoral tentacle is dark brown up to the bulb. The rhinophoral bases are close together.

The colour of the prey affects that of the digestive gland. This dietary factor, combined with a variable distribution of superficial white and brown pigment, makes this a difficult species to describe simply; in addition it should be noted that both bodily coloration and rhinophoral bulbs may be lacking in juveniles.

ANATOMY (Fig. 33e–g). The radula is uniseriate and a dental metamorphosis occurs at a body-length of approximately 7 mm. Before that size the radula contains teeth which possess an elongated median cusp flanked on either side by up to 6 fine denticles. After that size, the teeth are smooth. The radular formula of a 4 mm specimen was $20 \times 0.1.0$; 4·5 mm, $20 \times 0.1.0$; 6·5 mm, $23 \times 0.1.0$; 9·0 mm, $25 \times 0.1.0$ (Miller, 1958). In the largest specimens the number of teeth recorded was 35 (Labbé, 1929), and each tooth exhibited up to 12 denticles on either side. Denticles are present along the cutting edges of the jaw plates.

It is well established that cnidosacs are present at the tips of the cerata (Meyer & Möbius, 1865; Hecht, 1895; Vayssière, 1919; Henneguy, 1925). The central nervous system was described by Ceccatty & Planta (1954).

HABITS. *Favorinus branchialis* feeds on both calyptoblastic hydroids and on the egg masses of other opisthobranchs. Labbé (1931) claimed that his specimens would devour fragments of the actinians *Actinia equina* (L.) and *Anemonia sulcata* (Pennant), but this could not be repeated by Haefelfinger (1962b). It is clear, however, that *F. branchialis* feeds at certain times of the year on the hydroids *Obelia geniculata* (L.) and *Sertularia argentea* (L.) (Miller, 1961). The major intake is opisthobranch eggs. In the Mediterranean Haefelfinger (1962b) has reported feeding upon the spawn of *Aplysia fasciata* Poiret, *Placida dendritica* (Alder & Hancock), *Elysia viridis* (Montagu), *Tylodina perversa* (Gmelin), *Platydoris argo* (L.) and *Hypselodoris tricolor* (Cantraine). In Britain we have taken it on the eggs of *Archidoris pseudoargus* (Rapp) in Cornwall, and Renouf (1935) found it on the same prey in Lough Ine. Garstang (1890) reported finding it on the eggs of a *Polycera*, probably *quadrilineata* (Müller).

The spawn has been illustrated by Thompson (1976); it forms a delicate white spiral, containing up to 20,000 ova, each measuring 40 μm (Haefelfinger, 1962b) or 70 μm (Rasmussen, 1951). The larva has been figured by Trinchese (1881) and by Rasmussen (1951). Breeding was reported from July to September at Arcachon (Cuénot, 1927), and in November in the Plymouth list (M.B.A., 1957).

DISTRIBUTION (Map 9). There are not many British records, but they are widely scattered around our coasts (Seaward, 1982). Further records range from Murmansk and Norway to the Mediterranean Sea, to 20 m (Loyning, 1927; O'Donoghue, 1929; Rasmussen, 1951; Sordi & Majidi, 1957; Miller, 1958; Haefelfinger, 1962b).

Dicata Schmekel, 1967

(type *Dicata odhneri* Schmekel, 1967)

Favorinids having smooth rhinophores which lack bulbous swellings, and radular teeth which exhibit strong lateral denticles. The jaws have smooth cutting edges.

104. *Dicata odhneri* Schmekel, 1967

APPEARANCE IN LIFE (Plate 39). It was only in 1980 that we detected this rare visitor to our shores on a visit to S.W. Ireland, 3 specimens in all, 8–13 mm in length, somewhat larger than Schmekel's types from the vicinity of Naples. The body is translucent white with a dense frosting of superficial opaque white pigment, extending over the cerata. The distal portions of both the rhinophoral and oral tentacles are lemon-yellow.

The cerata are arranged in up to 9 pairs of groupings, but only the first three on each side are clear and separate; these three spring from horseshoe-shaped digestive gland branches, whereas the remainder spring from simple obliquely transverse rows. Each of the anterior groupings may consist of up to 20 of the bulbous cerata. The genital openings lie below the first horseshoe on the right side (this is derived from the right lobe of the digestive gland). The anus is situated in the middle of the 2nd ceratal horseshoe on the same side. Stout, recurved propodial tentacles are present; the oral tentacles are twice as long, approximately the same length as the smooth, tapering rhinophores.

ANATOMY (Fig. 33h). The uniseriate radula of a 9 mm Irish specimen consisted of 20 teeth (Picton & Brown, 1981). More recently, we have examined the radula of a 9 mm preserved specimen from Lough Ine; the formula was $26 \times 0.1.0$ and each tooth exhibited a prominent median cusp flanked on either side by up to 11 strong denticles. The jaws have a smooth cutting edge (Picton & Brown, 1981). The reproductive organs were described by Schmekel & Portmann (1982).

HABITS. The diet is unknown. The spawn takes the form of a spiral, and was deposited in the month of August at Lough Ine. The ova are white, measure 80 μm in diameter, and develop to hatching in 6–7 days at 16 °C.

DISTRIBUTION (Map 9). *Dicata* is known only from the Naples area and from Lough Ine (Schmekel, 1967; Picton & Brown, 1981; Seaward, 1982).

AEOLIDIIDAE

Slow-moving, aggressive aeolidaceans, having a cleioproctic anal position (behind the foremost ceratal rows derived on the right side from the left digestive gland) approximately half-way back along the right side of the body. All the cerata arise in regular rows from subdivisions of the digestive gland (the rows are not easy to discern in *Spurilla*, but clear in *Aeolidia* and *Aeolidiella*). Propodial and long, agile oral tentacles are present; in British species the rhinophores are smooth, lacking either bulbs or lamellae. The penis is unarmed. Two receptacula are present.

Smooth-edged jaws are present. The teeth of the uniseriate radula are broad and pectinate, with or without a median emargination. The larval shell is of type 1.

Aeolidia Cuvier, 1798

(type *Limax papillosus* L., 1761; validated in I.C.Z.N. Opinion number 779 (1966)).

The radular teeth each form a smoothly curved arch bearing regularly graded denticles, not emarginated in the mid-line. The nephroproct lies close to the anal opening, at about the same level along the body.

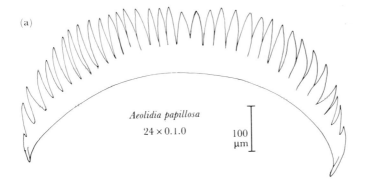

(a)

Aeolidia papillosa

24 × 0.1.0

100 μm

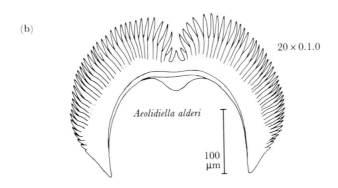

(b)

20 × 0.1.0

Aeolidiella alderi

100 μm

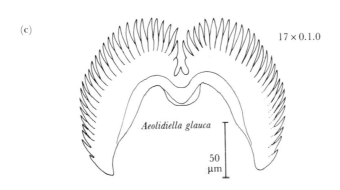

(c)

17 × 0.1.0

Aeolidiella glauca

50 μm

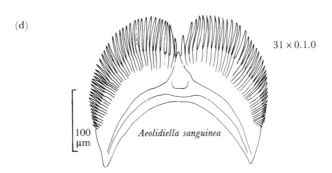

(d)

31 × 0.1.0

Aeolidiella sanguinea

100 μm

FIG. 34. Camera lucida drawings of aeolidacean radulae: *Aeolidia* and *Aeolidiella*.
 (a) *Aeolidia papillosa*, preserved length 40 mm, Helford Passage, Cornwall.
 (b) *Aeolidiella alderi*, preserved length 18 mm, Portland Harbour, Dorset.
 (c) *Aeolidiella glauca*, preserved length 14·5 mm, St Anthony, Cornwall.
 (d) *Aeolidiella sanguinea*, preserved length 22 mm, Donegal, Ireland.

105. *Aeolidia papillosa* (L. 1761)
Limax papillosus L., 1761
Eolida zetlandica Forbes & Goodsir, 1839
Eolis obtusalis Alder & Hancock, 1842
Eolis rosea Alder & Hancock, 1842
Aeolis lesliana MacGillivray, 1843
Aeolis murrayana MacGillivray, 1843

APPEARANCE IN LIFE (Plate 41). This is the largest British aeolidacean nudibranch, reaching 120 mm, and moreover it is without doubt one of the world's largest aeolids. The depressed body is broad in comparison with other British aeolids, and bears up to 25 obliquely transverse rows of elongated, somewhat flattened cerata (Thompson & Brown, 1976), up to 24 per half-row (Miller, 1958). The cerata leave bare a substantial median zone of the dorsum, behind the closely adjacent rhinophore bases. This bare area characteristically exhibits white pigment which forms a diamond or crescent shape; this white pigment often links with similar material covering the upper surfaces of the oral tentacles. The ground colour of the body is extremely variable, ranging from pale grey through shades of brown to rich purple-brown; superimposed upon the ground colour may be freckles of white, grey, brown or violet. The smooth rhinophores are usually darker than the rest of the body, especially at their bases, but the tips are pale. The oral tentacles are slightly longer than the rhinophores; propodial tentacles are conspicuous.

ANATOMY (Plate 35; Fig. 34a). Scanning E.M. photographs have been published by Thompson & Hinton (1968) and, using Californian material, by Bertsch (1974). The denticles are usually simple, and arranged in a smoothly graded series over each dental arch, although Colgan (1914) described bifid and trifid denticles in one individual. Miller (1958) examined a considerable number of radulae of Manx specimens. An 8 mm specimen had 12 teeth (with 19 denticles on each tooth); 15 mm, 19 (20); 46 mm, 27 (29); 56 mm, 27 (23); 61 mm, 24 (26); 77 mm, 31 (33). The greatest number of denticles on a tooth was 45 (Thompson & Hinton, 1968).

Many authors have investigated the general anatomy of *Aeolidia*, and illustrated accounts have been published by Alder & Hancock (1845–55), Meyer & Möbius (1865), Pelseneer (1894), Hecht (1895), Odhner (1939); Russell (1929) published a specialized description of the central nervous system.

HABITS. *Aeolidia papillosa* has been a favourite among ecologists, probably because its large size makes it easy to find and to experiment with. It also makes it an easier target for its natural enemies, both parasites and predators. Homans & Needler (1944) report finding adults in the stomachs of haddock *Gadus aeglefinus* L., casting doubt upon the fancied efficacy of the ceratal cnidosacs of the nudibranch. Leigh-Sharpe (1935) reports that *A. papillosa* acts as a host for the copepod parasites *Lichomolgus agilis* (Leydig) and *Splanchnotrophus*. We have observed *Aeolidia* spawn to be eaten by the dogwhelk *Nucella lapillus* (L.) in the Menai Straits. Yet *A. papillosa* remains one of the most common and successful shallow-water nudibranchs, because it has evolved exquisite mechanisms by which it can detect, overcome and consume one of the most abundant marine life-forms, actinian sea-anemones. Braams & Geelen (1953) demonstrated that *Aeolidia* was attracted by the scent of actinians, especially *Metridium* and *Actinia* (but not by hydroids). Wolter (1967) concluded that *Aeolidia* uses the rhinophores, oral tentacles and other chemoreceptors on the head to detect and orientate towards the prey. Edmunds, Potts, Swinfen & Waters (1974) quantified the preference shown for different anemones in laboratory tests and proposed a rank order of decreasing attractiveness in the series *Actinia equina* L., *Anemonia sulcata* (Pennant), *Anthopleura elegantissima* (Brandt). These authors then extended Waters' (1973) work on attractiveness of Californian actinians to Californian *Aeolidia papillosa*; *Anthopleura elegantissima* ranked high in the order of preference, while *Metridium senile* (L.) ranked low. This indicates that both British and Californian *Aeolidia* have similar food preferences. How far these laboratory studies can be taken to predict field behaviour is questionable, in the light of

Harris's (1967b) observation that the supposedly low-precedence (measured by laboratory ranking tests) *Metridium senile* forms a major part of the diet of *Aeolidia papillosa* on both the Atlantic and the Pacific coasts of the U.S.A.

Both Tardy & Bordes (1978) and Todd (1981) have published full lists of the species of actinians taken by *Aeolidia* in European waters. It is probably true to infer that any actinian will be attacked if the nudibranch is starved.

Edmunds, Potts, Swinfen & Waters (1976) went on to investigate the defensive behaviour of the actinian when attacked by *Aeolidia papillosa*. *Actinia equina*, for example, responds first by tentacle and column retraction, then by inflation of the column, active pedal locomotion and eventual detachment from the substratum. *Anemonia sulcata*, however, having much longer, non-retractile tentacles, uses these in active defence; it also crawls away but it does not detach. Actinians which possess acontia normally eject these when attacked; it appears probable that these do have a deterrent effect on *Aeolidia*. In the Danish Sound, Robson (1961) found a novel response in the actinian *Stomphia coccinea*; on attack this anemone detached and swam away by whole-body contorsions.

The nutritive value of these actinians is probably different. Swennen (1961) showed that Netherlands *Aeolidia* could be reared from juvenile to adult on a diet of *Actinia* or *Metridium* but not on other anemones. Another factor to be borne in mind is that *Aeolidia* will occasionally ingest other kinds of food, for example, *Mytilus* (Alder & Hancock, 1845–55) or *Tubularia indivisa* L. (Miller, 1961).

Most authors agree that *A. papillosa* is an annual organism, spawning in the spring months (e.g. January–June in the Clyde Sea area, Elmhirst, 1922; February–August in the Plymouth area, M.B.A., 1957), and dying after a total life span of 12–16 months (Hecht, 1895; Miller, 1962). It appears that the juveniles spend the first months of their lives sublittorally, moving into the shallowest waters to mate and oviposit. The spawn forms a spiral, attached to rocks; it may contain up to 30,000 ova (Miller, 1958), 83–88 μm (Miller, 1958) or 112–115 μm (personal) in diameter, arranged up to 12 per capsule (Colgan, 1914), exceptionally 19 (Roginsky, 1962, who records up to half a million ova in a White Sea spawn mass). Swennen (1961) believes that the precise nature of the diet influences the colour (pink or white) of the egg jelly, but McMillan (1942b) described a spawn mass which was at first white, but turned pink after 18 hours. The planktotrophic larvae have been described by Williams (1980) in California.

Aeolidia papillosa may attain considerable abundance; near den Helder Swennen (1959) counted 10/m² on a 20 m² patch; he reported this species from waters of salinity as low as 20‰.

DISTRIBUTION (Map 10). This world-wide species has been recorded from shores and shallow waters all around the British Isles (Seaward, 1982). Elsewhere, it has been recorded from the White Sea, Iceland, Baffin Land, New England, Faeroes, Kiel, Arcachon and Asturias (Meyer & Möbius, 1865; Cuénot, 1927; Lemche, 1938; Roginsky, 1962; Ortea, 1980); further afield there are reports from the Falkland Islands, California, Vancouver I. and Alaska (O'Donoghue, 1927; Lance, 1961; Bertsch, 1974). The majority of records are from the intertidal or shallow sublittoral, but some have come from considerable depths, to 800 m.

Aeolidiella Bergh, 1867
(type *Eolida soemmerringii* Leuckart, 1828)
= *Eolidina* Quatrefages, 1843
(type *Eolidina paradoxum* Quatrefages, 1843)

The radular teeth each form a curved arch which is markedly emarginate in the mid-line. A robust median denticle occupies the centre of this emargination, and smoothly graded regular denticles are present on either side of the mid-line. The nephroproct lies some way anterior to the anal opening.

The European species of *Aeolidiella* have been reviewed by Tardy (1969). The genus has attracted some controversy because Leuckart's *soemmeringii* is untraceable. A proposal to clarify this difficulty has been placed before the I.C.Z.N. (Brown, 1981b).

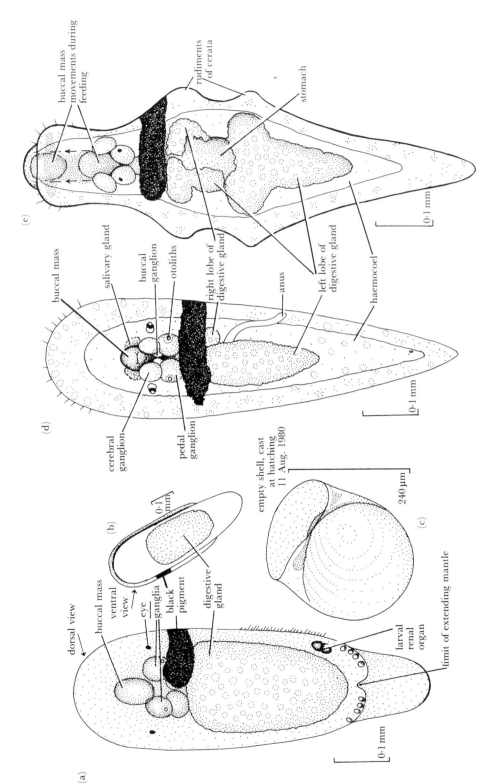

FIG. 35. Post-veliger development in *Aeolidiella alderi*, hatched in August 1980 at the Luc-sur-Mer Marine Laboratory in Normandy, France.
(a) length 0·60 mm, dorsal view.
(b) ventral view of the same.
(c) empty veliger shell, left behind in the spawn jelly at hatching, dorsal view.
(d) length 0·70 mm, 1 day after hatching, dorsal view.
(e) length 0·90 mm, 4 days after hatching, dorsal view.

106. **Aeolidiella alderi** (Cocks, 1852)
 Eolis alderi Cocks, 1852
 ?*Eolida soemmerringii* Leuckart, 1828
 Aeolidiella soemmeringii; Schmekel & Portmann, 1982

APPEARANCE IN LIFE (Plate 41). The body may measure up to 37 mm in length, pale cream or fawn with superficial orange pigment blushing parts of the dorsum. The colour of the digestive gland is closely governed by the diet and may vary from pale green-brown to pink or dark brown. The tips of the tentacles and of the cerata are white or pale yellow. The shape of the body is more slender than that of *Aeolidia papillosa*, but it is none the less robust.

Up to 16 obliquely transverse rows of cerata are present on either side of the dorsum (Garstang, 1890), each half-row consisting of up to 10 elongated, mobile cerata. All the cerata contain white cnidosacs near the distal extremity. This is especially conspicuous anteriorly, where the first one or two rows of cerata are unpigmented so that they exhibit the slightly larger cnidosacs. The digestive gland lobules reach only about half way up the cerata which constitute this anterior collar or "ruff". During feeding, Tardy (1964b) has observed that these cerata greatly elongate and shoot nematocysts into the stipe of the actinian prey, helping to overpower it before the nudibranch commences ingestion.

The oral tentacles are slightly longer than the smooth or slightly wrinkled rhinophores; conspicuous propodial tentacles are present.

ANATOMY (Fig. 34b). Colgan (1914) described the radula of a 30 mm specimen from Ireland. The formula was 15 × 0.1.0, and there were 22–25 denticles on each side of the recessed median cusp. We have examined a number of radulae from other parts of the geographical range. Three Adriatic specimens had formulae as follows: length 14 mm, 14 × 0.1.0 (with 19 denticles each side); 20 mm, 15 × 0.1.0 (29). A Dorset specimen 18 mm in preserved length: 20 × 0.1.0 (27).

The reproductive organs have been described from their earliest beginnings in development by Tardy (1970; 1971). The same author has published detailed observations on embryonic and larval development (1970b), and more specifically on the central nervous system (Tardy, 1974). Riva & Vicente (1976) confirmed Naville's (1926) claim that members of this genus maintained zooxanthellae (derived from the cnidarian prey) in the cells of the digestive gland.

HABITS. Tardy (1962) has given a good description of the main features of early development. The 280–300 µm ova develop to hatching in 13 days (25 °C) or 20 days (18 °C); each spawn mass may contain 200–1200 eggs. Development is of type 3 (Thompson, 1967). We succeeded in rearing juveniles at the Luc-sur-Mer laboratory feeding them upon *Actinothoe sphyrodeta* (Gosse); stages of early benthic development are illustrated in Figs 35 & 36. Tardy (1962) and Tardy & Bordes (1978) have given a substantial list of acceptable actinians: *Cereus pedunculatus* (Pennant), *Metridium senile* (L.), *Diadumene cincta* Stephenson, *Sagartia troglodytes* (Price) and *Sagartiogeton lacerata* (Dalyell). Miller (1961) contributed *Sagartia elegans* (Dalyell) and Riva & Vicente (1978a) described feeding upon the Mediterranean actinian *Parastephanauge pauxi*.

DISTRIBUTION (Map 10). We have taken this species in shallow water in both the Adriatic and Aegean Seas, and records exist from Naples (Schmekel & Portmann, 1982, as *A. soemmerringii*) and Morocco (Pruvot-Fol, 1954). Tardy (1969) reports it from the Île de Ré and from Arcachon, while we have taken it on the Cotentin Peninsula in Normandy. There are very few records in the British Isles (Seaward, 1982), all on western coasts.

107. **Aeolidiella glauca** (Alder & Hancock, 1845a)
 Eolis glauca Alder & Hancock, 1845a
 Eolis angulata Alder & Hancock, 1844a
 Eolis pallidula Lafont, 1871

APPEARANCE IN LIFE (Plate 41). This is a slightly larger species (up to 40 mm long) than *A. alderi*, but it takes a practised eye to distinguish between them. Despite this difficulty, there is

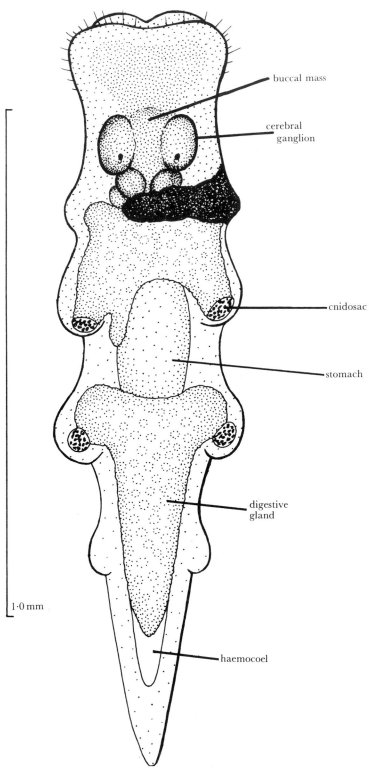

FIG. 36. Post-veliger development in *Aeolidiella alderi*, showing the development at a body-length of 1·5 mm (7 days after hatching) of paired ceratal processes containing rudimentary cnidosacs; dorsal view.

no doubt at all of their specific distinctness because their reproductive biology is so very different. *Aeolidiella glauca* has planktotrophic larval development, whereas *A. alderi* has direct development, lacking the planktonic veliger phase.

The body colour is pale yellow or grey-brown; the cerata have a pinkish tinge below the tip and the distal extremity of each is tipped by a cream-white zone. An important distinguishing feature is the presence of white spots, mottling or blotching over the whole dorsum, including the cerata and the tentacles. The pale brown or greenish brown digestive gland extends to within $\frac{1}{4}$ and $\frac{1}{3}$ of the tip of each ceras, including the most anterior rows. Consequently there is no pale "ruff" behind the rhinophores, of the type described for *Aeolidiella alderi*. All the cnidosacs are of equivalent size in *A.glauca*; there are no "giant" cnidosacs.

ANATOMY (Fig. 34c; Plate 35). The radula of a 14·5 mm preserved specimen from Cornwall has been examined; the formula was $17 \times 0.1.0$. Colgan (1913) reported on a 22 mm specimen which exhibited the radular formula $20 \times 0.1.0$ (with 33–46 denticles on either side of the recessed median cusp).

HABITS. Hadfield (1963) and Tardy (1969) agree that this species lays egg masses which each contain up to 100,000 ova; these develop to planktotrophic veliger larvae in 11 days at 10–12 °C.

Tardy & Bordes (1978) list the actinians on which this species has been seen to feed: *Sagartia troglodytes* (Price), *Sagartia elegans* (Dalyell), *Cereus pedunculatus* (Pennant), *Diadumene cincta* Stephenson, *Sagartiogeton undata* (Müller) and *Sagartiogeton lacerata* (Dalyell).

DISTRIBUTION (Map 10). This species has been recorded from localities all around the British Isles. Some field workers believe that this species is usually found in deeper water than the intertidal and shallow sublittoral regions occupied by *A. alderi*. Elsewhere, Tardy has sifted the published records and established that *A. glauca* is restricted to a short stretch of the European coast from Denmark to Arcachon on the French Biscay coast. It probably does not occur in the Mediterranean Sea (Tardy, 1969).

108. *Aeolidiella sanguinea* (Norman, 1877)
Eolis sanguinea Norman, 1877
Aeolis croisicensis Labbé, 1923

APPEARANCE IN LIFE (Plate 41). This is the largest and the most rare of the British species of *Aeolidiella*, up to 46 mm in a Donegal specimen. The pedal sole is white, but the rest of the body, including the cerata and the three pairs of tentacles, are pale yellow, orange or red (according to diet). Only the tips of the dorsal processes are white. There is no pale "ruff" behind the rhinophores, of the type described for *Aeolidiella alderi*.

ANATOMY (Fig. 34d). The radula of a 22 mm preserved specimen from Donegal was examined; it had the formula $31 \times 0.1.0$ (with up to 40 denticles on either side of the recessed median cusp).

HABITS. This species feeds, according to Tardy & Bordes (1978), on the actinians *Sagartia troglodytes* (Price), *S. elegans* (Dalyell), *Sagartiogeton undata* (Müller), *Diadumene cincta* Stephenson, *Cereus pedunctulatus* (Pennant) and *Aiptasia couchi* (Cocks). Tardy (1969) illustrates the planktotrophic veliger larva.

DISTRIBUTION (Map 10). There are only three localities in the British Isles, all on the western coasts of Ireland: Connemara (Norman, 1877), Donegal (personal) and Lough Ine (Wilson & Picton, 1983).

Elsewhere it has been recorded only from three sites on the French Atlantic coast: Île de Ré, Croisic and Fouras (Tardy, 1969).

EPILOGUE: SURVEY OF THE LITERATURE ON OPISTHOBRANCHIA SINCE THE PUBLICATION OF VOLUME I IN 1976

The years since volume I went to press have been years of intense activity for many enthusiastic students of the Opisthobranchia. Much of the resulting published material has dealt with the order Nudibranchia and it has been possible to incorporate a great amount of this into the systematic sections of the present volume. A substantial quantity remains, however, and the ensuing short review outlines progress in some topics of concern.

REPRODUCTION

One of the most important original papers to have been published is that by Lalli & Wells (1978) on the reproductive biology of the world's 7 species of the thecosome genus *Limacina* (*Spiratella* of some authorities). They are all protandrous hermaphrodites living pelagically (*L. helicoides* is bathypelagic), typically having an external autospermal groove and internal fertilization, like the benthic bullomorphs from which they are descended. Five species have a free-swimming larval phase, hatching from free-floating egg rafts. *Limacina helicoides* is ovoviviparous, however, with up to 10 encapsulated embryos held in the mucous gland until they attain a shell diameter of 5 mm and are expelled. *Limacina inflata* is unique in lacking mucous and albumen glands and is also aphallic, so that sperm transfer is by way of a spermatophore elaborated in the male tract by the prostate gland; this species has mantle cavity brood protection of unencapsulated embryos. A no-less-interesting account of the larval biology of the dorid nudibranch *Aegires punctilucens* has appeared, by Thiriot-Quiévreux (1977), distinguished by excellent photomicrographs from life, showing the unusual 2-stage metamorphosis of the veliger larvae, which possess greatly enlarged velar lobes (as big as those of some prosobranch veligers), which are resorbed long after the shell has been discarded (the operculum is said to be lacking). Thiriot-Quiévreux generously draws attention to a little known contribution by Garstang (1894), who studied the same planktonic larval form and stated that "the animal had assumed the form of a dorid when free-swimming and still retaining its velum".

A number of papers have attempted to relate zoogeography of opisthobranch species to reproductive strategy. The most successful investigations have resulted from comparative studies of numerous species from a restricted sea area. Clark (1975) for the New England coast, Clark & Goetzfried (1978) for Florida and Eyster (1980) for the S. Carolina coast set themselves such a task. The proportions of the 3 principal development-types (defined in volume I) are as follows, for two of these areas:

	type 1	type 2	type 3
New England (Clark, 1975)	90%	10%	0%
S. Carolina (Eyster, 1980)	44%	40%	16%

Clark & Goetzfried (1978) find these figures disturbing when combined with their own observations on the Florida coast, altogether indicating a surprisingly high proportion of types 2 and 3 towards the southern extremity of the eastern seaboard of north America. In short, the picture that emerges is opposite to expectation according to Thorson's "rule" that polar life (not tropical life) should impose such a change in reproductive mode. Clark & Goetzfried believe that direct (non-pelagic) development is an adaptation in tropical waters (like their area of study) to high climatic and/or trophic stability.

The description by Bridges (1975) and by Clark & Goetzfried (1978) of a number of opisthobranchs possessing extra-zygotic food reserves has opened up a new field of research, reviewed by Clark & Jensen (1981). In the aplysiid *Phyllaplysia taylori* (which has type 3 development), Bridges described a unique nutriment body (about 50 μm in diameter) in each egg capsule; this body breaks up later and is ingested orally by the developing embryo. Clark & Goetzfried have discovered spiralling masses of yolk granules situated outside the egg capsules, in the sacoglossans *Elysia cauze*, *E. papillosa* and *E. tuca*. This extra-zygotic yolk is utilized by the embryos as development proceeds. In another sacoglossan, *Costasiella lilianae*, the wall of each capsule is lined by a layer of vesicles which later break free and are "swallowed by the embryo, which grows to fill the capsule completely". In one of these sacoglossans, *Elysia cauze*, development is seasonally variable, a phenomenon referred to as poecilogony. Clark, Busacca & Stirts (1979) have shown that in Florida *E. cauze* exhibits type 1 development in early spring, type 2 in summer and type 3 (direct development) in autumn and early winter. These seasonal differences depend to a great extent on the ratio of egg numbers to the amount of extra-zygotic yolk. This variability is not phenotypic in one sense, because an individual adult does not reproduce by more than one mode; rather it is true that adults ripening at different seasons come to have different relative admixtures of deutoplasm and extra-zygotic yolk.

Another well-documented example of poecilogony has been provided by Eyster (1979) who studied developmental variability in the aeolid *Tenellia adspersa*. Adults in a single population laid (at 15–25 °C) either A. 70 μm ova which hatched as planktonic veligers in 68–122 hours; or B. 100 μm ova which underwent direct development in 140–190 hours. Each adult produced only one type of eggs throughout the period of observation, yet adults of the two strains were observed to mate indiscriminately.

Rivest (1978) has described what might well prove to be another case of poeciligony in *Cuthona nana*; it is already known for *Haminea hydatis* (Berrill, 1931; Thompson, 1981) and *Cylichnella* (= *Acteocina*) *canaliculata* (Franz, 1970a, 1971). Rivest found type 3 development (i.e. non-pelagic, although in *C. nana* a velum and shell are present at hatching) in direct contrast to earlier results (Harris, Wright & Rivest, 1975) in which type 1 development was noted.

The most courageous worker in this field is undoubtedly Todd (1978a, b, 1979), who has endeavoured to deepen our understanding of reproductive modes by his careful comparative investigations of the relationship between body-size, age, and the histology of the ovotestis and digestive gland in British dorid nudibranchs. Todd concluded that *Onchidoris muricata* (which has type 1 development) makes a greater relative reproductive effort that *Adalaria proxima* (type 2 development), but that, nevertheless, the latter species apportions a greater number of calories to reproduction. A particular tour de force was the publication (Todd, 1978a) of a photomicrograph showing the result of post-reproductive atrophy of the digestive gland in *Onchidoris muricata*, confirming that senescence and death are innately determined, in his words "primarily related to visceral consumption . . . not necessarily .... to gonadal exhaustion".

There has been a veritable explosion of publications dealing with the laboratory culture of opisthobranch veliger larvae, especially of the planktotrophic development-type 1. These achievements have been the result of patient efforts to understand all the metabolic requirements of the species selected, and the most recent papers have shown such technical mastery that comparative studies have for the first time become possible, such as those of Switzer-Dunlap & Hadfield (1977, 1979). The key factors in successful culture of the veliger larvae are the use of small vessels (about 1 litre), regular changes of culture medium, the availability of controlled quantities of algal cultures, preferably mixed, and the provision to the searching larvae of the species-specific trigger to progressive metamorphosis.

The trigger to metamorphosis is usually to be found in some element of the nutriment of the adult. For example, Harris (1975) showed that the late larvae of the aeolid nudibranch *Phestilla melanobranchia* required the 'smell' of living tissues of a dendrophylliid coral (but not actual contact with them), while *P. sibogae* required a chemical trigger from another coral, *Porites*. As Harris said: "Knowledge of the biology of the prey greatly facilitates the study of any nudibranch species". Following this dictum, other investigators have reported similar successes, with a variety of triggers. Some of these are listed below, but publications already mentioned in volume I are largely omitted here.

Species of opisthobranch	Trigger to metamorphosis	Authority
Aplysia dactylomela	Laurencia (Rhodophyta)	Switzer-Dunlap, 1978
A. juliana	Ulva fasciata & U. reticulata (Chlorophyta)	Do.
A. parvula	Chondrococcus hornemanni (Rhodophyta)	Do.
A. brasiliana	Callithamnion halliae & Polysiphonia (Rhodophyta)	Strenth & Blankenship, 1978
Stylocheilus longicaudus	Lyngbya majuscula (Cyanophyta)	Switzer-Dunlap, 1978
Dolabella auricularia	Unidentified blue-green	Do.
Tritonia diomedea	Virgularia (Cnidaria)	Kempf & Willows, 1977
Corambella obscura	Electra crustulenta (Polyzoa)	Perron & Turner, 1977
C. steinbergae	Membranipora villosa (Polyzoa)	Bickell & Chia, 1979
Rostanga pulchra	Ophlitaspongia pennata (Porifera)	Chia & Koss, 1978
Onchidoris bilamellata	living barnacles	Todd, per. comm.

Such laboratory studies on natural induction of metamorphosis have led to biochemical studies of artificial or counterfeit induction. Harrigan & Alkon (1978) have reported success with competent larvae of *Elysia chlorotica*, using a technique employed first by Bonar (1976) in his studies of *Phestilla sibogae*; this involved the use of an 0·1% seawater solution of succinylcholine chloride. Harrigan & Alkon reported also that the larvae of *Elysia chlorotica* (and those of the infaunal shelled bullomorph *Haminea solitaria*) can be stimulated to metamorphose in the laboratory by exposure to glass surfaces covered by a bacterial film derived from biotope sediments.

Hadfield (1978) alone has tried to characterize the chemicals involved in the natural induction or triggering of metamorphosis. He prepared lyophilized aqueous extracts of the coral *Porites compressa* which retained the metamorphosis-inducing capability of living coral when its effect was tested upon competent late veligers of *Phestilla sibogae*. Coral water (sea water in which living coral had stood for 24 hours) was also effective. Furthermore, Hadfield noted that a transient exposure to the trigger was sufficient to stimulate the onset of metamorphosis in "clean" sea water some hours later. Such experimental results led him (Hadfield, 1978), in a review of factors governing metamorphosis in marine molluscan larvae, to embark upon a theoretical comparison between larval metamorphosis in opisthobranchs and the vertebrate Anura. Of course, good comparisons can be made of the morphological aspects of this change of life in the two groups, as larval organs are rapidly lost and adult structures replace them. But the physiological control must be different because the key event in *Phestilla* (the example chosen by Hadfield) is a once-for-all triggering of the competent larvae by the inducer (Hadfield's term) from *Porites*, unlike the sustained requirement for thyroxin governing the transformation of the anuran tadpole into a frog. Certainly, metamorphosis in *Phestilla* does not need the intrinsic synthesis of new macro-molecules, because Hadfield has shown it is not halted by exposure to Actinomycin D, which classically blocks RNA production.

This consideration of such exciting new approaches must not detract from the importance of many recent papers which are helping to consolidate our understanding of the morphology of opisthobranch larvae and metamorphosing stages (a field of study which Bonar (1978) terms Metamorphogenesis). The review by Bonar must be required reading for some time to come, for anyone wishing to undertake research involving opisthobranch larvae, because this author has succeeded in tabulating and critically analysing all publications which have ever dealt with the metamorphosis of opisthobranchs, from the early days when mutilated dying larvae were wistfully claimed as healthy post-larvae, to the present time with its spectacular successes, such as the culture of *Aplysia juliana* through 5 generations in the laboratory (Switzer-Dunlap &

Hadfield, 1977), the beautifully detailed description of the histogenesis of larval organs in *Corambella steinbergae* (Bickell & Chia, 1979; Bickell, Chia & Crawford, 1981) and the comprehensive ultrastructural study of the metamorphosing stages of *Phestilla sibogae* (Bonar & Hadfield, 1974).

SACOGLOSSA

Strides have been taken in recent years towards a better understanding of the north Atlantic Sacoglossa. On a broad canvas, Marcus (1980) has published a taxonomic review of the Elysiidae of the area, a family which includes not only *Elysia* but also the warm-water *Bosellia* and *Tridachia*. In the British Isles the distinguished labours of Thomas Gascoigne have come to fruition in a series of careful papers. Gascoigne & Sigurdsson (1977) reported for the first time in British waters the tiny *Calliopaea oophaga* Lemche, 1974, hitherto known only from the type locality (Kattegat), now found on the Northumberland coast. This remarkable animal feeds upon the eggs of the bullomorphs *Philine denticulata* and *Retusa truncatula*, and is considered by the authors to be separate and distinct from the oophagous Mediterranean sacoglossan *Stiliger vesiculosus* (Deshayes).

Gascoigne & Todd (1977) gave a full re-description of a Yorkshire collection of *Calliopaea bellula* Orbigny, 1837 (listed as *Stiliger bellulus* in volume I). This was a single specimen, 3 mm in length, but the authors are convinced of the separateness of *C. bellula* from *C. oophaga*, although Gascoigne later expressed some doubt in a personal communication. Gascoigne recommends several nomenclatural changes from the arrangement in volume I. First, he prefers the ordinal name Ascoglossa to Sacoglossa. Second, he replaces *Stiliger* by *Calliopaea* for the British stiligerids. Third, he removes the genus *Hermaea* to its own family Hermaeidae, while at the same time elevating the subgenus *Placida* to a full genus, retained in the Stiligeridae.

Gascoigne has also published detailed anatomical studies of the Mediterranean *Caliphylla mediterranea* (1979b), *Hermaea variopicta* (1979a) and *Placida viridis*, which lacks a heart in the adult (Gascoigne & Sordi, 1980). These papers have resulted in the main from a brilliant exploration of the fauna in the vicinity of the Centro Interuniversitario di Biologia Marina at Livorno in the northern Tyrrhenian Sea. Gascoigne's acumen is matched only by the discovery on the north American mainland of the bivalved sacoglossan *Berthelinia caribbea* (by Moore & Miller, 1979, on Plantation Key, Florida, July 1978).

Careful work on the physiological ecology of *Limapontia capitata* by Jensen (1975a & b; 1977) has been illuminating. Working on the north coast of Zealand in Denmark, she has attempted to quantify the importance of *L. capitata* as a primary consumer of *Cladophora*. Starvation tolerance was found to be 9–10 days at 20–23 °C; in favourable circumstances, between 1 and 10% of the total standing crop of *Cladophora* was destroyed (i.e. consumed) by this small sacoglossan. The mollusc was, however, restricted in its diet to monosiphonaceous or coenocytic algae of a diameter matching the width of its foot. It could not ingest *Enteromorpha*, *Ceramium* or *Polysiphonia*, but would take *Cladophora* spp., *Chaetomorpha linum* or *Bryopsis plumosa*, seemingly without favouring one more than another. The calorific value of mean food intake in these studies was 0·3–0·4 cal/day/individual. The optimal temperature for nutrition was 15 °C.

These studies by Jensen of the phytivorous *Limapontia* lead into a consideration of the symbiotic Sacoglossa. Associations between sacoglossans and the plastids of Chlorophyta (and some Xanthophyta, according to Graves, Gibson & Bleakney, 1979) have continued to hold the attention of research teams around the world. Elysiid Sacoglossa provide the best examples, although the hermaeid *Costasiella lilianae* has been shown to retain symbiotic chloroplasts for at least 65 days in the laboratory (Clark, Jensen, Stirts & Fermin, 1981). Such chloroplasts, concludes one team, are surprisingly robust and the conclusion is reached that there is "nothing particularly remarkable about the phenomenon of chloroplast symbiosis" (Gallop, Bartrop & Smith, 1980). This "unremarkable" phenomenon certainly plays an important role in the symbiotic Sacoglossa. Clark, Busacca & Stirts (1979) showed that the young of *Elysia cauze* grew after hatching much better in the light than in the dark; moreover, mortality was more severe in dark-reared cultures. Working with *Elysia tuca* and its food-organism, the green alga

Halimeda discoidea, Stirts & Clark (1980) have shown that the maximal net carbon fixation rate (when functioning autotrophically) was higher for *E. tuca* than for *H. discoidea*! Facts like these highlight the need for careful comparative studies in a range of sacoglossan/algal symbioses. At one extreme, we know of examples such as *Alderia modesta* (Graves, Gibson & Bleakney, 1979), where the residual photosynthetic activity of ingested chloroplasts is of negligble value to the mollusc. At the other end of the spectrum is the tropical *Tridachia crispata*, where adults over a length of 10 mm have outgrown the need to take regular meals, and subsist predominantly autotrophically (Thompson, 1977).

Clark & Busacca (1978) have placed the wealth of scattered observations in a proper perspective by their studies of chlorophyll turnover rates in a number of sacoglossans in the shallow waters of Florida. Measured plastid half-lives were:

> 5 days in *Elysia tuca*
> 15 days in *E. cauze*
> 15 days in *Oxynoe antillarum*
> 58 days in *Tridachia crispata*.

Tridachia was shown to be far the "most efficient retainer". Not only this, but they have shown that in all likelihood *Tridachia* can "support chloroplasts from a variety of algae, and maintain them efficiently so that feeding—and restocking of chloroplasts—is seldom required". But *Tridachia* is never wholly autotrophic and cannot (in laboratory culture, at any event) sustain positive growth of body-weight without feeding, as we have ourselves verified.

Clark & Busacca (1978) have also analysed feeding specificity which they found is highest among the shelled Sacoglossa and relatively low in the more advanced Elysiidae and Stiligeridae. *Caulerpa* represents the most "primitive" sacoglossan food, with the Siphonales, other Chlorophyta, and then other foods being used by successively more advanced taxa. Fig. 37 shows something of the complex dietary requirements of the abundant herbivorous opisthobranch fauna of Florida, based to a great extent upon personal observations, supplemented by guidance from Dr K. B. Clark, and Jensen's (1980) review. Jensen (1981) presents detailed observations on the feeding mechanisms of Floridan sacoglossans. She analysed the size and shape of the radular teeth relative to the dimensions of the food plant. New observations on *Elysia ornata*, *Placida kingstoni* and *Ercolania funerea* show that cell sap may be ingested then returned to the algal filament several times before swallowing. In addition, Jensen describes for the first time predation on diatoms by *Elysia evelinae*; this habit may prove to be widespread.

GENERAL ECOLOGY

Among the most important of the recent papers dealing with opisthobranchs has been Todd's comprehensive review (1981) of nudibranch ecology. Not unnaturally, some of his conclusions are different from our own. For example, he considers that *Tritonia hombergi* and *Archidoris pseudoargus* undergo biennial life cycles, whereas we were and remain convinced that they grow to maturity during the first year of post-larval life, spawn and die soon after. Todd provides an invaluable review of nudibranch diets, incorporating many new facts discovered since the previous compilation published nearly 20 years ago (Thompson, 1964). He also reviews defensive mechanisms and furnishes the first published account of predation by birds (eider duck, *Somateria mollissima*) on nudibranchs (*Onchidoris bilamellata*). The acid-secreting epithelium, believed to play an important part in defence against teleost fish, was ignored by the ducks who swallowed the nudibranchs whole. Todd's review is essential reading for any student undertaking research work on general or reproductive biology of opisthobranchs; it is a full and well-balanced compilation.

Rudman (1981a, c) investigated the functional morphology of tropical aeolidacean nudibranchs which feed upon alcyonarians and scleractinians, respectively. These aeolids may exhibit cryptic shape, markings and deportment which have evolved to a high state of perfection (Fig. 38). Within the genus *Phyllodesmium*, Rudman found an evolutionary series of

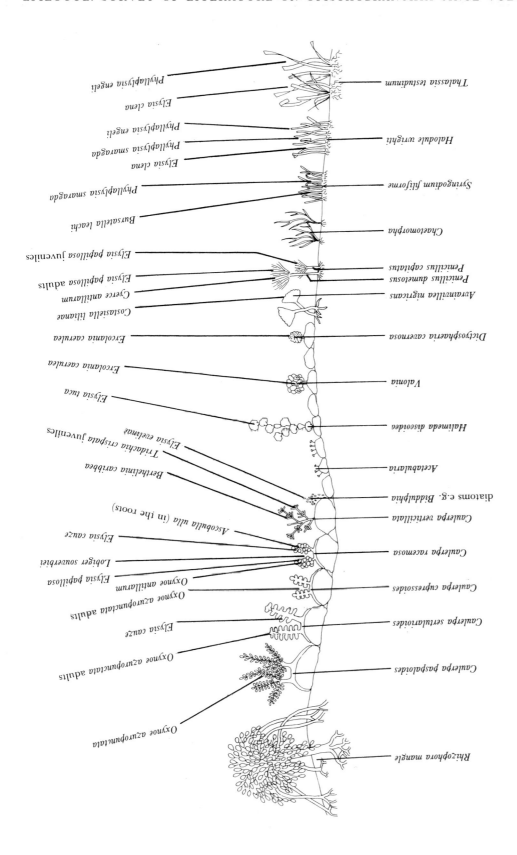

Fig. 37. Semi-diagrammatic representation of the dietary specificity of the herbivorous opisthobranchs of the Florida coast, U.S.A., compiled with the help of Drs K. B. Clark and K. R. Jensen. Opisthobranchia are listed above, plant-prey below.

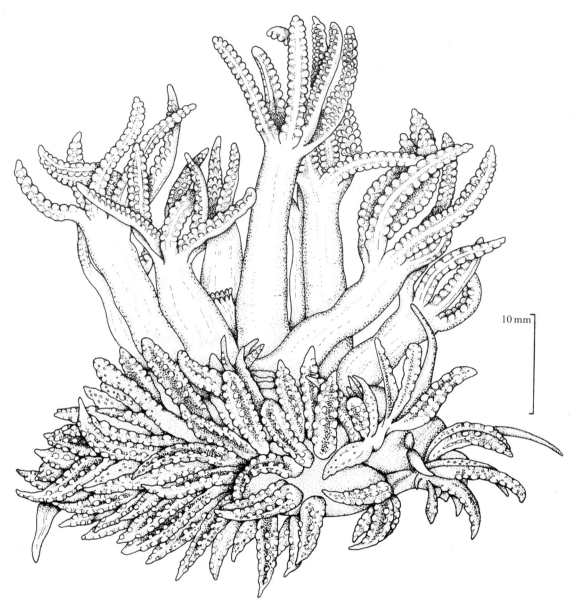

10 mm

FIG 38. The eastern Australian aeolidacean nudibranch *Phyllodesmium cryptica* Rudman on its prey, the alcyonarian soft-coral *Xenia*, showing near-perfect crypsis (after Rudman, 1981a).

increasingly intimate symbiosis between the aeolids and symbiotic dinoflagellates (zooxanthellae) derived from their alcyonarian prey. They all lack functional cnidosacs. Aeolidaceans which fed upon scleractinian corals of the genus *Porites*, in contrast, have well-developed cnidosacs and exhibit resource-partitioning such that each species utilizes a different part of the coral tissue as food. At least one of these aeolids has quite independently acquired the ability to transplant scleractinian zooxanthellae from the prey to its own tissues. Rudman (1982) elaborates on this discovery (first made by Hecht, 1895, Henneguy, 1925 and Naville, 1926), and provides a discussion of the evolution of symbiosis with zooxanthellae in the Nudibranchia; he believes that the relationship has evolved independently several times. Recent examples he has investigated include the giant aeolidacean *Pteraeolidia ianthina* (Angas) and the arminacean *Doridomorpha gardineri* Eliot (which feeds upon the blue coral *Heliopora*). In Europe, Riva & Vicente (1978b) confirmed the presence of zooxanthellae in the body and digestive gland lobules of the aeolidaceans *Spurilla neapolitana* (Chiaje) and *Aeolidiella alderi*. *Aeolidiella* also contains zoochlorellae. As Rudman (1982) observes, there are striking parallels between the symbiosis with algal cells of some carnivorous nudibranchs and the chloroplast symbiosis of many herbivorous sacoglossans reviewed above.

Earnest attempts to correlate radular structure with function in the Nudibranchia have not achieved the same success as noted for the Sacoglossa (Jensen, 1980). Bloom (1976) attempted such a correlation for doridaceans, and Nybakken & McDonald (1981) carried out a similar survey of those nudibranchs which prey upon Bryozoa, Tunicata, Cirripedia and Cnidaria. A complicating factor is that radular morphology, and sometimes diet, may alter significantly with age. Bloom paid attention too to the gastric caecum of some of the doridaceans. "A dorid with a caecum can handle large quantities of large and usually pointed spicules." The absence of such a caecum was associated with predation upon less harshly spiculose sponges, and accompanied in the alimentary canal by a more robust radula and a more muscular intestine. The author attempted to classify prey-sponges in a series of increasing difficulty of fragmentation. The easiest to masticate are the Halichondrida, followed by the Hadromerida, the Haplosclerida and the Poecilosclerida. Nybakken & McDonald (1981) sought and found similar functional correlations. For example, aeolidacean nudibranchs that consume hydroids have uniseriate or triseriate radulae. Those with uniseriate radulae effect ingress by puncturing the perisarc and ingesting the caenosarc. Those with triseriate radulae tend to feed directly on the polyps. Aeolidaceans which feed upon actinians tend to have uniseriate radulae consisting of broad teeth, each of which is strongly denticulate along the free border. Nudibranchs which attack members of the Pennatulacea, Gorgonacea or Alcyonacea have broad radulae; those which prey upon Stolonifera have narrow radulae. This approach has yielded many such interesting suggestions; much more undoubtedly remains to be discovered.

Among life-history studies, several must be mentioned. Protracted observations in arduous conditions led to Seager's (1979; 1982) analysis of population structure, growth, maturation, embryonic development and adult mortality of the Antarctic bullomorph *Philine gibba* Strebel. Development to hatching occupied 120 days at ambient $1 \cdot 1 \,°C$ and proved to be non-pelagic, unlike the pattern observed in the related *P. aperta*. Newly hatched juveniles of *P. gibba* resembled the adults in many respects but the shell remained uncovered by the mantle for some months. Sexual maturity was reached after approximately 3 years, spawning occurred at $3 \cdot 75$ years and death after a second period of spawning at $4 \cdot 75$ years. This slow ontogenetic rate contrasts vividly with Eyster's (1981) account of the life cycle of *Armina tigrina* Rafinesque in South Carolina, U.S.A. This developed to hatching in only 8 days at the ambient $23\,°C$. Sexual maturity was attained in about 80 days. Each individual laid 2–4 egg masses and then died. Eyster was able to calculate that between larval metamorphosis and the commencement of oviposition each individual *Armina* consumed approximately $1 \cdot 6\,g$ damp weight of the prey *Renilla reniformis* (Pallas). Furthermore, she estimated that the period of active spawning required the consumption of about $6 \cdot 1\,g$ damp weight of *Renilla* (two average colonies). Budgets like those measured by Eyster are all too rare. Eyster's observations are doubly important because so little had been published on the arminacean suborder.

Perhaps the most perfect case of crypsis in the nudibranchs is provided by Rudman's (1981b) observations on *Cuthona kuiteri*, an aeolidacean which is found on the coast of New South Wales, associated with the distinctive hydroid *Zyzzyzus spongicola* (Lendenfeld). The shape of the aeolid cerata (Fig. 39) is remarkably similar to that of the tubulariid polyp; in colour there is a perfect matching. Rudman also noted that the spawn masses of the *Cuthona* were attached in an ingenious way around the polyp stalks, thus gaining shelter from predators.

Finally, mention must be made of a recent discovery of the nudibranch epifauna associated with the gymnoblastic hydroid *Ectopleura dumortieri* (Beneden) in the south of the Gulf of Gascony (Lagardère & Tardy, 1980; Tardy & Gantès 1980). An illustration from this important study is reproduced in Fig. 40. Nudibranch enthusiasts will note that the *Ectopleura* epibiota include *Facelina bostoniensis*, *Polycera quadrilineata*, *Corambe testudinaria*, *Cuthona gymnota* and *Cumanotus cuenoti* (both swimming and creeping individuals are pictured). This undoubtedly represents the most complex nudibranch community yet discovered.

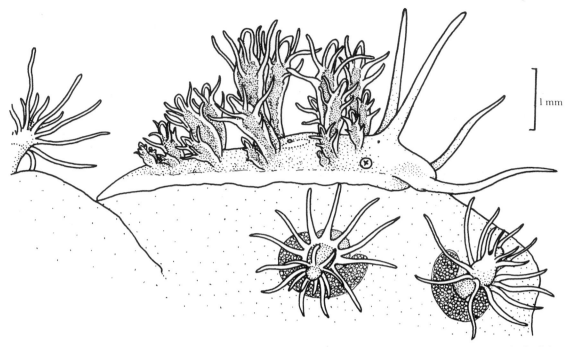

FIG. 39. The cryptic aeolidacean *Cuthona kuiteri* Rudman (New South Wales) on its prey, the tubulariid hydrozoan *Zyzzyzus*. Note the close resemblance of the nudibranch to the hydroid polyps, and the egg masses of *Cuthona* wrapped around the polyp stems (after Rudman, 1981b).

SYSTEMATIC ZOOLOGY OF BRITISH OPISTHOBRANCHIA

There have been new species, new records and some rearrangement of generic and supra-generic categories. Where these proposals relate to the true nudibranchs, they have been incorporated in the systematic sections of the present volume and need not be repeated here. However, a number of recent papers have consequences for the arrangement of the opisthobranch orders Bullomorpha, Aplysiomorpha, Pleurobranchomorpha, Acochlidiacea and Sacoglossa, which were dealt with in volume I (1976).

Kress (1977) described a new species of bullomorph, *Runcina ferruginea*, from south Devon, tabulating the distinguishing features of all the world's species of the suborder Runcinacea (later supplemented by Thompson, 1980). A reproduction of one of her drawings of *R. ferruginea* forms part of the Pictorial Synopsis in the present volume.

Wawra (1978) gave a detailed description of what is called in volume I (1976) *Microhedyle lactea* Hertling, 1930; he concluded that this species is conspecific with the senior *M. glandulifera* (Kowalevsky, 1901). The former name therefore falls into synonymy. Wawra's (1979) studies of acochlidiaceans have also led him to the far-reaching conclusion that certain species, such as *Platyhedyle denudata* Salvini-Plawen, 1973 from the Mediterranean, possess an ascus sac like the sacoglossans and accordingly should be moved to their own family, Platyhedylidae, close to the Limapontiidae within the order Sacoglossa. On the other side of the Atlantic, Rankin (1979) embarked on a wider ranging revision of the Acochlidiacea. Unfortunately, this was based upon the study of a single species found in the mud of a spring-fed mountain marsh, *Tantulum elegans*, together with a literature-survey. This was inadequate preparation for the author to propose no less than 5 new suborders in a re-defined Acochlidiacea, and large numbers of new families and genera. Progress in taxonomy is better brought about by carefully prepared steps, rather than by such impulsive saltations.

Poizat, using the meticulous meiobenthos-sampling and sorting methods described in his earlier papers (1927a & b), has published drawings and observations on many semi-microscopic opisthobranchs (and other molluscs) from Robin Hood's Bay and Northern

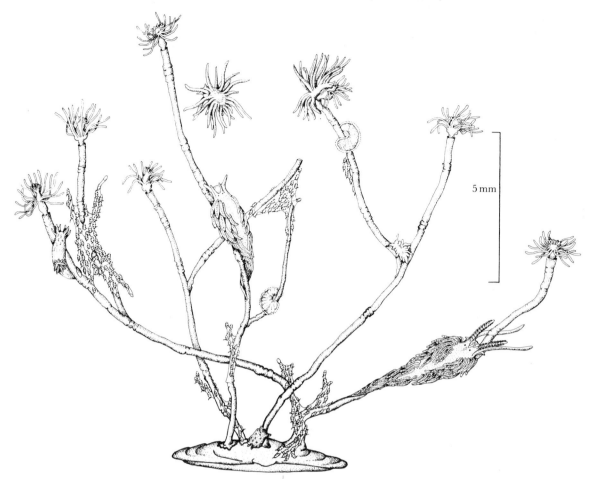

5 mm

FIG. 40. Epibionts associated with the tubularian *Ectopleura dumortieri* in the Gulf of Gascony. Many nudibranch species are illustrated: *Facelina bostoniensis* (juveniles and an adult), *Polycera quadrilineata*, *Corambe testudinaria* (both these doridaceans feed upon the epizoitic *Electra pilosa* (L.)), *Cuthona gymnota* and *Cumanotus cuenoti* (both pelagic and feeding individuals of *C. cuenoti* are pictured). Spawn masses of the last two species are present. (Modified after Lagardère & Tardy, 1980.)

Ireland in the British Isles, with comparative material from Marseilles and Var in France, and from the Skagerrak. Excellent photographs are given in his 1978 paper, and many drawings of British species in his most recent publication (1981), in which he suggests that these animals may undergo significant vertical migrations.

The discovery of abundant material of *Colpodaspis pusilla* off Lundy Island in the approaches to the Bristol Channel enabled a full re-description of this bullomorph (Brown, 1979), hitherto known only from a sketchy description of the Norwegian type specimen, augmented by Garstang's collection from red sandstone at 20 m near Plymouth (see volume I for details). His single specimen remained the sole British record until 1975 when another live individual was captured near Lundy I. Many more specimens were found in subsequent years, with a range-extension to the Donegal coast. The animal can easily be seen by a diver as a white speck, 2·5–5·0 mm in length; the delicate, acuminate internal shell was up to 3·2 mm long. These specimens were sexually mature, proving that *Colpodaspis* was not simply a juvenile form of another opisthobranch, but is a valid genus which probably belongs in the Diaphanidae. Brown (1979) gives a full description of the two known species of *Colpodaspis*, with a colour plate.

The most penetrating systematic study has been Morton's (1972) paper on the pallial organs of *Akera bullata* and of other opisthobranchs, in which he shows convincingly that Guiart (1901) had been correct in linking the Aplysiidae with the Akeridae. Taken to its logical conclusion, this means that the Akeridae must be transferred from the Bullomorpha to the Aplysiomorpha. We fully concur with this, and our own studies of the spermatozoa of *Akera* and *Aplysia* (which have certain unique features in common, not found in other opisthobranchs) prove its correctness. But the re-defined Aplysiomorpha now becomes an unwieldy order, difficult to delineate, because *Akera* has a large external shell, parapodia which are continuous with the pedal sole, a non-tentaculate cephalic shield, organs of Hancock, posterior pallial lobe and long visceral connectives, all features in which *Aplysia* differs. Nevertheless, the resemblances must be held to be more significant: the plicate gill, armoured gizzard, radula *n*.1.*n*, phytivorous habit, intermittent swimming by parapodial lobes, defensive purple gland, and spermatozoon with two mitochondrial spirals and a nucleus which is wound helically around the anterior extremity of the flagellum.

NUDIBRANCH DISTRIBUTION MAPS

These maps have been adapted with permission from the *Sea Area Atlas of the Marine Molluscs of Britain and Ireland* (Seaward, 1982), published for the Conchological Society by the Nature Conservancy Council. Some records which have come to us since the appearance of that publication have been added.

Each black circle denotes a locality report, without distinction between old and new records (the *Sea Area Atlas* made a distinction between pre 1951 and post 1951 live records).

MAP 1

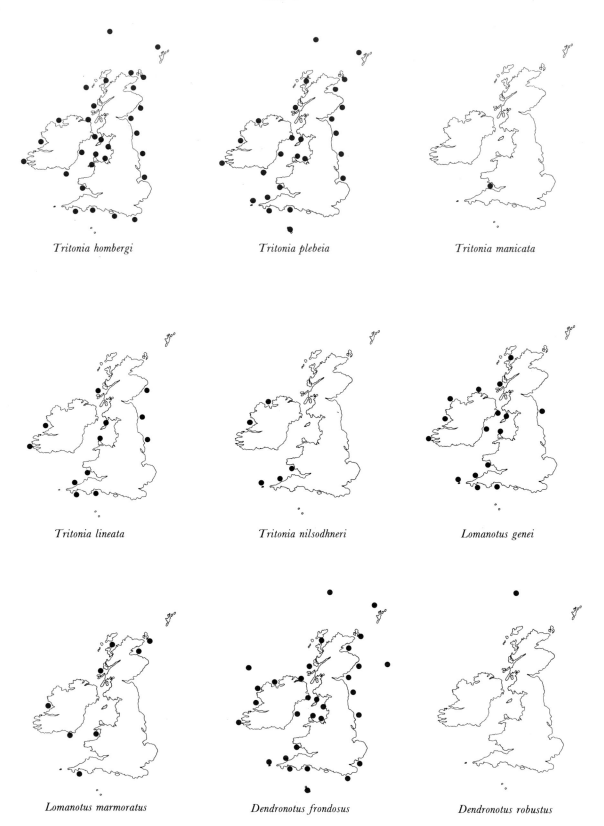

Tritonia hombergi   *Tritonia plebeia*   *Tritonia manicata*

Tritonia lineata   *Tritonia nilsodhneri*   *Lomanotus genei*

Lomanotus marmoratus   *Dendronotus frondosus*   *Dendronotus robustus*

Map 2

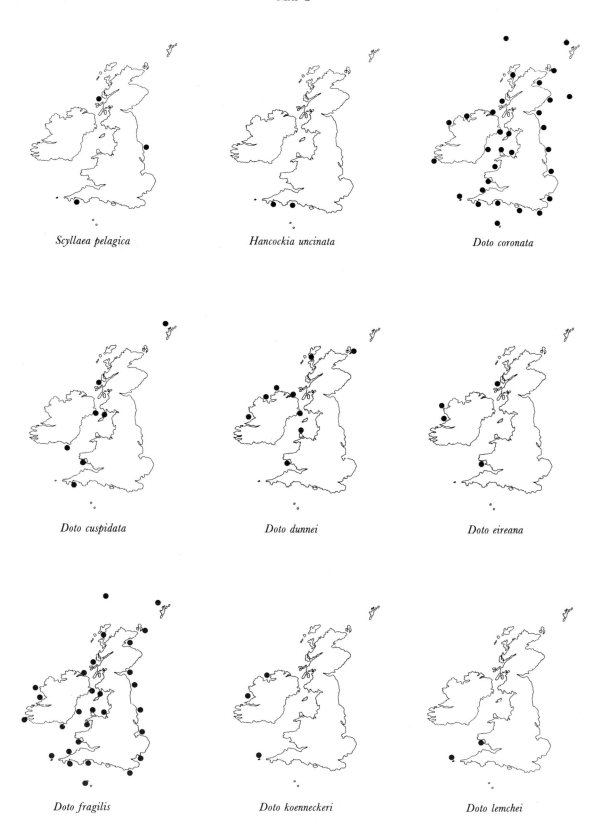

Scyllaea pelagica

Hancockia uncinata

Doto coronata

Doto cuspidata

Doto dunnei

Doto eireana

Doto fragilis

Doto koenneckeri

Doto lemchei

MAP 3

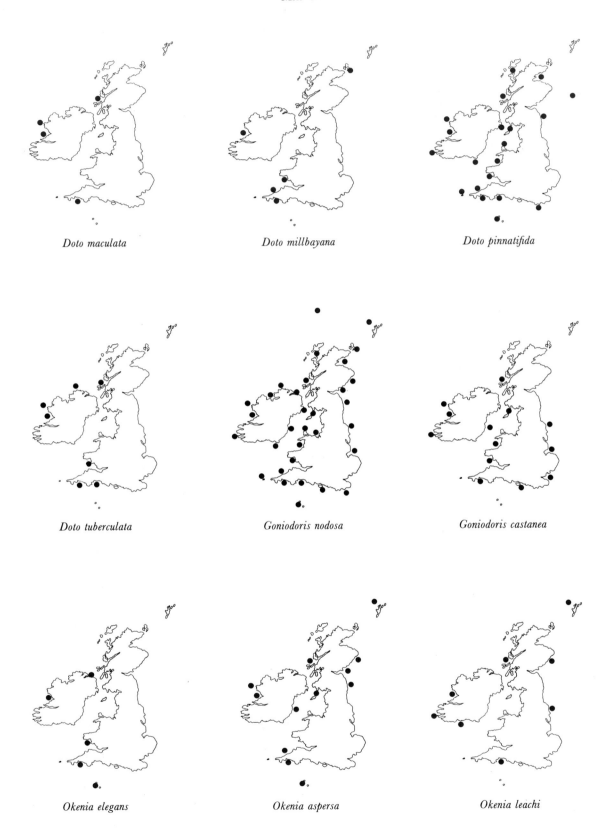

Doto maculata

Doto millbayana

Doto pinnatifida

Doto tuberculata

Goniodoris nodosa

Goniodoris castanea

Okenia elegans

Okenia aspersa

Okenia leachi

MAP 4

Okenia pulchella

Ancula gibbosa

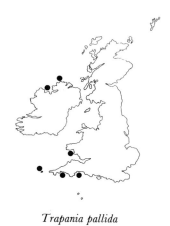

Trapania pallida

Trapania maculata

Acanthodoris pilosa

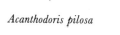

Adalaria proxima

Adalaria loveni

Onchidoris bilamellata

Onchidoris muricata

179

MAP 5

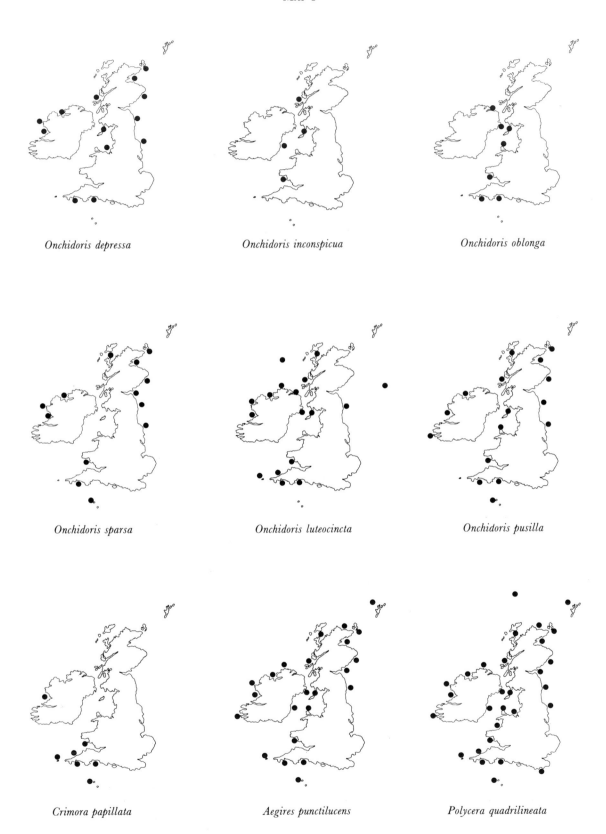

Onchidoris depressa   Onchidoris inconspicua   Onchidoris oblonga

Onchidoris sparsa   Onchidoris luteocincta   Onchidoris pusilla

Crimora papillata   Aegires punctilucens   Polycera quadrilineata

MAP 6

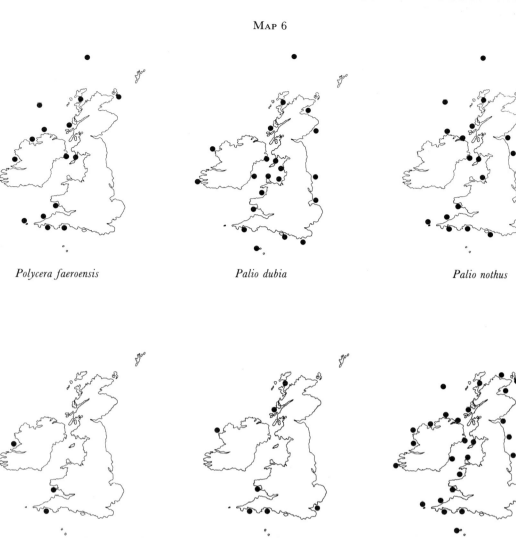

Polycera faeroensis

Palio dubia

Palio nothus

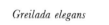

Greilada elegans

Thecacera pennigera

Limacia clavigera

Cadlina laevis

Aldisa zetlandica

Discodoris millegrana

Map 7

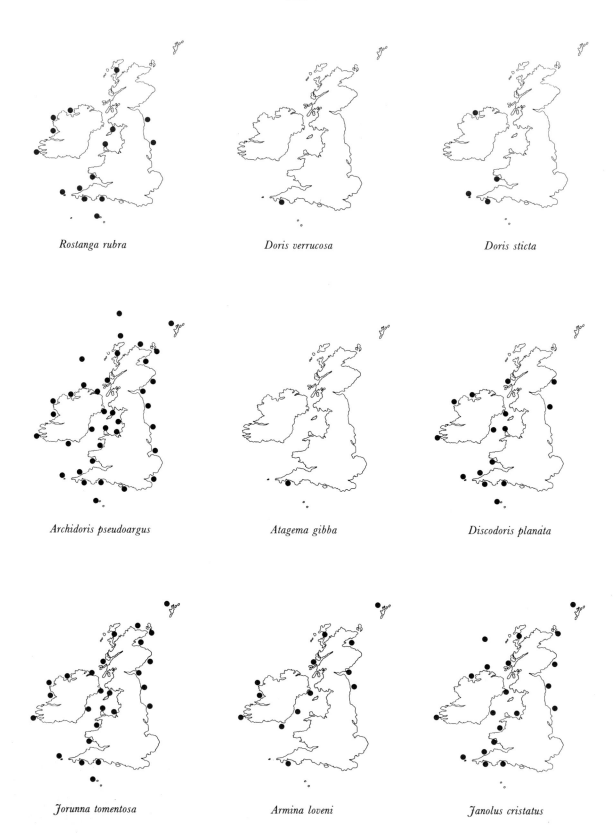

Rostanga rubra

Doris verrucosa

Doris sticta

Archidoris pseudoargus

Atagema gibba

Discodoris planata

Jorunna tomentosa

Armina loveni

Janolus cristatus

MAP 8

Janolus hyalinus

Proctonotus mucroniferus

Hero formosa

Coryphella browni

Coryphella gracilis

Coryphella verrucosa

Coryphella lineata

Coryphella pedata

Coryphella pellucida

MAP 9

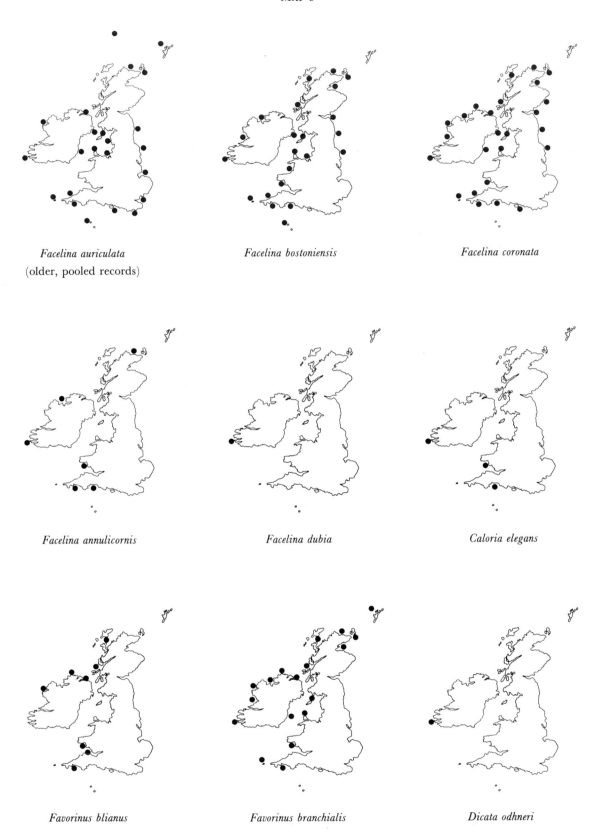

Map 10

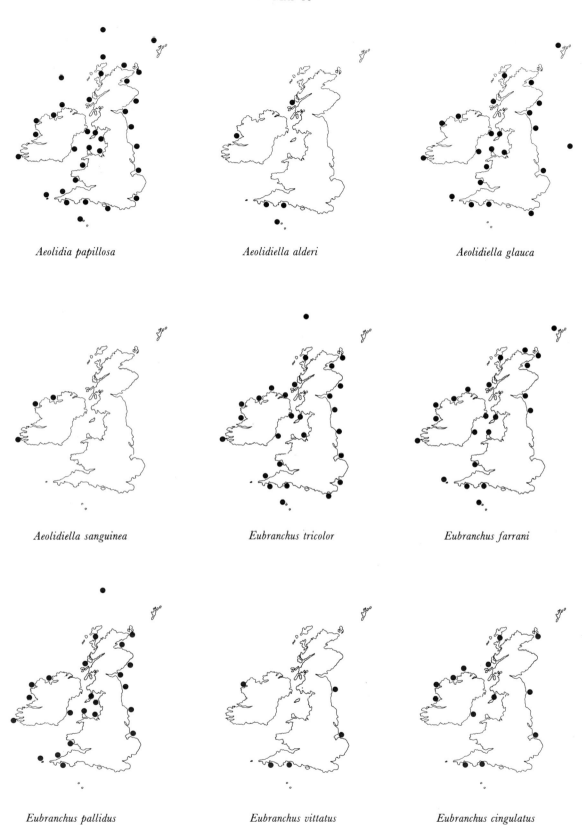

Aeolidia papillosa

Aeolidiella alderi

Aeolidiella glauca

Aeolidiella sanguinea

Eubranchus tricolor

Eubranchus farrani

Eubranchus pallidus

Eubranchus vittatus

Eubranchus cingulatus

MAP 11

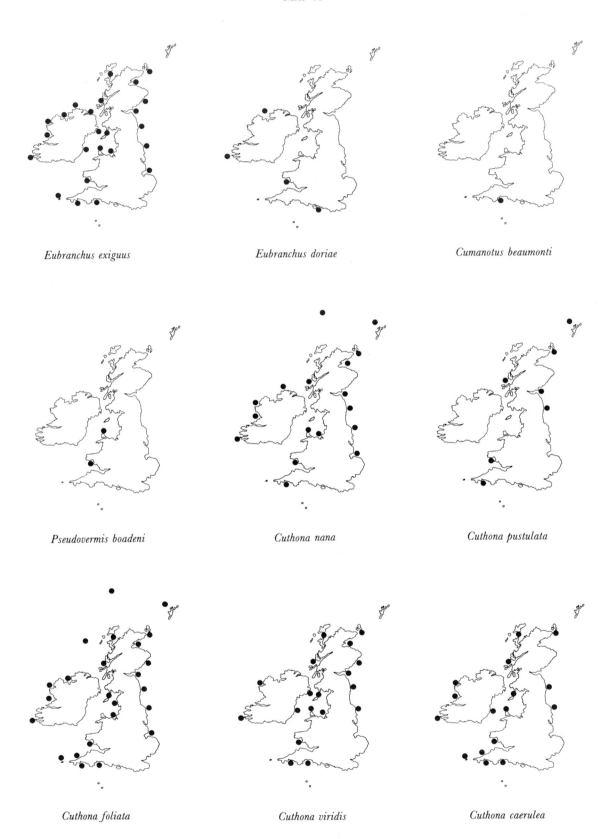

Eubranchus exiguus

Eubranchus doriae

Cumanotus beaumonti

Pseudovermis boadeni

Cuthona nana

Cuthona pustulata

Cuthona foliata

Cuthona viridis

Cuthona caerulea

MAP 12

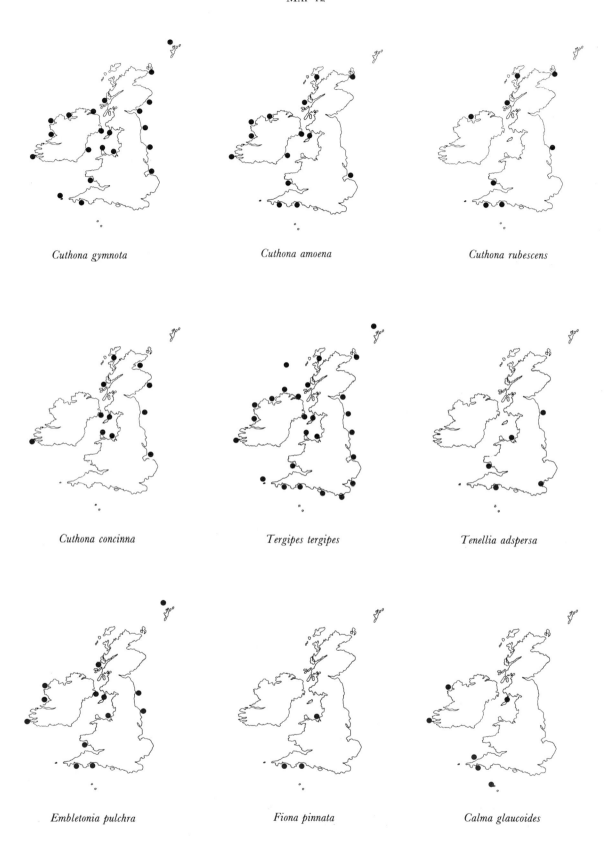

Cuthona gymnota

Cuthona amoena

Cuthona rubescens

Cuthona concinna

Tergipes tergipes

Tenellia adspersa

Embletonia pulchra

Fiona pinnata

Calma glaucoides

REFERENCES

ADAMS, A. & REEVE, L. 1850. *The zoology of the voyage of HMS Samarang.* London, Reeve & Benham. Pp. 1–66.

AGASSIZ, L. 1850. Minutes of 7 November 1849 meeting. *Proc. Boston Soc. nat. Hist.* **3**: 191.

ALDER, J. 1841. Observations on the genus *Polycera* of Cuvier, with descriptions of two new British species. *Ann. & Mag. nat. Hist.* **6**: 337–342.

ALDER, J. 1844. On the genus *Venilia. Ann. & Mag. nat. Hist.* **13**: 407.

ALDER, J. 1869. Account of the British Nudibranchiata. In Jeffreys (1863–69).

ALDER, J. & HANCOCK, A. 1842. Descriptions of several new species of nudibranchous Mollusca found on the coast of Northumberland. *Ann. & Mag. nat. Hist.* **9**: 31–36.

ALDER, J. & HANCOCK, A. 1843. Notice of a British species of *Calliopaea*, D'Orbigny, and of four new species of *Eolis*, with observations on the development and structure of the nudibranchiate Mollusca. *Ann. Mag. nat. Hist.* **12**: 233–238.

ALDER, J. & HANCOCK, A. 1844a. Description of a new genus of nudibranchiate Mollusca, with some new species of *Eolis. Ann. Mag. nat. Hist.* **13**: 161–167.

ALDER, J. & HANCOCK, A. 1844b. Descriptions of *Pterochilus*, a new genus of nudibranchiate Mollusca, and two new species of *Doris. Ann. & Mag. nat. Hist.* **14**: 329–331.

ALDER, J. & HANCOCK, A. 1845a. Notice of a new genus and several new species of nudibranchiate Mollusca. *Ann. & Mag. nat. Hist.* **16**: 311–316.

ALDER, J. & HANCOCK, A. 1845b. In *A monograph of the British nudibranchiate Mollusca*, part 1. London, Ray Society.

ALDER, J. & HANCOCK, A. 1845–55. *A monograph of the British nudibranchiate Mollusca.* London, Ray Society. Part 1 (1845); Part 2 (1846); Part 3 (1847); Part 4 (1948); Part 5 (1851); Part 6 (1854); Part 7 (1855).

ALDER, J. & HANCOCK, A. 1846. Notices of some new and rare British species of naked Mollusca. *Ann. & Mag. nat. Hist.* **18**: 289–294.

ALDER, J. & HANCOCK, A. 1847. In *A monograph of the British nudibranchiate Mollusca*, part 3. London, Ray Society.

ALDER, J. & HANCOCK, A. 1848a. Additions to the British species of nudibranchiate Mollusca. *Ann. Mag. nat. Hist.* **1**: 189–192.

ALDER, J. & HANCOCK, A. 1848b. In *A monograph of the British nudibranchiate Mollusca*, part 4. London, Ray Society.

ALDER, J. & HANCOCK, A. 1851. In *A monograph of the British nudibranchiate Mollusca*, part 5. London, Ray Society.

ALDER, J. & HANCOCK, A. 1854a. Notice of some new species of British Nudibranchiata. *Ann. Mag. nat. Hist.* **14**: 102–105.

ALDER, J. & HANCOCK, A. 1854b. In *A monograph of the British nudibranchiate Mollusca*, part 6. London, Ray Society.

ALDER, J. & HANCOCK, A. 1855. In *A monograph of the British nudibranchiate Mollusca*, part 7. London, Ray Society.

ALDER, J. & HANCOCK, A. 1862. Description of a new genus and some new species of naked Mollusca. *Ann. Mag. nat. Hist.* **10**: 261–265.

ALLAN, J. K. 1957. Some Opisthobranchia (Class Gastropoda) new to Australia or otherwise of interest. *J. malac. Soc. Australia* **1**: 3–7.

ALLEN, J. A. 1962. Mollusca. *Fauna of the Clyde Sea Area.* Scottish Marine Biological Association. Pp. 88.

ASCANIUS, P. 1774. Beskrivelse over en Norske sneppe og et sødyr. *Kgl. Norske Vidensk. Selsk. Skrift.* **5**: 153–158.

BABA, K. 1949. *Opisthobranchia of Sagami Bay collected by His Majesty, the Emperor of Japan.* Tokyo, Iwanami Shoten. Pp. 1–194.

BABA, K. 1955. *Opisthobranchia of Sagami Bay Supplement.* Tokyo, Iwanami Shoten. Pp. 1–59.

BABA, K. 1960. The genera *Polycera, Palio, Greilada* and *Thecacera* from Japan (Nudibranchia–Polyceridae). *Publs Seto mar. biol. Lab.* **8**: 75–78.

BABA, K. 1961. Three new species of the genus *Catriona* from Japan (Nudibranchia–Eolidacea). *Publs Seto mar. biol. Lab.* **9**: 367–372.

BABA, K. 1971. Description of *Doto (Doto) fragilis nipponensis* subsp. nov. from Sagami Bay, Japan (Nudibranchia: Dendronotoidea: Dotoidae). *The Veliger* **14**: 153–154.

BABA, K. 1974. New distributional record of *Aegires punctilucens* (d'Orbigny, 1837) from Sado Island, Japan. *The Veliger* **17**: 11–12.

BABA, K. & HAMATANI, I. 1963. A short account of the species, *Tenellia pallida* (A. & H.) taken from Mukaishima, Japan. *Publs Seto mar. biol. Lab.* **11**: 337–338.

BALCH, F. N. 1909. A spring collecting trip. Notes on New England nudibranchs. 2. *Nautilus* **23**: 35–38.

BARBOUR, M. A. 1979. A note on the distribution and food preferences of *Cadlina laevis. Nautilus* **93**: 61–62.

BARDARSON, G. G. 1919. Om den marine molluskfauna ved vestkysten af Island. *Det. Kgl. Danske Vidensk. Selsk. Biol. Meddel.* **2**: 1–139.

BARNARD, K. H. 1927. South African nudibranch Mollusca, with descriptions of new species, and a note on some specimens from Tristan d'Acunha. *Ann. S. Afr. Mus.* **25**: 171–215.

BARNARD, K. H. 1933. Description of a new species of *Thecacera. J. Conch., Lond.* **19**: 294–295.

BARNES, H. & POWELL 1954. *Onchidoris fusca* (Müller) a predator of barnacles. *J. anim. Ecol.* **23**: 361–363.

BAYER, F. M. 1963. Observations on pelagic mollusks associated with the siphonophores *Velella* and *Physalia. Bull. mar. Sci. Gulf Caribb.* **13**: 456–466.

BEAUMONT, W. I. 1900. The fauna and flora of Valencia harbour on the west coast of Ireland. XII. The opisthobranchiate Mollusca. *Proc. roy. Irish Acad.* (3) **5**: 832–854.

BEHRENTZ, A. 1931. Trekk av *Lamellidoris muricatas* biologi og av dens generationsorganers bygning. *Nyt. Mag. Naturvidesnk.* **70**: 1–26.

BERGH, R. 1857. Anatomisk Untersögelse af *Fiona atlantica*, Bgh. *Vidensk. Meddel. Naturh. Foren.* 273–335.

BERGH, R. 1860. Om Forekomsten af Neldefiim hos Mollusker. *Vidensk. Meddel. Naturh. Foren. Kjöbenhavn* (1860): 309–331.

BERGH, R. 1863. *Campaspe pusilla*, en ny Slaegstform af Dendronotidernes Gruppe, samst Bemaerkniger om Dotidernes Familie. *Naturh. Tidsskr.* 3, rzekke **1** (3): 471–483.

BERGH, R. 1864. Anatomiske bidrag til kundskab om Aeolidierne. *Kgl. Danske Vidensk. Selsk. Skrift. Naturvid. og Math. (5)* **7**: 139–316.

BERGH, R. 1866. Bidrag til en monographi ag Pleurophyllidierne, en Familie af de Gastraeopode Mollusker. *Naturh. Tidsskr. 3, raekke* **4** (2): 207–380.

BERGH, R. 1870–1904. Malacologische Untersuchungen. In Semper's *Reisen im Archipel der Philippinen.* Wiesbaden. **1** (1,): 1–30 (1870); (5): 205–246 (1873); (6): 247–285 (1874); (7): 287–314 (1874); (8): 315–344 (1875); (9): 345–376 (1875); **2** (10): 377–428 (1876); (12): 495–546 (1877); (13): 547–601 (1878c); (14): 603–645 (1878b); **4** (2): 79–128 (1881); **6** (1): 1–56 (1904).

BERGH, R. 1871a. Beiträge zur Kenntniss der Mollusken des Sargassomeeres. *Verhandl. k. k. zool.-bot. Gesell. Wien* **21**: 1273–1308.

BERGH, R. 1871b. In: Synopsis molluscorum marinum Daniae. Fortegnelse over de i de Danske have Forkommende Blöddyr. (By O. A. L. Mörch.) *Vidensk. Meddel. Naturh. Foren. Kjöbenhavn* (1871): 157–225.

BERGH, R. 1873. Beiträge zur Kenntnis der Aeolidiaden. 1. *Verhandl. k. k. zool.-bot. Gesell. Wien* **23**: 597–628.

BERGH, R. 1876. Malacologische Untersuchungen 1870–1904, **2** (10): 377–428.

BERGH, R. 1877. Malacologische Untersuchungen 1870–1904, **2** (12): 495–546.

BERGH, R. 1878a. Beiträge zur Kenntniss der Aeolidiaden. 6. *Verhandl. k. k. zool.-bot. Gesell. Wien* **28**: 553–584.

BERGH, R. 1878b. Malacologische Untersuchungen 1870–1904, **2** (14): 603–645.

BERGH, R. 1878c. Malacologische Untersuchungen 1870–1904, **2** (13): 547–601.

BERGH, R. 1879a. Beiträge zu einer Monographie der Polyceraden. 1. *Verhandl. k. k. zool.-bot. Gesell. Wien* **29**: 599–652.

BERGH, R. 1879b. On the nudibranchiate gasteropod Mollusca of the north Pacific Ocean, with special reference to those of Alaska. Part I. *Proc. Acad. nat. Sci. Philad.* (1879): 71–132.

BERGH, R. 1879c. Gattungen nordischer Doriden. *Arch. f. Naturg.* **45** (1): 340–369.

BERGH, R. 1880. On the nudibranchiate gasteropod Mollusca of the north Pacific Ocean, with special reference to those of Alaska. Part II. *Proc. Acad. nat. Sci. Philad.* (1880): 40–127.

BERGH, R. 1881. Malacologische Untersuchungen 1870–1904, **4** (2): 79–128.

BERGH, R. 1882. Beiträge zur Kenntniss der japanischen Nudibranchien. 2. *Verhandl. k. k. zool.-bot. Gesell. Wien* **31**: 219–254.

BERGH, R. 1884. Report on the Nudibranchiata dredged by H.M.S. Challenger during the years 1873–1876. *Rep. Sci. Results Voy. H.M.S. Challenger, Zoology* **10**: 1–154.

BERGH, R. 1885. Beiträge zur Kenntniss der Aeolidiaden. 8. *Verhandl. zool.-bot. Gesell. Wien* **35**: 1–60.

BERGH, R. 1886. Die Nudibranchien gesammelt wahrend der Fahrten des "Willem-Barents" in das nordliche Eismeer. *Bijd. tot de Dierkunde* **13**: 1–37.

BERGH, R. 1888. Beiträge zur Kenntniss der Aeolidiaden. 9. *Verhandl. zool.-bot. Gesell. Wien* **38**: 673–706.

BERGH, R. 1894. Reports on the dredging operations off the west coast of central America to the Galapagos, to the west coast of Mexico, and in the Gulf of California, in charge of Alexander Agassiz, carried on by the U.S. Fish Commission steamer "Albatross", during 1891, Lieut. Commander Z. L. Tanner, U.S.N., commanding. XIII. Die Opisthobranchien. *Bull. Mus. comp. Zool. Harvard Coll.* **25**: 125–233.

BERGH, R. 1900. Nudibranchiate Gasteropoda. *Danish Ingolf-Expedition, Copenhagen* **2**: 1–49.

BERGH, R. 1904. Malacologische Untersuchungen 1870–1904, **6** (1): 1–56.

BERRILL, N. J. 1931. The natural history of *Bulla hydatis* Linn. *J. mar. biol. Ass., U.K.* **17**: 567–571.

BERTSCH, H. 1968. Effects of feeding by *Armina californica* on the bioluminescence of *Renilla koellikeri*. *The Veliger* **10**: 440–441.

BERTSCH, H. 1974. Descriptive study of *Aeolidia papillosa* with scanning electron micrographs of the radula. *The Tabulata* (Jan. 1974): 3–6.

BICKELL, L. R. & CHIA, F. S. 1979. Organogenesis and histogenesis in the planktonic veliger of *Doridella steinbergae* (Opisthobranchia: Nudibranchia). *Mar. Biol.* **52**: 291–313.

BICKELL, L. R., CHIA, F. S. & CRAWFORD, B. J. 1981. Morphogenesis of the digestive system during metamorphosis of the nudibranch *Doridella steinbergae* (Gastropoda): conversion from phytoplanktivore to carnivore. *Mar. Biol.* **62**: 1–16.

BLAINVILLE, H. M. D. 1816. Quatrième mémoire sur les mollusques, de l'ordre des cyclobranches. *Bull. Sci. Soc. Philom.* (1816): 93–97.

BLAINVILLE, H. M. D. 1824. *Dictionnaire des sciences naturelles* **32**. Paris. Pp. 276–282.

BLEAKNEY, J. S. & SAUNDERS, C. L. 1978. Life history observations on the nudibranch mollusc *Onchidoris bilamellata* in the intertidal zone of Nova Scotia. *Canad. Field-Nat.* **92**: 82–85.

BLOOM, S. A. 1976. Morphological correlations between dorid nudibranch predators and sponge prey. *The Veliger* **18**: 289–301.

BOADEN, P. J. S. 1961. Littoral interstitial species from Anglesey representing three families new to Britain. *Nature, Lond.* **191**: 512.

BOADEN, P. J. S. 1963. The interstitial fauna of some north Wales beaches. *J. mar. biol. Ass., U.K.* **43**: 79–96.

BOHADSCH, J. B. 1761. *De quibusdam animalibus marinis, eorumque proprietatibus, orbi litterario vel nondum vel minus notis, liber.* Dresden.

BOLOT, E. 1886. Sur la ponte des *Doris. C. R. Acad. Sci. Paris* **102**: 829–831.

BONAR, D. B. 1876. Molluscan metamorphosis: a study in tissue transformation. *Amer. Zool.* **16**: 573–592.

BONAR, D. B. 1978. Morphogenesis at metamorphosis in opisthobranch molluscs. (In) *Settlement and Metamorphosis of Marine Invertebrate Larvae*, eds F. S. Chia & M. E. Rice, pp. 177–196. New York, Elsevier.

BONAR, D. B. & HADFIELD, M. G. 1974. Metamorphosis of the marine gastropod *Phestilla sibogae* Bergh (Nudibranchia: Aeolidacea). I. Light and electron microscopic analysis of larval and metamorphic stages. *J. exp. mar. Biol. Ecol.* **16**: 227–255.

BOUCHARD-CHANTEREAUX, N. R. 1836. Catalogue des mollusques marins observés, jusqu'a ce jour, à l'état vivant, sur les côtes du Boulonnais. *Mem. Soc. Agric. Boulogne-sur-Mer, Hist. nat. Zool.* **3**: 101–168.

BOUCHET, P. 1976. *Trinchesia genovae* (O'Donoghue, 1926) éolidien méconnu du littoral Méditerranéen. *Vie Millieu* **26**: 235–242.

BOUCHET, P. & MORETEAU, J.-C. 1976. Additions à l'inventaire des Mollusques de Roscoff: Gastéropodes Opisthobranches. *Trav. Stat. biol. Roscoff* **23**: 1–8.

BOUCHET, P. & TARDY, J. 1976. Faunistique et biogéographie des nudibranches des côtes françaises de l'Atlantique et de la Manche. *Ann. Inst. océanogr.* **52**: 205–213.

BRAAMS, W. G. & GEELEN, F. M. 1953. The preference of some nudibranchs for certain coelenterates. *Arch. Néerl. Zool.* **10**: 241–264.

BRIDGES, C. B. 1975. Larval development of *Phyllaplysia taylori* Dall, with a discussion of development in the Anaspidea (Opisthobranchiata: Anaspidea). *Ophelia* **14**: 161–184.

BROWN, G. H. 1978. On *Tritonia manicata* Deshayes, 1853, a dendronotacean nudibranch (Gastropoda, Opisthobranchia) new to the British fauna. *J. Conch., Lond.* **29**: 305–308.

BROWN, G. H. 1979a. Studies on the taxonomy, ecology and zoogeography of opisthobranch molluscs in European coastal waters. Ph.D. thesis, University of Bristol.

BROWN, G. H. 1979b. An investigation of the anatomy of *Colpodaspis pusilla* (Mollusca: Opisthobranchia) and a description of a new species of *Colpodaspis* from Tanzanian coastal waters. *J. Zool., Lond.* **187**: 201–221.

BROWN, G. H. 1980. The British species of the aeolidacean family Tergipedidae (Gastropoda: Opisthobranchia) with a discussion of the genera. *Zool. J. Linn. Soc.* **69**: 225–255.

BROWN, G. H. 1981a. The taxonomy of the British species of the genus *Facelina* Alder & Hancock (Opisthobranchia: Nudibranchia). *J. moll. Stud.* **47**: 334–336.

BROWN, G. H. 1981b. *Aeolidiella* Bergh, 1867 (Gastropoda, Opisthobranchia): proposals to clarify the type species of the genus. *Bull. zool. Nomencl.* **38**: 294–296.

BROWN, G. H. & PICTON, B. E. 1976. *Trapania maculata* Haefelfinger, a doridacean nudibranch new to the British fauna. *J. Conch., Lond.* **29**: 63–65.

BRUCE, J. R., COLMAN, J. S. & JONES, N. S. 1963. *Marine Fauna of the Isle of Man and its surrounding seas.* Liverpool University Press. Pp. 307.

BÜRGIN-WYSS, U. 1961. Die rückenanhänge von *Trinchesia coerulea* (Montagu). *Rev. Suisse Zool.* **68**: 461–582.

BURN, R. 1958. Further Victorian Opisthobranchia. *J. malac. Soc. Australia* number 2: 20–36.

BURN, R. 1961. Two new species and a new genus of opisthobranch molluscs from New South Wales. *Proc. roy. zool. Soc., N.S.W.* (1958–9): 131–135.

BURN, R. 1962. Descriptions of Victorian nudibranchiate Mollusca, with a comprehensive review of the Eolidacea. *Mem. nat. Mus. Melbourne* number 25: 95–128.

BURN, R. 1964. Descriptions of Australian Eolidacea (Mollusca: Opisthobranchia) 2. The genera *Nossis*, *Eubranchus*, *Trinchesia* and *Toorna*. *J. malac. Soc. Australia* number 8: 10–22.

BURN, R. 1966. Descriptions of Australian Eolidacea (Mollusca: Opisthobranchia). 4. The genera *Pleurolidia*, *Fiona*, *Learchis* and *Cerberilla* from Lord Howe Island. *J. malac. Soc. Australia* number 10: 21–34.

BURN, R. 1978. A new record of *Thecacera pennigera* (Montagu, 1815) (Opisthobranchia: Polyceridae) from New South Wales. *J. malac. Soc. Australia* **4**: 22.

BYNE, L. St G. 1893. A contribution towards a list of the marine Mollusca of Teignmouth. *J. Conch., Lond.* **7**: 175–188.

CASTEEL, D. B. 1904. The cell-lineage and early development of *Fiona marina*, a nudibranch mollusk. *Proc. Acad. nat. Sci. Philadelphia* **56**: 325–405.

CECCATTI, M. P. & PLANTA, O. 1954. Note sur le système nerveux central des Eolidiens (Mollusques Nudibranches). *Bull. Soc. Zool. France* **79**: 152–158.

CHALLIS, D. A. 1969. An ecological account of the marine interstitial opisthobranchs of the British Solomon Islands Protectorate. *Phil. Trans. Roy. Soc. Ser. B* **255**: 527–539.

CHAMISSO, A. de & EYSENHARDT, C. G. 1821. De animalibus quibusdam e classe vermium Linneana, in circumnavigatione terrae, auspicante Comite N. Romanzoff, duce Ottone de Kotzebue, annis 1815–1818 peracta *observatis. Fasc. II, reliquos Vermes continentur. Nova Acta Physico-Medica Academiae Caesareae Leopolino-Carolinae* **10**: 343–373.

CHIA, F. S. & KOSS, R. 1978. Development and metamorphosis of the planktotrophic larvae of *Rostanga pulchra* (Mollusca: Nudibranchia). *Mar. Biol.* **46**: 109–119.

CHIAJE, S. DELLE. 1824. Memorie sulla storia e notomia degli animali senza vertebre del regno di Napoli. Molluschi. Napoli, **1**: 1–184.

CHIAJE, S. DELLE. 1828. Memorie sulla storia e notomia degli animali senza vertebre de regno di Napoli. Molluschi. Napoli, **3**: 1–232.

CHIAJE, S. DELLE. 1841. Descrizione e notomia degli animali invertebrati della Sicilia citeriore osservati vivi negli anni 1822–1830. Napoli, **7**: 86–173.

CHRISTENSEN, H. 1977. Feeding and reproduction in *Precuthona peachi* (Mollusca, Nudibranchia). *Ophelia* **16**: 131–142.

CLARK, K. B. 1975. Nudibranch life cycles in the northwest Atlantic and their relationship to the ecology of fouling communities. *Helg. wiss. Meeres.* **27**: 28–69.

CLARK, K. B. & BUSACCA, M. 1978. Feeding specificity and chloroplast retention in four tropical Ascoglossa, with a discussion of the extent of chloroplast symbiosis and the evolution of the order. *J. moll. Stud.* **44**: 272–282.

CLARK, K. B., BUSACCA, M. & STIRTS, H. 1979. Nutritional aspects of development of the ascoglossan, *Elysia cauze.* (In) *Reproductive Ecology of Marine Invertebrates,* ed. S. E. Stancyk, pp. 11–23. Columbia, University of South Carolina Press.

CLARK, K. B. & GOETZFRIED, A. 1976. *Lomanotus stauberi*, a new dendronotacean nudibranch from central Florida (Mollusca: Opisthobranchia). *Bull. mar. Sci.* **26**: 474–478.

CLARK, K. B. & GOETZFRIED, A. 1978. Zoogeographic influences on development patterns of north Atlantic Ascoglossa and Nudibranchia, with a discussion of factors affecting egg size and number. *J. moll. Stud.* **44**: 283–294.

CLARK, K. B. & JENSEN, K. R. 1981. A comparison of egg size, capsule size, and development patterns in the order Ascoglossa (Sacoglossa) (Mollusca: Opisthobranchia). *Intern. J. Invert. Reprod.* **3**: 57–64.

CLARK, K. B., JENSEN, K. R., STIRTS, H. M. & FERMIN, C. 1981. Chloroplast symbiosis in a non-elysiid mollusc, *Costasiella lilianae* (Marcus) (Hermaeidae: Ascoglossa (= Sacoglossa)): effects of temperature, light intensity, and starvation on carbon fixation rate. *Biol. Bull. Woods Hole* **160**: 43–54.

COCKS, W. P. 1852. New species of Mollusca. *Naturalist* **2**: 1.

COLGAN, N. 1908. Contributions towards a revision of the genus *Lomanotus*. *Ann. Mag. nat. Hist.* **2**: 205–218.

COLGAN, N. 1911. Marine Mollusca. Clare Island survey, part 22. *Proc. roy. Irish Acad.* **31**: 1–36.

COLGAN, N. 1913. Some additions to the nudibranch fauna of County Dublin. *Irish Nat. J.* **22**: 165–168.

COLGAN, N. 1914. The opisthobranch fauna of the shores and shallow waters of County Dublin. *Irish Nat.* **23**: 161–204.

COOMANS, A. & CONINCK, L. 1962. *Embletonia pallida*, nudibranche nouveau pour la faune Belge. *Bull. Inst. roy. Sci. nat. Belg.* **38**: 1–4.

CORNET, R. & MARCHE-MARCHAD, I. 1951. Inventaire de la faune marine de Roscoff, Mollusques. *Trav. Stat. biol. Roscoff* supp. 5: 1–80.

COSTA, A. 1866. Saggio sui molluschi eolididei del Golfo di Napoli. *Ann. Mus. zool. Napoli* **3**: 59–60.

COSTA, A. 1867. Illustrazione di due generi di molluschi nudibranchi. *Rend. Accad. Sci. Fis. Mat. Soc. r. Napoli* **6**: 136–137.

COUTHOUY, J. P. 1838. Descriptions of new species of Mollusca and shells, and remarks on several polypi found in Massachusetts Bay. *Boston J. nat. Hist.* **2**: 53–111.

COUTHOUY, J. P. 1839. Monograph on the family Osteodesmacea of Deshayes, with remarks on two species of Patelloidea, and descriptions of new species of marine shells, a species of *Anculotus*, and one of *Eolis*. *Boston J. nat. Hist.* **2**: 129–189.

CRAMPTON, D. M. 1977. Functional anatomy of the buccal apparatus of *Onchidoris bilamellata* (Mollusca: Opisthobranchia). *Trans. zool. Soc. Lond.* **34**: 45–86.

CUÉNOT, L. 1903. Contributions à la faune du Bassin d'Arcachon III.—Doridiens. *Trav. Soc. scient. Arcachon Stat. biol.* (1903) 1–22.

CUÉNOT, L. 1927a. Contributions à la faune du Bassin d'Arcachon IX.—Revue générale de la faune et bibliographie. *Bull. Stat. biol. Arcachon* **24**: 229–308.

CUÉNOT, L. 1927b. Recherches sur la valeur protrectrice de l'homochromie chez quelques animaux aquatiques. *Ann. Sci. nat. Zool.* **10**: 123–150.

CUVIER, G. L. C. F. D. 1798. *Tableau elémentaire de l'histoire naturelle des animaux.* Paris, Baudouin. Pp. 1–710.

CUVIER, G. L. C. F. D. 1803. Mémoire sur le genre *Tritonia*, avec la description et l'anatomie d'une espèce nouvelle, *Tritonia hombergii*. *Ann. Mus. Hist. nat. Paris* **1**: 480–496.

CUVIER, G. L. C. F. D. 1804. Mémoire sur le genre *Doris*. *Ann. Mus. Hist. nat. Paris* **4**: 447–473.

CUVIER, G. L. C. F. D. 1805. Mémoire sur la Scyllée, l'Eolide et le *Glaucus* avec des additions au mémoire sur la Tritonie. *Ann. Mus. Hist. nat. Paris* **6**: 416–436.

CUVIER, G. L. C. F. D. 1817. *Le règne animale.* Paris, Déterville. **2**: 389–395.

CUVIER, G. L. C. F. D. 1836. *Le règne animale*, 3rd ed. Bruxelles, Hauman. Pp. 266.

DALYELL, J. G. 1853. *The powers of the creator displayed in the creation.* London, Van Voorst. **2**, Pp. 359.

DAUTZENBERG, P. & DUROUCHOUX, P. 1913. Les mollusques de la Baie de Saint-Malo. *Feuilles Jeunes Nat.* (4) **36**: 1–8.

DESHAYES, G. P. 1839–53. *Traité élémentaire de conchyliologie avec les applications de cette science à la geologie*. Paris, Masson. Pp. 384.

DICQUEMARE, J. F. 1785. Suite des extraits du porte-feuille. Limaces de Mer. *Observ. Mém. Phys. Hist. nat. Paris* **27** (2): 262–264.

DREYER, T. F. 1912. A contribution to our knowledge of the reproductive organs of the Nudibranchiata. *Rep. S. African Assoc. Adv. Sci.* (1911): 340–349.

EDMUNDS, M. 1961. *Polycera elegans* Bergh: a new species to Britain and discussion of its taxonomy. *Proc. malac. Soc. Lond.* **34**: 316–322.

EDMUNDS, M. 1964. Eolid Mollusca from Jamaica, with descriptions of two new genera and three new species. *Bull. mar. Sci. Gulf Caribb.* **14**: 1–32.

EDMUNDS, M. 1966. Protective mechanisms in the Eolidacea (Mollusca Nudibranchia). *J. Linn. Soc. (Zool.)* **47**: 27–71.

EDMUNDS, M. 1968. Acid secretion in some species of Doridacea (Mollusca, Nudibranchia). *Proc. malac. Soc. Lond.* **38**: 121–133.

EDMUNDS, M. 1970. Opisthobranchiate Mollusca from Tanzania II. Eolidacea (Cuthonidae, Piseinotecidae and Facelinidae). *Proc. malac. Soc. Lond.* **39**: 15–57.

EDMUNDS, M. 1971. Opisthobranchiate Mollusca from Tanzania (Suborder: Doridacea). *Zool. J. Linn. Soc.* **50**: 339–396.

EDMUNDS, M. & KRESS, A. 1969. On the European species of *Eubranchus* (Mollusca Opisthobranchia). *J. mar. biol. Ass., U.K.* **49**: 879–912.

EDMUNDS, M., POTTS, G. W., SWINFEN, R. C. & WATERS, V. L. 1974. The feeding preferences of *Aeolidia papillosa* (L.) (Mollusca, Nudibranchia). *J. mar. biol. Ass., U.K.* **54**: 939–947.

ELIOT, C. N. E. 1902. On some nudibranchs from Zanzibar. *Proc. zool. Soc. Lond.* **2**: 62–72.

ELIOT, C. N. E. 1905. Notes on two rare British nudibranchs, *Hero formosa* var. *arborescens*, and *Staurodoris maculata*. *Proc. malac. Soc. Lond.* **6**: 239–243.

ELIOT, C. N. E. 1906. Notes on some British nudibranchs. *J. mar. biol. Ass. U.K.* **7**: 333–382.

ELIOT, C. N. E. 1908. Reports on the marine biology of the Sudanese Red Sea.—XI. Notes on a collection of nudibranchs from the Red Sea. *J. Linn. Soc. Lond. Zool.* **31**: 86–122.

ELIOT, C. N. E. 1910. *A monograph of the British nudibranchiate Mollusca*. London, Ray Society. Supplementary Volume. Pp. 198.

ELMHIRST, R. 1922. Notes on the breeding and growth of marine animals in the Clyde Sea area. *Ann. Rep. Scottish mar. biol. Ass.* (1922): 19–43.

ESCHSCHOLTZ, J. F. 1831. Zoologischer Atlas-Beschreibungen neuer Thierarten—Zweiter Reise um die Welt. Pt 4, 1–19.

EVANS, T. J. 1922. *Calma glaucoides*: a study in adaptation. *Quart. J. micr. Sci.* **66**: 439–455.

EVANS, W. & EVANS, W. E. 1917. Some nudibranchs, including *Hermaea dendritica*, and *Lamellidoris aspera* from the Forth area. *Scott. Nat.* (1917): 105–110.

EYSTER, L. 1979. Reproduction and developmental variability in the opisthobranch *Tenellia pallida*. *Mar. Biol.* **51**: 133–140.

EYSTER, L. S. 1980. Distribution and reproduction of shell-less opisthobranchs from South Carolina. *Bull. mar. Sci.* **30**: 580–599.

EYSTER, L. S. 1981. Observations on the growth, reproduction and feeding of the nudibranch *Armina tigrina*. *J. moll. Stud.* **47**: 171–181.

FABRICIUS, O. 1780. *Fauna Groenlandica*. Hafniae: Impensis Ioannis Gottlob Rothe. Pp. 1–452.

FABRICIUS, O. 1797. Om tvende Faeröiste Blöddyr, en Doride og en Söe-Nelde. *Naturhist. Selsk. Skr.* **4** (1): 38–55.

FARRAN, G. P. 1903. The nudibranchiate molluscs of Ballynakill and Boffin Harbours, Co. Galway. *Rep. Sea & Island Fish. Ireland* (1901) part 2, app. VIII: 123–132.

FARRAN, G. P. 1909. Nudibranchiate Mollusca of the trawling grounds off the east and south coasts of Ireland. *Fish. Ireland sci. Invest.* (1907) **6**: 1–18.

FISCHER, P. 1867. Catalogue des nudibranches et céphalopodes des côtes océaniques de la France. *J. Conchyl.* (3) **15**: 5–15.

FISCHER, P. 1872. Catalogue des nudibranches et céphalopodes des côtes océaniques de la France (2e supplément). *J. Conchyl.* (3) **20**: 5–26.

FISCHER, H. 1892. Recherches sur la morphologie du foie des gastéropodes. *J. Conchyliol., Paris* **40**: 380–382.

FISHER, N. 1936. Notes on a rare British nudibranch. *Proc. malac. Soc. Lond.* **22**: 73–74.

FIZE, A. 1961. Note préliminaire sur *Pseudovermis setensis* n. sp., mollusque opisthobranche éolidien mésopsammique de la côte languedocienne. *Bull. Soc. zool. France* **86**: 253.

FLEMING, J. 1820. Entry in Brewster's *Edinburgh Encyclopaedia*. **14** (2): 618.

FLEMING, J. 1822. *Encyclopaedia Britannica*. Mollusca. **5**: 575.

FLEMING, J. 1828. *A history of British animals, exhibiting the descriptive characters and systematical arrangement of the genera and species of quadrupeds, birds, reptiles, fishes, Mollusca and Radiata of the United Kingdom*. Edinburgh, Bell & Bradfute. Pp. 1–565.

FORBES, E. 1838. *Malacologia monensis, a catalogue of the Mollusca inhabiting the Isle of Man and the neighbouring sea*. Edinburgh, Carfrae. Pp. 1–63.

FORBES, E. 1840. On some new and rare British Mollusca. *Ann. Mag. nat. Hist.* **5**: 102–108.

FORBES, E. & GOODSIR, J. 1839. Notice of zoological researches in Orkney and Shetland during the month of June 1839. *Athenaeum* no. 618 (August): 647.

FORBES, E. & HANLEY, S. 1848–53. *A history of British Mollusca, and their shells.* London, Van Voorst. **1**. Pp. 486 (1848); **3**. Pp. 556 (1849); **3**. Pp. 616 (1850–1); **4**. Pp. 301 (1852–53).

FORREST, J. E. 1953. On the feeding habits and the morphology and mode of functioning of the alimentary canal in some littoral dorid nudibranchiate Mollusca. *Proc. Linn. Soc. Lond.* **164**: 225–235.

FORSKÅL, P. 1775. *Vermes.* **1**. *Mollusca.* (In) *Descriptiones animalium, avium, amphibiorum, piscium, insectorum, vermum quae in itinere orientali observavit Petrus Forskål.* Hauniae. Pp. 99–140.

FRANZ, D. R. 1970a. Zoogeography of northwest Atlantic opisthobranch molluscs. *Marine Biology* **7**: 171–180.

FRANZ, D. R. 1970b. The distribution of the nudibranch *Doris verrucosa* Linne in the northwest Atlantic. *Nautilus* **83**: 80–85.

FRANZ, D. R. 1971. Development and metamorphosis of the gastropod *Acteocina canaliculata* (Say). *Trans. Amer. microsc. Soc.* **90**: 174–182.

FRIELE, H. 1879. Catalog der auf der norwegischen Nordmeerexpedition bei Spitzbergen gefundenen Mollusken. *Jahrb. Deutsch. Malakoz. Gesell.* **6**: 264–286.

FRIELE, H. & GRIEG, J. A. 1901. *Den Norske Nordhaus-Expedition, 1876–1878.* 16. Zoologi. Mollusca **3**: 1–129. Christiana.

FRIELE, H. & HANSEN, G. A. 1876. Bidrag til kundskaben om de Norske nudibranchier. *Vidensk. Selsk., Forhandl.* (1876): 69–80.

GADZIKIEWICZ, W. 1907. Das plötzliche aufreten einer vergleichweise grossen Zahl von Dorididae Cryptobranch-eatae (*Staurodoris bobretzii* n. sp.) in dem Meeresbuchten bei Sebastopol. *Biol. Centralb.* **27**: 508–510.

GALLOP, A., BARTROP, J. & SMITH, D. C. 1980. The biology of chloroplast acquisition by *Elysia viridis. Proc. roy. Soc. Lond.* B **207**: 335–349.

GANTÈS, H. 1956. Complément à l'étude des opisthobranches des côtes du Maroc. *Bull. Soc. Sci. nat. phys. Maroc* **36**: 257–263.

GARSTANG, W. 1889. Report on the nudibranchiate Mollusca of Plymouth Sound. *J. mar. biol. Ass., U.K.* **1**: 173–198.

GARSTANG, W. 1890. A complete list of the opisthobranchiate Mollusca found at Plymouth, with further observations on their morphology, colours, and natural history. *J. mar. biol. Ass., U.K.* **1**: 399–457.

GARSTANG, W. 1894. Faunistic notes at Plymouth during 1893–4. *J. mar. biol. Ass., U.K.* **3**: 210–235.

GARSTANG, W. 1895. On *Doris maculata*, a new species of nudibranchiate mollusk found at Plymouth. *J. mar. biol. Ass., U.K.* **4**: 167–168.

GASCOIGNE, T. 1979a. Observations on the anatomy of *Hermaea variopicta* (Opisthobranchia: Ascoglossa). *J. Zool., Lond.* **187**: 223–233.

GASCOIGNE, T. 1979b. A redescription of *Caliphylla mediterranea* Costa, 1867. *J. moll. Stud.* **45**: 300–311.

GASCOIGNE, T. 1980. A redescription of *Placida viridis* Trinchese, 1873 (Gastropoda: Ascoglossa). *J. Conch.* **30**: 167–179.

GASCOIGNE, T. & SIGURDSSON, J. B. 1977. *Calliopaea oophaga* Lemche, 1974, a species new to the British fauna (Opisthobranchia: Sacoglossa). *J. moll. Stud.* **43**: 286–289.

GASCOIGNE, T. & TODD, C. D. 1977. A description of a specimen of *Calliopaea bellula* d'Orbigny, 1857 found at Robin Hood's Bay, North Yorkshire (Opisthobranchia: Sacoglossa). *J. moll. Stud.* **43**: 290–295.

GMELIN, J. F. 1791. (In) Linnaeus' *Systema naturae*, ed. 13. Lipsiae. Pp. 1–3103.

GOSLINER, T. M. 1980. The systematics of the Aeolidacea (Nudibranchia: Mollusca) of the Hawaiian Islands, with descriptions of two new species. *Pacific Science* **33** (1979): 37–77.

GOSLINER, T. M. 1981a. The South African Janolidae (Mollusca, Nudibranchia) with a description of a new genus and two new species. *Ann. S. Afr. Mus.* **86**: 1–42.

GOSLINER, T. M. 1981b. A new species of tergipedid nudibranch from the coast of California. *J. moll. Stud.* **47**: 200–205.

GOSLINER T. M. & GRIFFITHS, R. J. 1981. Description and revision of some South African aeolidacean Nudibranchia (Mollusca, Gastropoda). *Ann. S. African Mus.* **84**: 105–150.

GOSSE, P. H. 1877. On *Hancockia eudactylota*, a genus and species of mollusk supposed to be new. *Ann. Mag. nat. Hist.* **20**: 316–319.

GOULD, A. A. 1841. *Report on the Invertebrata of Massachusetts.* Cambridge, Massachusetts. Pp. 1–373.

GOULD, A. A. 1862. Otia Conchologica: Descriptions of shells and mollusks from 1839 to 1862. Boston. Pp. 1–256.

GOULD, A. A. & BINNEY, W. G. 1870. *Report on the Invertebrata of Massachusetts.* Cambridge, Massachusetts. Pp. 1–524.

GRAHAM, A. 1955. Molluscan diets. *Proc. malac. Soc. Lond.* **31**: 144–159.

GRAVES, D. A., GIBSON, M. A. & BLEAKNEY, J. S. 1979. The digestive diverticula of *Alderia modesta* and *Elysia chlorotica. The Veliger* **21**: 415–422.

GRAY, J. E. 1857. *Guide to the systematic distribution of Mollusca in the British Museum.* London, British Museum (Natural History). Pp. 1–230.

GRAY, M. E. 1850. Figures of molluscous animals, selected from various authors. London, **2** (pls. 79–199); **3** (pls 200–312); **4**(pls 1–124).

GRIEG, M. J. 1907. Invertébrés du fond. In: Duc d'Orleans Croisière Océanographique dans la Mer du Grönland en 1905, pp. 503–569.

GUIART, J. 1901. Contribution à l'étude des gastéropodes opisthobranches et en particulier des céphalaspides. *Mém. Soc. zool. France* **14**: 5–219.

HADDON, A. C. 1882. Notes on the development of Mollusca. *Quart. J. micr. Sci.* **22**: 367–370.

HADFIELD, M. G. 1963. The biology of nudibranch larvae. *Oikos* **14**: 85–95.

HADFIELD, M. G. 1978. Metamorphosis in marine molluscan larvae: an analysis of stimulus and response. (In) *Settlement and Metamorphosis of Marine Invertebrate Larvae*, eds F. S. Chia & M. E. Rice, pp. 165–175. New York, Elsevier.

HAEFELFINGER, H.-R. 1960a. Catalogue des opisthobranches de la rade de Villefranche-sur-Mer et ses environs (Alpes Maritimes). *Rev. Suisse Zool.* **67**: 323–351.

HAEFELFINGER, H.-R. 1960b. Neue und wenig bekannte Opisthobranchier der gattungen *Trapania* und *Caloria* aus de bucht von Villefranche-sur-Mer. *Rev. Suisse Zool.* **67**: 226–238.

HAEFELFINGER, H.-R. 1960c. Beobachtungen an *Polycera quadrilineata* (Müller) (Moll., Nudibr.). *Rev. Suisse Zool.* **67**: 101–117.

HAEFELFINGER, H.-R. 1962a. *Crimora papillata* Alder 1862, opisthobranche nouveau pour la Méditerranée. *Vie et Milieu* **13**: 161–165.

HAEFELFINGER, H.-R. 1962b. Quelques faits concernant le nutrition chez *Favorinus branchialis* (Rathke 1806) et *Stiliger vesiculosus* (Deshayes 1864), deux Mollusques Opisthobranches. *Comm. Ass. gén. Soc. Suisse Zool.* (1962): 311–316.

HAEFELFINGER, H.-R. 1963. Remarques biologiques et systématiques au sujet de quelques Tritoniidae de la Méditerranée (Moll. Opisthobranchia). *Rev. Suisse Zool.* **70**: 61–76.

HAEFELFINGER, H.-R. 1968. Zur taxonomischen Problematik der Spezies *Aegires Leuckarti* Verany und *Aegires punctilucens* (d'Orbigny) (Mollusca, Gastropoda, Opisthobranchia). *Rev. Suisse Zool.* **75**: 575–583.

HAMOND, R. 1972. The marine Mollusca of Norfolk. *Trans. Norfolk Norwich nat. Soc.* (1972): 271–306.

HANCOCK, A. 1865. On the structure and homologies of the renal organ in the nudibranchiate Mollusca. *Trans. Linn. Soc. Lond.* **24**: 511–530.

HANCOCK, A. & EMBLETON, D. 1852. On the anatomy of *Doris. Phil. Trans. R. Soc. Ser. B* **142**: 207–252.

HANCOCK, A. & NORMAN, A. M. 1863. On *Splanchnotrophus*, an undescribed genus of Crustacea, parasitic in nudibranchiate molluscs. *Trans. Linn. Soc. Lond.* **24**: 49–60.

HARRIGAN, J. F. & ALKON, D. L. 1978. Laboratory cultivation of *Haminea solitaria* (Say, 1822) and *Elysia chlorotica* (Gould, 1870). *The Veliger* **21**: 299–305.

HARRIS, L. G. 1973. Nudibranch associations. *Current Topics comp. Pathobiol.* **2**: 213–315.

HARRIS, L. G. 1975. Studies on the life history of two coral-eating nudibranchs of the genus *Phestilla*. *Biol. Bull. Woods Hole* **149**: 539–550.

HARRIS, L. G. 1976. Comparative ecological studies of the nudibranch *Aeolidia papillosa* and its anemone prey *Metridium senile* along the Atlantic and Pacific coasts of the United States. *J. moll. Stud.* **42**: 301.

HARRIS, L. G., WRIGHT, L. W. & RIVEST, B. R. 1975. Observations on the occurrence and biology of the aeolid nudibranch *Cuthona nana* in New England waters. *The Veliger* **17**: 267–268.

HARTLEY, J. P. 1979. On the offshore Mollusca of the Celtic Sea. *J. Conch., Lond.* **30**: 81–92.

HECHT, E. 1895. Contribution à l'étude des nudibranches. *Mém. Soc. zool. France* **8**: 539–711.

HEINCKE, F. 1897. Nachtrage zur Fisch- und Molluskenfauna Helgolands 1 (Nr 7 der Beiträge zur Meeresfauna von Helgoland). *Wiss. Meeresunters.* **2** (1): 233–252.

HENNEGUY, L.-F. 1925. Contribution à l'histologie des nudibranches. *Archs Anat. microsc. Morph. exp.* **21**: 400–468.

HERDMAN, W. A. 1881. Additional notes to the invertebrate fauna of Lamlash Bay. *Proc. Roy. phys. Soc. Edinburgh* **6**: 17–30.

HERDMAN, W. A. & CLUBB, J. A. 1889. Second report on the Nudibranchiata of the L.M.B.C. district. *Trans. Liverpool biol. Soc.* **3**: 225–239.

HERDMAN, W. A. & CLUBB, J. A. 1892. Third report upon the Nudibranchiata of the L.M.B.C. district. *Trans. Liverpool biol. Soc.* **4**: 131–169.

HERTLING, H. 1930. Über eine Hedylide von Helgoland und Bemerkungen zur Systematik der Hedyliden. *Wiss. Meeresunters. Helgoland* **18**, Heft 1 (5): 1–10.

HESSE, M. 1872. Diagnose de nudibranches nouveaux des côtes de Bretagne. *J. Conchyl.* **20**: 345–348.

HOLLEMAN, J. J. 1972. Observations on growth, feeding, reproduction, and development in the opisthobranch, *Fiona pinnata* (Eschscholtz). *The Veliger* **15**: 142–146.

HOMANS, R. E. S. & NEEDLER, A. W. H. 1944. Food of the haddock. *Proc. Nova Scotia Inst. Sci.* **21**: 1–35.

HUNNAM, P. J. & BROWN, G. H. 1975. Sublittoral nudibranch Mollusca (sea slugs) in Pembrokeshire waters. *Field Studies* **4**: 131–159.

HURST, A. 1957. The egg masses and veligers of thirty northeast Pacific opisthobranchs. *The Veliger* **9**: 255–288.

IHERING, H. 1879. Einiges neue über Mollusken. *Zool. Anz.* **2**: 136–138.

IHERING, H. 1885. Beiträge zur Kenntniss der Nudibranchien des Mittelmeeres. 2. *Malakoz. Blätt.* **8**, 12–48.

IHERING, H. 1886. Zur Kenntniss der Nudibranchien der brasilianischen Küste. *Jahrb. Deutsch. Malakoz. Gesell.* **13** (3): 223–240.

IREDALE T. & O'DONOGHUE, C. H. 1923. List of British nudibranchiate Mollusca. *Proc. malac. Soc. Lond.* **15**: 195–233.

JAECKEL, S. 1952. Zur Verbreiting und Lebensweise der Opisthobranchier in der Nordsee. *Kieler Meeresforsch.* **8**: 249–259.

JEFFREYS, J. G. 1863–69. *British Conchology.* (5 vols.) **1**. Pp. 341 (1863); **2**. Pp. 465 (1863); **3**. Pp. 393 (1865); **4**. Pp. 486 (1867); **5**. Pp. 258 (1869). London, Van Voorst.

JENSEN, K. R. 1975a. The importance of *Limapontia capitata* (Mueller) (Opisthobranchia, Sacoglossa) as a primary consumer in the Cladophora-belt. *10th Europ. Symp. mar. Biol.* **2**: 339–350.

JENSEN, K. R. 1975b. Food preferences and food consumption in relation to growth of *Limpontia capitata*. *Ophelia* **14**: 1–14.

JENSEN, K. R. 1977. Optimal salinity and temperature intervals of *Limapontia capitata* (Opisthobranchia, Sacoglossa) determined by growth and heart rate measurements. *Ophelia* **16**: 175–185.

JENSEN, K. R. 1980. A review of sacoglossan diets, with comparative notes on radular and buccal anatomy. *Malacol. Rev.* **13**: 55–77.

JENSEN, K. R. 1981. Observations on feeding methods in some Florida ascoglossans. *J. moll. Stud.* **47**: 190–199.

JOHNSON, C. W. 1934. List of marine Mollusca of the Atlantic coast from Labrador to Texas. *Proc. Boston Soc. nat. Hist.* **40**: 1–203.

JOHNSTON, G. 1828. A few remarks on the class Mollusca, in Dr. Fleming's work on British animals; with descriptions of some new species. *Edinburgh New philos. J.* **5**: 74–81.

JOHNSTON, G. 1832. Illustrations in British zoology. *Loudon Mag. nat. Hist.* **5**: 428–429.

JOHNSTON, G. 1834. Illustrations in British zoology. *Loudon Mag. nat. Hist.* **7**: 490–492.

JOHNSTON, G. 1835. Illustrations in British zoology. *Loudon Mag. nat. Hist.* **8**: 376–380.

JOHNSTON, G. 1838a. Miscellanea zoologica. *Ann. Mag. nat. Hist.* **1**: 44–56.

JOHNSTON, G. 1838b. Miscellanea zoologica. *Ann. Mag. nat. Hist.* **1**: 114–125.

KAY, E. A. & YOUNG, D. K. 1969. The Doridacea (Opisthobranchia; Mollusca) of the Hawaiian Islands. *Pacific Science* **23**: 172–231.

KEMPF, S. C. & WILLOWS, A. O. D. 1977. Laboratory culture of the nudibranch *Tritonia diomedea* Bergh (Tritoniidae: Opisthobranchia) and some aspects of its behavioural development. *J. exp. mar. Biol. Ecol.* **30**: 261–276.

KENT, W. S. 1869. On a new British nudibranch (*Embletonia grayi*). *Proc. zool. Soc. Lond.* pp. 109–111.

KOWALEVSKY, A. 1901. Les hedylides, étude anatomique. *Mém. Acad. Sci. St Pétersb.* (Sci. math. phys. nat.) **12** (6): 1–32.

KRAUSE, A. 1885. Ein Beitrag zur Kenntniss der Mollusken-fauna des Beringmeeres. *Arch. für Naturg.* **51**: 256–302.

KRESS, A. 1968a. *Trapania pallida* sp.nov. (Opisthobranchia, Gastropoda), a genus new to Britain. *Proc. malac. Soc. Lond.* **38**: 161–165.

KRESS, A. 1968b. Untersuchungen zur Histologie, Autotomie und Regeneration dreier *Doto*-Arten *Doto coronata, D. pinnatifida, D. fragilis* (Gastropoda, Opisthobranchiata). *Rev. Suisse Zool.* **75**: 235–303.

KRESS, A. 1970. A new record of *Trapania pallida* (Opisthobranchia, Gastropoda) with a description of its reproductive system and a comparison with *T. fusca*. *Proc. malac. Soc. Lond.* **39**: 111–116.

KRESS, A. 1971. Über die Entwicklung der Eikapselvolumina bei verschiedenen Opisthobranchier-Arten (Mollusca, Gastropoda). *Helg. wiss. Meeres.* **22**: 326–349.

KRESS, A. 1972. Veränderungen der Eikapselvolumina während der Entwicklung verschiedener Opisthobranchier-Arten (Mollusca, Gastropoda). *Marine Biology* **16**: 236–252.

KRESS, A. 1975. Observations during embryonic development in the genus *Doto* (Gastropoda, Opisthobranchia). *J. mar. biol. Ass., U.K.* **55**: 691–701.

KRESS, A. 1977. *Runcina ferruginea* n.sp. (Cephalaspidea: Opisthobranchia: Gastropoda), a new runcinid from Great Britain. *J. mar. biol. Ass., U.K.* **57**: 201–211.

KRESS, A. 1981. A scanning electron microscope study of notum structures in some dorid nudibranchs. *J. mar. biol. Ass., U.K.* **61**: 177–191.

KROPP, B. 1931. The pigment of *Velella spirans* and *Fiona marina*. *Biol. Bull. Woods Hole* **60**: 120–123.

KUZIRIAN, A. M. 1977. The rediscovery and biology of *Coryphella nobilis* Verrill, 1880 in New England. *J. moll. Stud.* **43**: 230–240.

KUZIRIAN, A. M. 1979. Taxonomy and biology of four New England coryphellid nudibranchs. *J. moll. Stud.* **45**: 239–261.

LABBÉ, A. 1922. Contribution à la faune du Croisic. 2. *Pleurophyllidia henneguyi* n.sp. *Arch. Zool. exp. gén.* **61** (3): 51–58.

LABBÉ, A. 1923. Note préliminaire sur cinq espèces nouvelles d'Éolidiens de la Station de Croisic. *Bull. Soc. zool. France* **48**: 265–268.

LABBÉ, A. 1931. Les polyceradés de la Station du Croisic et description sommaire d'une espèce nouvelle: *Polycera salamandra*, nov. sp. *Bull. Soc. zool. France* **56**: 19–24.

LAFONT, A. 1871–72. Note pour servir à la faune de la Gironde contenant la liste des animaux marins dont la présence a été constatée à Arcachon pendant les années 1869–1970. *Actes. Soc. Linn. Bordeaux* (3) **28** (1): 237–280.

LAFONT, A. 1874. Description d'un nouveau genre de nudibranche des côtes de France. *J. Conchyl.* (3) **22**: 369–370.

LAGARDÈRE, F. & TARDY, J. 1980. Un faciès d'épifaune nouveau: le faciès a *Ectopleura dumortieri* (van Beneden) et *Electra pilosa* (Linné). Faune associée, cartographie et évolution saisonnière. *Cah. Biol. mar.* **21**: 265–278.

LALLI, C. M. & WELLS, F. E. 1978. Reproduction in the genus *Limacina* (Opisthobranchia: Thecosomata). *J. Zool., Lond.* **186**: 95–108.

LANCE, J. R. 1961. A distributional list of southern Californian opisthobranchs. *The Veliger* **4**: 64–69.

LANCE, J. R. 1962. Two new opisthobranch mollusks from southern California. *The Veliger* **4**: 155–159.

LEACH, W. E. 1847. The classification of the British Mollusca. *Ann. Mag. nat. Hist.* **20**: 267–273.

LEIGH-SHARPE, W. 1935. A list of British invertebrates with their characteristic parasitic and commensal Copepoda. *J. mar. biol. Ass. U.K.* **20**: 47–48.

LEMCHE, H. 1929. Gastropoda Opisthobranchiata. *Zoology Faroes* No. 53: 1–35.

LEMCHE, H. 1936. On some nudibranchiate gastropods from the northern Atlantic. *Vidensk. Meddel. Naturh. Foren.* **99**: 131–148.

LEMCHE, H. 1938. Gastropoda Opisthobranchiata. *Zoology Iceland* **4** (61): 1–54.

LEMCHE, H. 1941. Gastropoda Opisthobranchiata. Zoology of E. Greenland. *Meddr Grønland* **121** (7): 1–50.

LEMCHE, H. 1964. Proposed stabilisation of the generic name *Trinchesia* Ihering, 1879, and suppression under the plenary powers of *Diaphoreolis* Iredale & O'Donoghue, 1923 (Class Gastropoda). *Bull. zool. Nomencl.* **21**: 52–55.

LEMCHE, H. 1976. New British species of *Doto* Oken, 1815 (Mollusca: Opisthobranchia). *J. mar. biol. Ass., U. K.* **56**: 691–706.

LEMCHE, H. & THOMPSON, T. E. 1974. Three opisthobranch gastropods new to the British fauna. *Proc. malac. Soc. Lond.* **41**: 185–193.

LEUCKART, F. S. 1828. *Breves animalium quorundum maxima ex parte marinorum descriptiones.* Heidelberg, Osswaldi. Pp. 1–24.

LINNAEUS, C. 1758. *Systema Naturae,* ed. 10. Holmiae, Laurentii Salvii. Pp. 1–823.

LINNAEUS, C. 1761. *Fauna Svecica sistens animalia Sveciae regni* (ed. 2): 1–578.

LINNAEUS, C. 1767. *Systema Naturae,* ed. 12, tome 1. Holmiae, Laurentii Salvii. Pp. 1–1327.

LLOYD, H. M. 1952. A study of the reproductive systems of some opisthobranchiate molluscs. *Ph.D. Thesis, University of London.*

LOMAN, J. C. C. 1893. Anteekening over twee voor de Nederlandsche fauna niewe Nudibranchiata. *Tidschr. Nederlandsche Dierk.* (2) **4**: 35–37.

LOVELAND, R. E., HENDLER, G. & NEWKIRK, G. 1969. New records of nudibranchs from New Jersey. *The Veliger* **11**: 418–420.

LOVÉN, S. L. 1841. Bidrag till Kännedomen af Molluskernas utveckling. *K. Vet. Acad. Handl. Stockholm* (1839): 227–241.

LOVÉN, S. L. 1844. Om nordiska hafs-mollusker. *Ofvers. K. Vet. Akad. Forh. Stockholm* **1** (3): 48–53.

LOVÉN, S. L. 1846. Index molluscorum litora Scandinaviae Occidentalia habitantium. *Ofvers. K. Vet. Akad. Forh. Stockholm* **3**: 135–160.

LOYNING, P. 1927. Nudibranchs from Bergen, collected in the neighbourhood of the Biological Station at Herdla. *Nyt. Mag. f. Naturvid.* **65**: 243–264.

McDONALD, G. R. & NYBAKKEN, J. W. 1978. Additional notes on the food of some California nudibranchs, with a summary of known food habits of California species. *The Veliger* **21**: 110–119.

MacFARLAND, F. M. 1905. A preliminary account of the Dorididae of Monterey Bay, California. *Proc. biol. Soc. Washington* **18**: 35–54.

MacFARLAND, F. M. 1966. Studies of opisthobranchiate mollusks of the Pacific coast of north America. *Mem. Calif. Acad. Sci.* **6**: 1–546.

MacGILLIVRAY, W. 1843. A history of the molluscous animals of the counties of Aberdeen, Kincardine, and Banff. London, Bohn. Pp. 1–372.

MacGINITIE, N. 1959. Marine Mollusca of Point Barrow, Alaska. *Proc. U.S. nat. Mus.* **109**: 59–208.

McKAY, D. W. & SMITH, S. M. 1979. *Marine Mollusca of east Scotland.* Edinburgh, Royal Scottish Museum. Pp. 1–185.

M'INTOSH, W. C. 1863. On the nudibranchiate Mollusca of St Andrews. *Proc. Roy. Soc. Edinburgh* **5**: 387–393.

M'INTOSH, W. C. 1865. On the nudibranchiate Mollusca of St Andrews; *Edwardsia;* and the polyps of *Alcyonium digitatum. Proc. Roy. Soc. Edinburgh,* 1864–65: 1–8.

M'INTOSH, W. C. 1874. On the invertebrate marine fauna and fishes of St Andrews. *Ann. Mag. nat. Hist.* (4) **13**: 420–432.

McMILLAN, N. F. 1942a. Food of nudibranchs. *J. Conch., Lond.* **21**: 327.

McMILLAN, N. F. 1942b. Spawn of *Aeolidia papillosa* L. *J. Conch., Lond.* **21**: 324.

McMILLAN, N. F. 1944. The marine Mollusca of Greenisland, Co. Antrim. *Irish Nat. J.* **8**: 158–164.

MacNAE, W. 1957. The families Polyceridae and Goniodorididae (Mollusca, Nudibranchiata) in southern Africa. *Trans. Roy. Soc. S. Africa* **35**: 341–372.

MANUEL, R. L. 1980. *The Anthozoa of the British Isles—a colour guide.* Manchester, Underwater Conservation Society. Pp. 1–69.

MARCUS, E. 1955. Opisthobranchia from Brazil (1). *Bolm Fac. Fil. Ciênc. Univ. S. Paulo, Zool.* **164**: 165–203.

MARCUS, E. 1957. On Opisthobranchia from Brazil (2). *J. Linn. Soc. Lond. (Zool.)* **43**: 390–486.

MARCUS, E. 1958. On western Atlantic opisthobranchiate gastropods. *Amer. Mus. Novit.* number 1906: 1–82.

MARCUS, E. 1959. Lamellariacea und Opisthobranchia. *Lunds Univ. Arsskrift* **55** (9): 1–133.

MARCUS, E. 1961. Opisthobranch mollusks from California. *The Veliger* **3** (suppl.): 1–85.

MARCUS, E. 1965. Some Opisthobranchia from Micronesia. *Malacologia* **3** (2): 263–286.

MARCUS, E. 1976. On *Kentrodoris* and *Jorunna* (Gastropoda Opisthobranchia). *Bolm Zool. Univ. S. Paulo* **1**: 11–68.

MARCUS, E. 1977. An annotated checklist of the western Atlantic warm water opisthobranchs. *J. moll. Stud.* suppl. 4: 1–22.

MARCUS, E. 1980. Review of western Atlantic Elysiidae (Opisthobranchia Ascoglossa) with a description of a new *Elysia* species. *Bull. mar. Sci.* **30**: 54–79.

MARCUS, E. 1983. The western Atlantic Tritoniidae. *Bolm Zool. Univ. S. Paulo* **7**: 177–214.

MARCUS, E. & MARCUS, E. 1955. Über Sand-Opisthobranchier. *Kieler Meeresforsch.* **11**: 230–243.

MARCUS, E. & MARCUS, E. 1958. Opisthobranchia aus dem Schill von Helgoland. *Kieler Meeresforsch.* **14**: 91–96.

MARCUS, E. & MARCUS, E. 1960. Opisthobranchia aus dem Roten Meer und von den Malediven. *Akad. Wiss. Lit. Abh. Math. Naturwiss.* number 12: 873–933.

MARINE BIOLOGICAL ASSOCIATION. 1957. *Plymouth marine fauna*, 3rd ed. M.B.A., Plymouth. Pp. 457.

MAZZARELLI, G. F. 1903. Note biologische sugli opisthobranchi del Golfo di Napoli. *Atti Soc. Ital. Sci. nat.* **42** (3): 280–296.

MAZZARELLI, G. F. 1904. Contributo alla conoscenza larve libere degli opistobranchi. *Archo zool. Ital.* **2**: 19–78.

MENKE, K. T. 1830. *Synopsis methodica molluscorum generum omnium et specierum earum, quae in Museo Menkeano adservantur.* Pyrmonti, Uslar. Pp. 1–168.

MEYER, H. A. & MÖBIUS, K. 1865. *Fauna der Kieler Bucht. Vol. I Die Hinterkiemer oder Opisthobranchia.* Leipzig, Engelmann. Pp. 1–88.

MEYER, H. A. & MÖBIUS, K. 1872. *Fauna der Kieler Bucht. Vol. II Die Prosobranchia und Lamellibranchia nebst einem Supplement zu den Opisthobranchia.* Leipzig, Engelmann. Pp. 1–139.

MEYER, K. B. 1971. Distribution and zoogeography of fourteen species of nudibranchs of northern New England and Nova Scotia. *The Veliger* **14**: 137–152.

MILLER, M. C. 1958. Studies on the nudibranchiate Mollusca of the Isle of Man. Ph.D. Thesis, University of Liverpool.

MILLER, M. C. 1961. Distribution and food of the nudibranchiate Mollusca of the south of the Isle of Man. *J. anim. Ecol.* **30**: 95–116.

MILLER, M. C. 1962. Annual cycles of some Manx nudibranchs, with a discussion of the problem of migration. *J. anim. Ecol.* **31**: 545–569.

MILLER, M. C. 1971. Aeolid nudibranchs (Gastropoda: Opisthobranchia) of the families Flabellinidae and Eubranchidae from New Zealand waters. *Zool. J. Linn. Soc.* **50**: 311–337.

MILLER, M. C. 1974. Aeolid nudibranchs (Gastropoda: Opisthobranchia) of the family Glaucidae from New Zealand waters. *Zool. J. Linn. Soc.* **54**: 31–61.

MILLER, M. C. 1977. Aeolid nudibranchs (Gastropoda: Opisthobranchia) of the family Tergipedidae from New Zealand waters. *Zool. J. Linn. Soc.* **60**: 197–222.

MILLOTT, N. 1937. On the morphology of the alimentary canal, process of feeding, and physiology of digestion of the nudibranch mollusc, *Jorunna tomentosa* (Cuvier). *Phil. Trans. R. Soc. Ser. B* **228**: 173–217.

MÖLLER, H. P. C. 1842. Index molluscorum Groenlandiae. *Naturhist. Tidskr. Kjöbenhavn* **4** (1): 76–97.

MONTAGU, G. 1804. Description of several marine animals found on the south coast of Devonshire. *Trans. Linn. Soc. Lond.* **7**: 61–85.

MONTAGU, G. 1808. Description of several marine animals found on the south coast of Devonshire. *Trans. Linn. Soc. Lond.* **9**: 81–114.

MONTAGU, G. 1815. Descriptions of several new or rare animals, principally marine, discovered on the south coast of Devonshire. *Trans. Linn. Soc. Lond.* **11**: 1–26.

MOORE, D. R. & MILLER, M. F. 1979. Discovery of living bivalved gastropods in the Florida Keys. *Nautilus* **93**: 106.

MÖRCH, O. A. L. 1857. (In) RINK, H. J., Grönland geographisk og statistik beskrevet. Pt. 4, Blöddyr (molluscs). *Prodromus faunae moluscorum Grölandiae. Fortegnelse over Grönlands Blöddyr*, pp. 1–28.

MÖRCH, O. A. L. 1868. Faunula molluscorum Islandiae. *Vidensk. Meddel. Naturh. Foren.* (1868): 185–227.

MÖRCH, O. A. L. 1877. Synopsis of the Greenland fauna. (In) RINK, H. J., *Danish Greenland its people and its products*, pp. 429–463.

MORSE, M. P. 1968. Functional morphology of the digestive system of the nudibranch mollusc *Acanthodoris pilosa*. *Biol. Bull.* **134**: 305–319.

MORTON, J. E. 1972. The form and functioning of the pallial organs in the opisthobranch *Akera bullata* with a discussion on the nature of the gill in Notaspidea and other tectibranchs. *The Veliger* **14**: 337–349.

MÜLLER, O. F. 1776. *Zoologiae Danicae. Prodromus su animalium Daniae et Norvegiae ingenarum characteres, nomina, et synonyma imprimis populatium* (1–32): 1–282.

MÜLLER, O. F. 1778. Molluscorum marinorum Norvegiae. *Nova Acta Acad. Caes. Leopold Nat. Cur.* **6**: 48–54.

MÜLLER, O. F. 1789. *Zoologica Danica seu animalium Daniae et Norvegiae rariorum ac minus notorum descriptiones et historia*, ed. 3. Hauniae, Möller. **3**: 1–71.

NAVILLE, A. 1926. Notes sur les eolidiens. Un eolidien d'eau saumâtre. Origine des nématocystes. Zooxanthelles et homochromie. *Rev. Suisse Zool.* **33**: 251–289.

NOBRE, A. 1896. Mollusques et brachiopodes du Portugal. *Ann. Sci. Nat.* **3**: 97–108.

NOBRE, A. 1905. Mollusques et brachiopodes du Portugal. *Ann. Sci. Nat.* **8**: 1–147.

NORDMANN, A. 1845. Versuch einer Monographie des *Tergipes edwardsi*, in Beiträg zur Natur und Entwicklungsgeschichte der Nacktkiemer. *Bull. Phys. Math. Acad. Imp. Sci. St Pétersbourg* **3**: 270–271.

NORDSIECK, F. 1972. *Die europäischen Meeresschnecken (Opisthobranchia mit Pyramidellidae; Rissoacea.)* Stuttgart, Gustav Fischer. Pp. 1–327.

NORMAN, A. M. 1877. On two new British nudibranchiate Mollusca. *Ann. Mag. nat. Hist.* **20**: 517–519.

NORMAN, A. M. 1890. Revision of British Mollusca. *Ann. Mag. nat. Hist.* (6) **6**: 60–91.

NYBAKKEN, J. & MCDONALD, G. 1981. Feeding mechanisms of west American nudibranchs feeding on Bryozoa, Cnidaria and Ascidiacea, with special respect to the radula. *Malacologia* **20**: 439–449.

ODHNER, N. H. 1907. Northern and Arctic invertebrates in the collection of the Swedish State Museum (Riksmuseum). III. Opisthobranchia and Pteropoda. *K. Svenska Vetensk. Handl.* **41** (4): 1–118.

ODHNER, N. H. 1914. Beiträge zur Kenntnis der marinen Molluskenfauna von Rovigno in Istrien. *Zool. Anz.* **44**: 156–170.

ODHNER, N. H. 1922. Norwegian opisthobranchiate Mollusca in the collections of the Zoological Museum of Kristiana. *Nyt Mag. Naturv.* **60**: 1–47.

ODHNER, N. H. 1926. Nudibranchs and lamellariids from the Trondhjem Fjord. *Kgl. Norske Vidensk. Selsk. Skrift.* (1926) No. 2: 1–36.

ODHNER, N. H. 1929. Aeolidiiden aus dem Nördlichen Norwegen. *Tromsø Museums Arsh.* **50**: 1–22.

ODHNER, N. H. 1934. The Nudibranchiata. *Br. Antarctic (Terra Nova) Exped.* 1910. *Nat. Hist. Rep. Zool.* **7** (5): 229–310.

ODHNER, N. H. 1936. Nudibranchia Dendronotacea a revision of the system. *Mém. Mus. Roy. Hist. nat. Belg.* 2 sér, fasc. 3: 1057–1128.

ODHNER, N. H. 1939. Opisthobranchiate Mollusca from the western and northern coasts of Norway. *Kgl. Norske Vidensk. Selsk. Skrift.* (1939) No. 1: 1–92.

ODHNER, N. H. 1940. Eine neue Nacktschnecke *Xenocratena suecica* n.gen. n.sp., und ihre Verwandtschaft. *Ark. Zool.* **32**B (2): 1–8.

ODHNER, N. H. 1941. New polycerid nudibranchiate Mollusca and remarks on this family. *Göteborgs Vetensk.-Samh. Handl.* ser. B, **1** (11): 1–20.

ODHNER, N. H. 1963. On the taxonomy of the family Tritoniidae (Mollusca: Opisthobranchia). *The Veliger* **6**: 48–52.

O'DONOGHUE, C. H. 1924. Notes on the nudibranchiate Mollusca from the Vancouver Island region IV. Additional species and records. *Trans. Roy. Can. Inst.* **15**: 1–33.

O'DONOGHUE, C. H. 1926. A list of the nudibranchiate Mollusca recorded from the Pacific coast of north America, with notes on their distribution. *Trans. Roy. Can. Inst.* **15**: 199–247.

O'DONOGHUE, C. H. 1927. Notes on a collection of nudibranchs from Laguna Beach, California. *J. Entomol. Zool. Pomona College* **19**: 77–119.

O'DONOGHUE, C. H. 1929. Report on the Opisthobranchiata with a comparison of the nudibranch fauna of the Red Sea and of the Mediterranean (Cambridge Expedition to the Suez Canal). *Trans. zool. Soc. Lond.* **22**: 713–841.

OERSTED, A. S. 1844. *De regionibus marinis. Elementa topographiae historiconaturalis freti Oresund.* Hauniae. Pp. 1–88 (doctoral thesis).

OKEN, L. 1815. *Lehrbuch der Naturgeschichte.* Jena: Schmid. Pt 3, **1** Zoologie: 278–286, 328–330.

OLIVEIRA, M. P. D'. 1895. Opisthobranches du Portugal de la collection de M. Paulino d'Oliveira. *Inst. Coimbra* **42**: 574–592.

ORBIGNY, A. D'. 1837. Mémoire sur les espèces et sur les genres nouveaux de l'ordre des nudibranches, observés sur les côtes de France. *Mag. de Zool.* **7** (5): 1–16.

ORTEA, J. A. 1976a. Catalogo brevemente comentado de la fauna de moluscos marinos gasteropodos y bivalves existentes en el estuario de Villaviciosa. *Asturnatura* **3**: 109–120.

ORTEA, J. A. 1976b. *Eubranchus exiguus* (Alder & Hancock, 1848) un opistobranquio nuevo para le fauna Iberica. *Asturnatura* **3**: 159–162.

ORTEA, J. A. 1977. Un molusco poco conocido: *Duvaucelia manicata. Vida Silvestre* No. 24: 237–241.

ORTEA, J. A. 1978. Cinco opistobranquios nuevos para la fauna Iberica (Gastropoda: Opisthobranchia) colectados en Asturias. *Supl. Cien. Bol. Idea* No. 23: 107–120.

ORTEA, J. A. 1979a. Dos nuevas especies Ibericas de *Onchidoris* colectadas en Asturias. *Supl. Cien. Bol. Idea* No. 24: 167–175.

ORTEA, J. A. 1979b. *Onchidoris sparsa* (Alder & Hancock, 1846) in Asturias, northern Spain. *The Veliger* **22**: 45–48.

ORTEA, J. A. 1980. Sobre la biologia de *Aeolidia papillosa* (L.) (Mollusca, Nudibranchia) en Asturias. *Bol. Cienc. Natur.* number 25: 73–76.

ORTEA, J. A. 1982. A new *Favorinus* (Nudibranchia: Aeolidoidea) from the Canary Islands. *Nautilus* **96**: 45–48.

ORTEA, J. A. & URGORRI, V. 1978. El genero *Doto* (Oken, 1815) en el norte y noroeste de España. *Bol. Estación Centr. Ecol.* **7**: 73–92.

ORTEA, J. A. & URGORRI, V. 1979. Primera cita de *Hancockia uncinata* (Hesse, 1872) (Gasteropoda; Nudibranchia) para el litoral ibérico. *Trabaj. Compostel. Biol.* **8**: 79–85.

ORTON, J. H. 1914. Preliminary account of a contribution to an evaluation of the sea: the life-history of *Galvina picta. J. mar. biol. Ass., U.K.* **10**: 323–324.

OTTO, A. W. 1820. Eine neue Roche und eine gleichfalls neue Molluske. *Nova Acta Acad. Caes. Leopold Nat. Cur.* **10** (1): 113–126.

PELSENEER, P. 1891. Sur quelques points d'organisation des nudibranches et sur leur phylogénie. *Ann. Soc. Roy. Belg.* **26**: 68–71.

PELSENEER, P. 1894. Recherches sur divers opisthobranches. *Mém. Acad. Roy. Belg. Cl. Sci.* **53**: 1–157.

PELSENEER, P. 1906. *Mollusca.* Part V of *A Treatise on Zoology* ed. Lankester. London, Black. Pp. 1–355.

PELSENEER, P. 1911. Recherches sur l'embryologie des gastropodes. *Mém. Acad. Roy Belg. Cl. Sci.* **3** (6): 1–167.

PELSENEER, P. 1922. Sur une habitude de *Doris bilamellata*. *Ann. Soc. Roy. Belg.* **53**: 28–32.

PELSENEER, P. 1935. Essai d'éthologie zoologique d'après l'ètude des Mollusques. *Mem. Acad. Roy. Belg. Cl. Sci.* No. 1: 1–622.

PENNANT, T. 1777. *British zoology*, ed. 4. London & Warrington, White. **4**. Pp. 1–379.

PÉRIASLAVZEV, S. 1891. Additions to the fauna of the Black Sea; with two tables. *Trudy Obschchestva ispitatelei prirody pri imperatorskum Kharkovskum Universiteté* **25**: 267.

PERRON, F. E. & TURNER, R. D. 1977. Development, metamorphosis, and natural history of the nudibranch *Doridella obscura* Verrill (Corambidae: Opisthobranchia). *J. exp. mar. Biol. Ecol.* **27**: 171–185.

PHILIPPI, R. A. 1841. Zoologische Bemerkengen. *Arch. für Naturg.* **7**: 42–59.

PHILIPPI, R. A. 1844. *Enumerato molluscorum Siciliae cum viventium tum in tellure tertiaria fossilium quae in itinere suo observavit.* **2**. Halis Saxonum. Pp. 1–303.

PICTON, B. E. 1979. *Caloria elegans* (Alder & Hancock) *comb. nov.* Gastropoda: Opisthobranchia, an interesting rediscovery from S.W. England. *J. moll. Stud.* **45**: 125–130.

PICTON, B. E. 1980. A new species of *Coryphella* (Gastropoda: Opisthobranchia) from the British Isles. *Irish Nat. J.* **20**: 15–19.

PICTON, B. E. & BROWN, G. H. 1978. A new species of *Cuthona* (Gastropoda: Opisthobranchia) from the British Isles. *J. Conch., Lond.* **29**: 345–348.

PICTON, B. E. & BROWN, G. H. 1981. Four nudibranch gastropods new to the fauna of Great Britain and Ireland including a description of a new species of *Doto* Oken. *Irish Nat. J.* **20**: 261–268.

POHL, H. 1905. Über den feinern bau des genitalsystems von *Polycera quadrilineata*. *Zool Jahrb.* **21**: 427–452.

POIZAT, C. 1972a. Étude preliminaire des gastéropodes opisthobranches de quelques sables marins du Golfe de Marseille. *Téthys* **3**: 875–896.

POIZAT, C. 1972b. Méthodes d'élevage des gastéropodes opisthobranches de petites et moyennes dimensions. Mise au point d'un circuit fermé en eau de mer. Premiers résultats. *Téthys* **4**: 251–268.

POIZAT, C. 1978. Gastéropodes mesopsammiques de fonds sableux du Golfe de Marseille: ecologie et reproduction. 2 volumes, D.Sc. thesis, Universite d'Aix-Marseille.

POIZAT, C. 1981. Gastéropodes mesopsammiques de la Mer du Nord (Robin Hood's Bay, U.K.) ecologie et distribution. *J. moll. Stud.* **47**: 1–10.

PORTMANN, A. & SANDMEIER, E. 1960. Zur kenntnis von *Diaphorodoris* (Gastr., Nudibranchia) und ihrer mediterranean formen. *Verh. Naturf. Ges. Basel* **71**: 174–183.

POTTS, G. W. 1970. The ecology of *Onchidoris fusca* (Nudibranchia). *J. mar. biol. Ass., U.K.* **50**: 269–292.

POTTS, G. W. 1981. The anatomy of respiratory structures in the dorid nudibranchs, *Onchidoris bilamellata* and *Archidoris pseudoargus*, with details of the epidermal glands. *J. mar. biol. Ass. U.K.* **61**: 959–982.

PRUVOT-FOL, A. 1931. Notes de systématique sur les opisthobranches. *Bull. Mus. Nation. Hist. nat. Paris* (2) **3** (3): 308–316.

PRUVOT-FOL, A. 1936. Note préliminaire sur les nudibranches de Risso. *Rev. Suisse Zool.* **43**: 531–533.

PRUVOT-FOL, A. 1948. Deux aeolidiens d'Arcachon. *J. Conchyliol., Paris* **88**: 97–100.

PRUVOT-FOL, A. 1951. Études des nudibranches de la Méditerranée. 2. *Arch. Zool. exp. gén.* **88**: 1–80.

PRUVOT-FOL, A. 1953. Étude de quelques opisthobranches de la côte Atlantique du Maroc et du Sénégal. *Trav. Inst. scient. Chérifien* No. 5, Zool. No. 2: 1–105.

PRUVOT-FOL, A. 1954. *Faune de France, Mollusques opisthobranches.* Paris, Lechevalier. Pp. 1–460.

PRUVOT-FOL, A. 1955a. Note sur deux nudibranches attribués à la famille des Polyceradae. *Bull. Soc. zool. France* **80**: 350–359.

PRUVOT-FOL, A. 1955b. Les Arminidae (Pleurophyllidiadae ou Diphyllidiadae des anciens auteurs). *Bull. Mus. Nation. Hist. nat. Paris* (2) **27** (6): 462–468.

QUATREFAGES, A. DE. 1843. Mémoire sur l'Eolidine paradoxale. *Ann. Sci. nat. Paris* (2) **19**: 274–312.

QUATREFAGES, A. DE. 1844. Sur les gastéropodes phlébentérés (Phlébentérata Nob.), ordre nouveau de la classe des gastéropodes, proposé d'après l'examen anatomique et physiologique des genres Zéphyrine (*Zephyrina* Nob.), Actéon (*Acteon* Oken), Actéonie (*Acteonie* Nob.), Amphorine (*Amphorina* Nob.), Pavois (*Pelta* Nob.), Chalide (*Chalidis* Nob.). *Annls. Sci. nat.* (3) (*Zool.*) **1**: 129–183.

QUOY, J. R. C. & GAIMARD, J. P. 1832. Voyage de découvertes de l'Astrolabe pendant les années 1826–1829 sous le commandement de M. J. Dumont d'Urville. *Zool.* **2**: 1–686.

RAFINESQUE, C. S. 1814. *Précis des découvertes somiologiques ou zoologiques et botaniques.* Palermo, published privately. Pp. 1 30.

RANKIN, J. J. 1979. A freshwater shell-less mollusc from the Caribbean: structure, biotics, and contribution to a new understanding of the Acochlidiacea. *Royal Ontario Museum Life Sciences Contribution* number **116**: 1–123.

RAPP, W. L. 1827. Ueber das Mollusken geschlecht *Doris*. *Nova Acta Acad. Leop. Carol. Nat. Curius.* **13** (2): 516–522.

RASMUSSEN, E. 1944. Faunistic and biological notes on marine invertebrates I. *Vidensk. Medd. Dansk Naturh. Foren. Kbh.* **107**: 207–233.

RASMUSSEN, E. 1951. Faunistic and biological notes on marine invertebrates. II. The eggs and larvae of some Danish marine gastropods. *Vidensk. Medd. Dansk Naturh. Foren. Kbh.* **113**: 202–249.

RASMUSSEN, E. 1973. Systematics and ecology of the Isefjord marine fauna (Denmark). *Ophelia* **11**: 1–495.

RATHKE, J. 1806. In MÜLLER, O. F. *Zoologiae Danica seu animalium Daniae et Norvegiae rariorum ac minus notorum descriptiones et historia*, ed. 3, **4**: 1–46.

RENOUF, L. P. W. 1915. Note on the occurrence of two generations of the nudibranchiate Mollusca *Lamellidoris bilamellata* (Linn.) and of *Archidoris tuberculata* (Cuvier) in the course of a year. *Proc. Roy. Phys. Soc. Edinb.* **20** (1): 12–15.

RENOUF, L. P. W. 1935. Observations on periodic, sporadic, occasional and unusual appearances of certain organisms. *Acta Phaenologica* **3**: 110–154.

RISBEC, J. 1928. Contribution à l'étude des nudibranches Néo-Calédoniens. *Faune Colon. Française* **2** (1): 1–328.

RISBEC, J. 1937. Note préliminaire au sujet de nudibranches Néo-Calédoniens. *Bull. Mus. Nation. Hist. Nat. Paris* (2) **9**: 159–164.

RISSO, A. 1818. Mémoire. Sur quelques Gastéropodes nouveaux, Nudibranches et Tectibranches observés dans la mer de Nice. *J. de Physique* **87**: 368–377.

RISSO, A. 1826. *Histoire naturelle des principales productions de l'Europe Méridionale et particulièrement de celles des environs de Nice et des Alpes Maritimes.* Paris, Levrault. **4**: 1–439.

RIVA, A. & VICENTE, N. 1978a. Rapports trophiques entre les nudibranches *Aeolidiella alderi*, *Spurilla neapolitana* et un anthozoaire *Parastephanauge pauxi*. *Haliotis* (1976) **7**: 112–115.

RIVA, A. & VINCENTE, N. 1978b. Observations d'algues symbiotiques dans l'organisme de *Aeolidia alderi*, *Spurilla neapolitana* et *Favorinus branchialis*. *Haliotis* (1976) **7**: 116–119.

RIVEST, B. R. 1978. Development of the eolid nudibranch *Cuthona nana* (Alder & Hancock, 1842), and its relationship with a hydroid and hermit crab. *Biol. Bull. Woods Hole* **154**: 157–175.

ROBILLIARD, G. A. 1970. The systematics and some aspects of the ecology of the genus *Dendronotus* (Gastropoda: Nudibranchia). *The Veliger* **12**: 433–479.

ROBILLIARD, G. A. 1971. Predation by the nudibranch *Dirona albolineata* on three species of prosobranchs. *Pacific Science* **25**: 429–435.

ROBILLIARD, G. A. 1972. A new species of *Dendronotus* from the northeastern Pacific with notes on *Dendronotus nanus* and *Dendronotus robustus* (Mollusca: Opisthobranchia). *Canad. J. Zool.* **50**: 421–432.

ROBSON, E. A. 1961. The swimming response and its pacemaker system in the anemone *Stomphia coccinea*. *J. exp. Biol.* **38**: 685–694.

ROGINSKY, I. S. 1962. The egg-masses of nudibranchs of the White Sea. *Biol. White Sea* **1**: 201–214.

ROGINSKY, I. S. 1964. A large-sized nudibranch mollusc *Coryphella fusca* O'Donoghue—a predator of small nudibranch molluscs—*Coryphella rufibranchialis* Johnston and *Cuthona* sp. *Zool Zhurn.* **43**: 1717–1719.

ROGINSKY, I. S. 1965. Data relating to the reproduction and development of the Nudibranchiata of the White and Barents Seas. *Vtore Soveshchania Po Izucheniya* (1965): 38–39.

ROGINSKY, I. S. 1970. *Tenellia adspersa*, a nudibranch new to the Azov Sea, with notes on its taxonomy and ecology. *Malacogical Review* **3**: 167–174.

ROLLER, R. A. & LONG, S. J. 1969. An annotated list of opisthobranchs from San Luis Obispo County, California. *The Veliger* **11**: 424–430.

Ros, J. 1975. Opisthobranquios (Gastropoda: Euthyneura) del litoral ibérico. *Inv. Pesq. Barcelona* **39**: 269–372.

ROSE, R. M. 1971. Functional morphology of the buccal mass of the nudibranch *Archidoris pseudoargus*. *J. Zool., Lond.* **165**: 317–336.

ROWETT, H. G. Q. 1946. A comparison of the feeding mechanisms of *Calma glaucoides* and *Nebaliopsis typica*. *J. mar. biol. Ass. U.K.* **26**: 352–357.

RUDMAN, W. B. 1981a. Further studies on the anatomy and ecology of opisthobranch molluscs feeding on the scleractinian coral *Porites*. *Zool. J. Linn. Soc.* **71**: 373–412.

RUDMAN, W. B. 1981b. Polyp mimicry in a new species of aeolid nudibranch mollusc. *J. Zool., Lond.* **193**: 421–427.

RUDMAN, W. B. 1981c. The anatomy and biology of alcyonarian-feeding aeolid opisthobranch molluscs and their development of symbiosis with zooxanthellae. *Zool. J. Linn. Soc.* **72**: 219–262.

RUDMAN, W. B. 1982. The taxonomy and biology of further aeolidacean and arminacean nudibranch molluscs with symbiotic zooxanthellae. *Zool. J. Linn. Soc.* **74**: 147–196.

RUNHAM, N. W. 1963. The histochemistry of the radulas of *Acanthochitona communis*, *Lymnaea stagnalis*, *Helix pomatia*, *Scaphander lignarius* and *Archidoris pseudoargus*. *Ann. Histochem.* **8**: 433–442.

RUNNSTRÖM, S. 1927. Über die Thermopathie der Fortpflanzung und Entwickelung mariner Tiers in Begichung zu ihrer geographischen Verbreitung. *Bergens Mus. Årbok, Natur. Rekke* No. 2: 1–67.

RUSSELL, H. D. 1971. *Index Nudibranchia a catalog of the literature* 1554–1965. Greenville, Delaware, Delaware Museum of Natural History. Pp. 1–141. ·

RUSSELL, L. 1929. The comparative morphology of the elysioid and aeolidioid types of the molluscan nervous system, and its bearing on the relationships of the ascoglossan nudibranchs. *Proc. Zool. Soc. Lond.* (1929): 197–233.

RYLAND, J. 1970. *Bryozoans.* London, Hutchinson. Pp. 175.

SALVINI-PLAWEN, L. 1973. Zur Kenntnis der Philinoglossacea und der Acochlidiacea mit Platyhedylidae fam. nov. (Gastropoda, Cephalaspidea). *Sond. Z. zool. Systematik Evolutionsforsch.* **11**: 110–133.

SALVINI-PLAWEN, L. & STERRER, W. 1968. Zur Kenntnis der mesopsammalen Gattung *Pseudovermis* (Gastropoda, Nudibranchia). *Helg. wiss. Meeres.* **18**: 69–77.

SARS, G. O. 1878. *Bidrag til Kundskaben om Norges Arktiske Fauna* I. *Mollusca regionis arcticae Norvegiae.* Christiana, Universitets. Pp. 1–466.

SARS, M. 1829. *Bidrag til söedyrenes naturhistorie.* Pt 1. Bergen. Pp. 1–59.

SARS, M. 1840. Beitrag zur Entwicklungsgeschichte der Mollusken und Zoophyten. *Arch. für Naturg.* **6** (1): 196–219.

SARS, M. 1850. Beretning om i Sommeren 1849 foretagen Zoologisk Reise i Lofoten og Finmarken. *Nyt. Mag. Naturv.* **6**: 121–211.

SARS, M. 1870. Bidrag til Kundskab om Christiana fjordens Fauna. 2. *Nyt. Mag. Naturv.* **17**: 113–232.

SAUVAGE, H. E. 1873. Catalogue des nudibranches des Côtes du Boulonnais, dressé d'après les notes de Bouchard-Chantereaux. *J. Conchyl.* (3) **21**: 25–36.

SCHMEKEL, L. 1967. *Dicata odhneri*, n.sp., n.gen., ein neuer Favorinide (Gastr. Opisthobranchia) aus dem Golf von Neapel. *Pubbl. Staz. zool. Napoli* **35**: 263–273.

SCHMEKEL, L. 1968. Ascoglossa, Notaspidea und Nudibranchia im Litoral des Golfes von Neapel. *Rev. Suisse Zool.* **75**: 103–155.

SCHMEKEL, L. 1970. Anatomie der Genitalorgane von Nudibranchiern (Gastropoda Euthyneura). *Pubbl. Staz. Zool. Napoli* **38**: 120–214.

SCHMEKEL, L. 1971. Histologie und Feinstruktur der Genitalorgane von Nudibranchiern (Gastropoda, Euthyneura). *Z. Morph. Tiere* **69**: 115–183.

SCHMEKEL, L. & KRESS, A. 1977. Die Gattung *Doto* (Gastropoda: Nudibranchia) in Mittelmeer und Armelkanal, mit Beschreibung von *Doto acuta*, n.sp. *Malacologia* **16**: 467–499.

SCHMEKEL, L. & PORTMANN, A. 1982. *Opisthobranchia des Mittelmeeres Nudibranchia und Saccoglossa.* Berlin, Springer-Verlag. Pp. 1–410.

SCHMEKEL, L. & WEISCHER, M. L. 1973. Die Blutdrüse der Doridoidea (Gastropoda, Opisthobranchia) als Ort möglicher Hämocyanin-Synthese. *Z. Morph. Tiere* **76**: 261–284.

SCHONENBERGER, M. & SCHONENBERGER, N. 1969. Zur Kenntnis von *Facelina dubia* Pruvot-Fol (Gastr. Opisthobranchia. *Pubbl. Staz. zool. Napoli* **37**: 293–302.

SCHULTZE, M. S. 1849. Ueber die Entwicklung des *Tergipes lacinulatus. Arch. für Naturg.* **15**: 268–279.

SEAGER, J. R. 1979. Reproductive biology of the Antarctic opisthobranch *Philine gibba* Strebel. *J. exp. mar. Biol. Ecol.* **41**: 51–74.

SEAGER, J. R. 1982. Population dynamics of the Antarctic opisthobranch *Philine gibba* Strebel. *J. exp. mar. Biol. Ecol.* **60**: 163–179.

SEAWARD, D. R. 1982. *Sea Area Atlas of the Marine Molluscs of Britain and Ireland.* Published for the Conchological Society of Great Britain and Ireland by the Nature Conservancy Council. Pp. 1–53, maps 1–746.

SNELI, J. A. & STEINNES, A. 1975. The marine Mollusca of Jan Mayen Island. *Astarte* **8**: 7–16.

SORDI, M. & MAJIDI, P. 1957. Osservazioni sui nudibranchi e gli ascoglossi (Gasteropodi Opistobranchi) del litorale livornese. *Boll. Pesca Piscicolt. Idrobiol.* **11**: 235–245.

STARMÜHLNER, F. 1955. Zur Molluskenfauna des Felslitorals und submariner Höhlen am Capo di Sorrento (I. Teil). Ergebnisse der Österreichischen Tyrrhenia-Expedition 1952, Teil IV. *Österreichische Zool. Zeitschr.* **6**: 147–249.

STEARNS, R. E. C. 1873. Description of a new genus and two new species of nudibranchiate mollusks from the coast of California. *Proc. Calif. Acad. Sci.* **5**: 77–78.

STIMPSON, W. 1854. Synopsis of the marine Invertebrata of Grand Manan: or the region about the mouth of the Bay of Fundy, New Brunswick. *Smithsonian Cont. Knowledge* (1854): 1–66.

STIRTS, H. M. & CLARK, K. B. 1980. Effects of temperature on products of symbiotic chloroplasts in *Elysia tuca* Marcus. *J. exp. mar. Biol. Ecol.* **43**: 39–47.

STORROW, B. 1911. Notes on nudibranchs. *Northumberland Sea Fish. Comp. Rep. Sci. Invest.* (1911): 28–29.

STRENTH, N. E. & BLANKENSHIP, J. E. 1978. Laboratory culture, metamorphosis and development of *Aplysia brasiliana* Rang, 1828 (Gastropoda: Opisthobranchia). *The Veliger* **21**: 99–103.

SUTHERLAND, P. C. 1890. Nudibranchiate Mollusca. *Trans. Inverness Sci. Soc.* **2**: 337–339.

SWENNEN, C. 1959. The Netherlands coastal waters as an environment for Nudibranchia. *Basteria* **23**: 56–62.

SWENNEN, C. 1961a. On a collection of Opisthobranchia from Turkey. *Zool. Medd. Riksmus. nat. Hist. Leiden* **38**: 41–75.

SWENNEN, C. 1961b. Data on distribution, reproduction and ecology of the nudibranchiate molluscs occurring in the Netherlands. *Netherlands J. Sea Res.* **1**: 191–240.

SWITZER-DUNLAP, M. 1978. Larval biology and metamorphosis of aplysiid gastropods. (In) *Settlement and Metamorphosis of Marine Invertebrate Larvae*, eds F. S. Chia & M. E. Rice, pp. 197–206. New York, Elsevier.

SWITZER-DUNLAP, M. & HADFIELD, M. G. 1977. Observations on development, larval growth and metamorphosis of four species of Aplysiidae (Gastropoda: Opisthobranchia) in laboratory culture. *J. exp. mar. Biol. Ecol.* **29**: 245–261.

SWITZER-DUNLAP, M. & HADFIELD, M. G. 1979. Reproductive patterns of Hawaiian aplysiid gastropods. (In) *Reproductive Biology of Marine Invertebrates*, ed. S. E. Stancyk, pp. 199–210. Belle Baruch Library in Marine Sciences, number 9.

TARDY, J. 1962. Première liste concernant la fauna des mollusques nudibranches et ascoglosses sur la côte nord-ouest de l'Ile de Ré (Charente-Maritime). *87e Congrès Soc. Sav.* (1962): 1217–1227.

TARDY, J. 1963. Description d'une nouvelle espèce de Tritoniidae: *Duvaucelia odhneri*, recoltée sur la côte atlantique française. *Bull. Inst. Océanog. Monaco* **60**: 1–10.

TARDY, J. 1964a. Observations sur le développement de *Tergipes despectus* (Gastéropodes Nudibranches). *C. R. Acad. Sci. Paris* **258**: 1635–1637.

TARDY, J. 1964b. Comportement prédateur de *Eolidiella alderi* (Mollusque Nudibranche). *C. R. Acad. Sci. Paris* **258**: 2190–2192.

TARDY, J. 1969. Étude systématique et biologique sur trois espèces d'Aeolidielles des côtes européennes (Gastéropodes Nudibranches). *Bull. Inst. Oceanogr. Monaco* **68**: 1–40.

TARDY, J. 1970. Contribution à la connaissance de la biologie chez les nudibranches: développement et métamorphose; vie prédatrice: I. *Facelina coronata* (Forbes) et *Aeolis* sp. *Bull. Soc. zool. France* **95**: 765–772.

TARDY, J. 1971. Étude expérimentale de la régénération germinale après castration chez les Aeolidiidae. *Ann. Sci. nat. Zool.* **13**: 91–147.

TARDY, J. & BORDES, M. 1978. Regime, preferendum et ethologie predatrice des Aeolidiidae des côtes de France. *Haliotis* **9**: 43–52.

TARDY, J. & DUFRENNE, M. 1978. Effets de groupement et du volume disponible sur la sexualité du mollusque nudibranche *Eubranchus doriae* (Trinchese, 1879). *Haliotis* **7**: 66–68.

TARDY, J. & GANTÈS, H. 1980. Un mollusque nudibranche peu connu: *Cumanotus cuenoti* A. Pruvot-Fol (1948); redescription, biologie. *Bull. Soc. zool. Fr.* **105**: 199–207.

TCHANG-SI 1931. Contribution à l'étude des mollusques opisthobranches de la côte provençale. Thèses de la Faculté des Sciences de l'Université de Lyon. Pp. 5–211.

TCHANG-SI 1934. Contribution à l'étude des opisthobranches de la côte de Tsingtao. *Contr. Inst. Zool. Peiping* **2** (2): 1–148.

THIELE, J. 1931. *Handbuch der Systematischen Weichtierkunde.* **1**, *Mollusca.* Amsterdam, Asher. Pp. 1–778.

THIRIOT-QUIÉVREUX, C. 1977. Véligère planctotrophe du doridien *Aegires punctilucens* (d'Orbigny) (Mollusca: Nudibranchia: Notodorididae): description et métamorphose. *J. exp. mar. Biol. Ecol.* **26**: 177–190.

THOMPSON, T. E. 1958a. Observations on the radula of *Adalaria proxima* (Alder & Hancock) (Gastropoda, Opisthobranchia). *Proc. malac. Soc. Lond.* **33**: 49–56.

THOMPSON, T. E. 1958b. The natural history, embryology, larval biology and post-larval development of *Adalaria proxima* (Alder & Hancock) (Gastropoda, Opisthobranchia). *Phil. Trans. Roy. Soc. Ser. B* **242**: 1–58.

THOMPSON, T. E. 1958c. The influence of temperature on spawning in *Adalaria proxima* (A. & H.) (Gastropoda, Nudibranchia). *Oikos* **9**: 246–252.

THOMPSON, T. E. 1959. Feeding in nudibranch larvae. *J. mar. biol. Ass., U.K.* **38**: 239–248.

THOMPSON, T. E. 1960a. On a disputed feature of the anatomy of the nudibranch *Dendronotus frondosus* Ascanius. *Proc. malac. Soc. Lond.* **34**: 24–26.

THOMPSON, T. E. 1960b. Defensive adaptations in opisthobranchs. *J. mar. biol. Ass., U.K.* **39**: 123–134.

THOMPSON, T. E. 1961a. The importance of the larval shell in the classification of the Sacoglossa and the Acoela (Gastropoda, Opisthobranchia). *Proc. malac. Soc. Lond.* **34**: 233–238.

THOMPSON, T. E. 1961b. Observations on the life history of the nudibranch *Onchidoris muricata* (Müller). *Proc. malac. Soc. Lond.* **34**: 239–242.

THOMPSON, T. E. 1961c. The structure and mode of functioning of the reproductive organs of *Tritonia hombergi* (Gastropoda, Opisthobranchia). *Q. Jl microsc. Sci.* **102**: 1–14.

THOMPSON, T. E. 1962. Studies on the ontogeny of *Tritonia hombergi* Cuvier (Gastropoda, Opisthobranchia). *Phil. Trans. Roy. Soc. Ser. B* **245**: 171–218.

THOMPSON, T. E. 1964. Grazing and the life cycles of British nudibranchs. *Brit. ecol. Soc. Symp.* No. 4: 275–297.

THOMPSON, T. E. 1966. Studies on the reproduction of *Archidoris pseudoargus* (Rapp) (Gastropoda, Opisthobranchia). *Phil. Trans. Roy. Soc. Ser. B* **250**: 343–375.

THOMPSON, T. E. 1967. Direct development in the nudibranch *Cadlina laevis*, with a discussion of developmental processes in Opisthobranchia. *J. mar. biol. Ass., U.K.* **47**: 1–22.

THOMPSON, T. E. 1972. Eastern Australian Dendronotoidea (Gastropoda, Opisthobranchia). *Zool. J. Linn. Soc.* **51**: 63–77.

THOMPSON, T. E. 1973a. Sacoglossan gastropod molluscs from eastern Australia. *Proc. malac. Soc. Lond.* **40**: 239–251.

THOMPSON, T. E. 1973b. Euthyneuran and other molluscan spermatozoa. *Malacologia* **14**: 167–206.

THOMPSON, T. E. 1975. Dorid nudibranchs from eastern Australia (Gastropoda, Opisthobranchia). *J. Zool., Lond.* **176**: 477–517.

THOMPSON, T. E. 1976. *Biology of opisthobranch molluscs I.* London, Ray Society. Pp. 1–206.

THOMPSON, T. E. 1977. Jamaican opisthobranch molluscs I. *J. moll. Stud.* **43**: 93–140.

THOMPSON, T. E. 1980a. Jamaican opisthobranch molluscs II. *J. moll. Stud.* **46**: 74–99.

THOMPSON, T. E. 1980b. New species of the bullomorph genus *Runcina* from the northern Adriatic Sea. *J. moll. Stud.* **46**: 154–157.

THOMPSON, T. E. 1981. Taxonomy of three misunderstood opisthobranchs from the northern Adriatic Sea. *J. moll. Stud.* **47**: 73–79.

THOMPSON, T. E. 1982. Sea slugs. (In) *Encyclopaedia of Marine Invertebrates*, ed. J. G. Walls, pp. 363–443. Neptune, N.J., T.F.H. Inc.

THOMPSON, T. E. & BEBBINGTON, A. 1973. Scanning electron microscope studies of gastropod radulae. *Malacologia* **14**: 147–165.

THOMPSON, T. E. & BROWN, G. H. 1974. *Atagema gibba* Pruvot-Fol, a doridacean nudibranch new to the British fauna. *J. Conch., Lond.* **28**: 233–237.

THOMPSON, T. E. & BROWN, G. H. 1976. *British opisthobranch molluscs.* London, Linnean Society & Academic Press. Pp. 1–203.

THOMPSON, T. E. & BROWN, G. H. 1981. Biology and relationships of the nudibranch mollusc *Notobryon wardi* in South Africa, with a review of the Scyllaeidae. *J. Zool., Lond.* **194**: 437–444.

THOMPSON, T. E. & HINTON, H. E. 1968. Stereoscan electron microscope observations on opisthobranch radulae and shell-sculpture. *Bijd. tot de Dierkunde* **38**: 91–96.

THOMPSON, W. 1840. Contributions towards a knowledge of the Mollusca Nudibranchia and Mollusca Tunicata of Ireland, with descriptions of some apparently new species of Invertebrata. *Ann. Mag. nat. Hist.* **5** (29): 84–102.

THOMPSON, W. 1843. Report on the fauna of Ireland, division Invertebrata. *Rep. Brit. Ass. Sci.* (1843): 245–291.

THOMPSON, W. 1845. Additions to the fauna of Ireland including descriptions of some apparently new species of Invertebrata. *Ann. Mag. nat. Hist.* **15**: 308–322.

THOMPSON, W. 1860. On a species of *Eolis*, and also a species of *Lomanotus* new to science; with a description of a specimen of *Eolis caerulea* of Montagu. *Ann. Mag. nat. Hist.* (3) **5**: 48–51.

THORSON, G. 1946. Reproduction and larval development of Danish marine bottom invertebrates, with special reference to the planktonic larvae in the Sound (Øresund). *Meddr Kommn Havunders*, Ser. Plankt. **4** (1): 1–523.

TODD, C. D. 1978a. Gonad development of *Onchidoris muricata* (Müller) in relation to size, age and spawning. *J. moll. Stud.* **44**: 190–199.

TODD, C. D. 1978b. Changes in spatial pattern of an intertidal population of the nudibranch mollusc *Onchidoris muricata* in relation to life-cycle, mortality and environmental heterogeneity. *J. anim. Ecol.* **47**: 189–203.

TODD, C. D. 1979. Reproductive energetics of two species of dorid nudibranchs with planktotrophic and lecithotrophic larval strategies. *Marine Biology* **53**: 57–68.

TODD, C. D. 1981. The ecology of nudibranch molluscs. *Oceanogr. mar. Biol. ann. Rev.* **19**: 141–234.

TRINCHESE, S. 1874. Descrizione di alcuni nuovi Eolididei del Porto di Genova. *Mem. R. Accad. Sci. Istit. Bologna* **4**: 197–203.

TRINCHESE, S. 1877. Descrizione del genere *Rizzolia*. *Rendic. Sess. Accad. Sci. Istit. Bologna* (1877): 147–150.

TRINCHESE, S. 1879. Aeolididae e famiglie affini del Porto di Genova. *Rendic. Sess. Com. Nom. Accad. Sci. Istit. Bologna* (1879): 47–52.

TRINCHESE, S. 1881. Aeolididae e famiglie affini. *Atti Accad. naz. Lincei Memorie* ser. 3 **11**: 3–142.

TRINCHESE, S. 1883. Di una nuova forma del genere *Lomanotus* e del suo sviluppo. *Rendic. Accad. Sci. Fis. Mat. Soc. Roy. Napoli* **22** (3): 92–94.

TRINCHESE, S. 1885. Diagnosi del nuovo genere *Govia*. *Rendic. Accad. Sci. Fis. Mat. Soc. Roy. Napoli* **24** (6): 179–180.

TRINCHESE, S. 1888. Descrizione de nuovo genere *Caloria*. *Mem. R. Accad. Sci. Inst. Bologna* **9**: 291–295.

VANNUCCI, M. & HOSOE, E. K. 1953. Sôbre *Embletonia mediterranea* (Costa) nudibrânquio de região lagunar de Cananéia. *Bol. Inst. Oceanograf.* **4**: 103–120.

VAYSSIÈRE, A. 1877. Sur un nouveau genre de la famille des Tritoniadés. *C. R. Acad. Sci., Paris* **85**: 299–301.

VAYSSIÈRE, A. 1888. Recherches zoologiques et anatomiques sur les mollusques opistobranches du Golfe de Marseille deuxième partie nudibranches (cirrobranches) et ascoglosses. *Ann. Mus. Hist. nat. Marseille, Zool.* **3** (4): 1–160.

VAYSSIÈRE, A. 1901. Recherches zoologiques et anatomiques sur les mollusques opistobranches du Golfe de Marseille (suite et fin). 3. *Ann. Mus. Hist. nat. Marseille Zool.* **6**: 1–130.

VAYSSIÈRE, A. 1904. Étude zoologique de l'*Archidoris stellifera* H. von Ihering. *J. Conchyl.* **52**: 123–131.

VAYSSIÈRE, A. 1913. Mollusques de la France et des régions voisines. In *Encycl. Scient.* 1. Paris, O. Doin. Pp. 1–420.

VAYSSIÈRE, A. 1917. Recherches zoologiques et anatomiques sur les mollusques amphineures et gastéropodes (opisthobranches et prosobranches). *2me Exped. Antarct. Francaise* (1908–1910): 1–44.

VAYSSIÈRE, A. 1919. Recherches zoologiques et anatomiques sur les mollusques opisthobranches du Golfe de Marseille. *Ann. Mus. Hist. nat. Marseille* **17**: 53–92.

VÉRANY, G. B. 1844. Description de deux genres nouveaux de mollusques nudibranches. *Rev. Zool. Soc. Cuvier* (1844): 302–303.

VÉRANY, G. B. 1846. Catalogo degli animali invertebrati marini del Golfo di Genova e Nizza. *Mus. Stor. Nat. Genova* (1846): 1–30.

VÉRANY, G. B. 1853. Catalogue des mollusques céphalopodes, pteropodes, gastéropodes nudibranches, etc., des environs de Nice. *J. Conchyl.* **4**: 375–392.

VERRILL, A. E. 1870. Contributions to zoology from the Museum of Yale College. No. 8. Descriptions of some New England Nudibranchiata. *Amer. J. Sci. Art* (2) **50**: 405–408.

VERRILL, A. E. 1873. Report upon the invertebrate animals of Vineyard Sound and the adjacent waters, with an account of the physical characters of the region. *New England Fish. Rep.* 1871–2, *U.S. Comm. Fish.*: 295–778.

VERRILL, A. E. 1875. Brief contributions to zoology from the Museum of Yale College. No. 33. Results of dredging expeditions off the New England coast in 1874. *Amer. J. Sci.* (3) **10**: 36–43.

VERRILL, A. E. 1878. Notice of recent additions to the marine fauna of the eastern coast of north America. *Amer. J. Sci.* (3) **16**: 207–214.

VERRILL, A. E. 1879. Notice of recent additions to the marine fauna of the east coast of north America. No. 4. *Amer. J. Sci.* (3) **17**: 309–315.

VERRILL, A. E. 1880. Notice of recent additions to the marine Invertebrata, of the northeastern coast of America, with descriptions of new genera and species and critical remarks on others. Pt. 2, Mollusca. *Proc. U.S. nat. Mus.* **3**: 356–405.

VESTERGAARD, K. & THORSON, G. 1938. Über den Laich und die Larven von *Duvaucelia plebeja*, *Polycera quadrilineata*, *Eubranchus pallidus* und *Limapontia capitata* (Gastropoda, Opisthobranchiata). *Zool. Anz.* **124**: 129–138.

VICENTE, N. 1963. Mollusques opisthobranches récoltés en plongée dans le Golfe de Marseille. *Rec. Trav. Sta. Mar. Endoume* **31**: 173–185.

VICENTE, N. 1967. Contribution a l'étude des gastéropodes opisthobranches du Golfe de Marseille. *Rec. Trav. Stat. mar. Endoume* **42**: 133–179.

VOLODCHENKO, N. I. 1955. (In) PAVLOVSKII, E. N. *Atlas bespozvonochnykh dal' nevostochnykh morei SSSR.* Moscow, Akademii Nauk SSSR. Pp. 1–240.

WALTON, C. L. 1908. Nudibranchiata collected in the North Sea by the s.s. "Huxley" during July and August, 1907. *J. mar. biol. Ass., U.K.* **8**: 227–240.

WATERS, V. L. 1973. Food-preference of the nudibranch *Aeolidia papillosa*, and the effect of the defenses of the prey on predation. *The Veliger* **15**: 174–192.

WAWRA, E. 1978. Zur identität von *Microhedyle glandulifera* (Kowalevsky, 1901) und *Microhedyle lactea* (Hertling, 1930). *Ann. Naturhistor. Mus. Wien* **81**: 607–617.

WAWRA, E. 1979. Zur systematischen Stellung von *Platyhedyle denudata* Salvini-Plawen, 1973. *Sond. Z. zool. Systematik Evolutionsforsch.* **17**: 221–225.

WILLAN, R. C. 1976. The opisthobranch *Thecacera pennigera* (Montagu) in New Zealand with a discussion on the genus. *The Veliger* **18**: 347–356.

WILLIAMS, G. C. & GOSLINER, T. M. 1979. Two new species of nudibranchiate molluscs from the west coast of north America, with a revision of the family Cuthonidae. *Zool. J. Linn. Soc.* **67**: 203–223.

WILLIAMS, L. G. 1971. Veliger development in *Dendronotus frondosus* (Ascanius, 1774) (Gastropoda: Nudibranchia). *The Veliger* **14**: 166–171.

WILSON, K. & PICTON, B. E. 1983. A list of the Opisthobranchia: Mollusca of Lough Hyne nature reserve, Co. Cork, with notes on distribution and nomenclature. *Irish nat. J.* **21**: 69–72.

WINCKWORTH, R. 1932. The British marine Mollusca. *J. Conch., Lond.* **19**: 211–252.

WINCKWORTH, R. 1941. The name *Cratena*. *Proc. malac. Soc. Lond.* **24**: 146–149.

WINCKWORTH, R. 1951. A list of the marine Mollusca of the British Isles, additions and corrections. *J. Conch., Lond.* **23**: 131–134.

WOLTER, H. 1967. Beiträge zur Biologie, Histologie und Sinnesphysiologie (insbesondere der Chemorezeption) einiger Nudibranchier (Mollusca, Opisthobranchia) der Nordsee. *Z. Morph. Ökol. Tiere* **60**: 275–337.

WOODLAND, W. 1907. Studies in spicule formation. VI. Scleroblastic development of the spicules in some Mollusca and in one genus of colonial ascidians. *Q. Jl microsc. Sci.* **51**: 45–53.

WRIGHT, P. 1859. Notes on the Irish Nudibranchiata. *Nat. Hist. Rev. Lond.* **6** (2): 86–88.

YONGE, C. M. & THOMPSON, T. E. 1976. *Living marine molluscs.* London, Collins. Pp. 1–288.

SYSTEMATIC INDEX
(Compiled by Joyce Ablett)

In this alphabetic index, orders and suborders are printed thus: DORIDACEA
Superfamilies, tribes, families and subfamilies are printed: **TRITONIINAE**
Current genera and species (i.e. senior synonyms): ***Archidoris pseudoargus***
Obsolete genera and species (i.e. junior synonyms): *Cuthona aurantia*
Page numbers printed in bold arabic numerals are principal entries.

PICTORIAL SYNOPSIS OF THE BRITISH NAKED OPISTHOBRANCH MOLLUSCS

For each species, a drawing is given, with a reference to a colour or black and white illustration. This may be in the present volume, or in volume I (Thompson, 1976), or elsewhere. Finally, the maximal recorded body-length in British waters is indicated.

BRITISH NUDIBRANCHIA 1

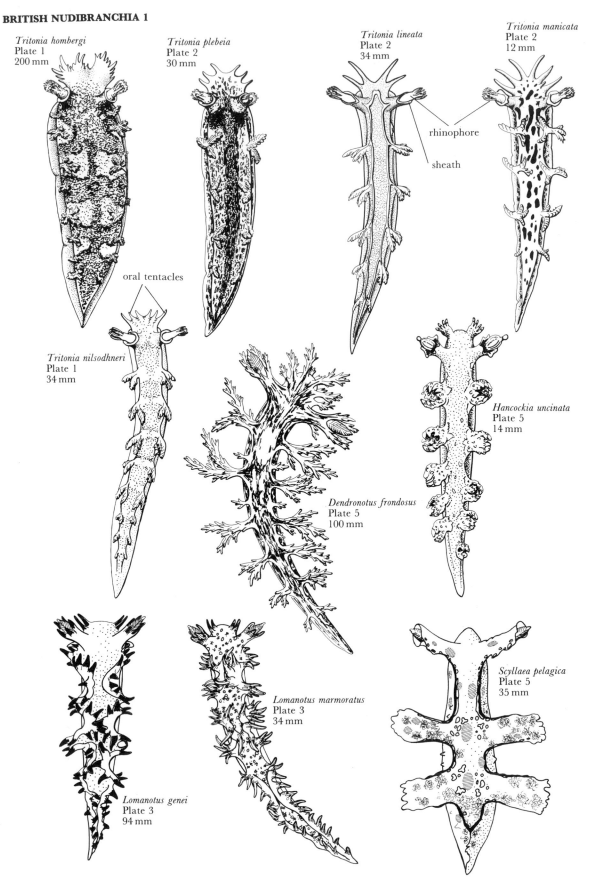

Tritonia hombergi
Plate 1
200 mm

Tritonia plebeia
Plate 2
30 mm

Tritonia lineata
Plate 2
34 mm

Tritonia manicata
Plate 2
12 mm

rhinophore

sheath

oral tentacles

Tritonia nilsodhneri
Plate 1
34 mm

Dendronotus frondosus
Plate 5
100 mm

Hancockia uncinata
Plate 5
14 mm

Lomanotus marmoratus
Plate 3
34 mm

Scyllaea pelagica
Plate 5
35 mm

Lomanotus genei
Plate 3
94 mm

BRITISH NUDIBRANCHIA 2

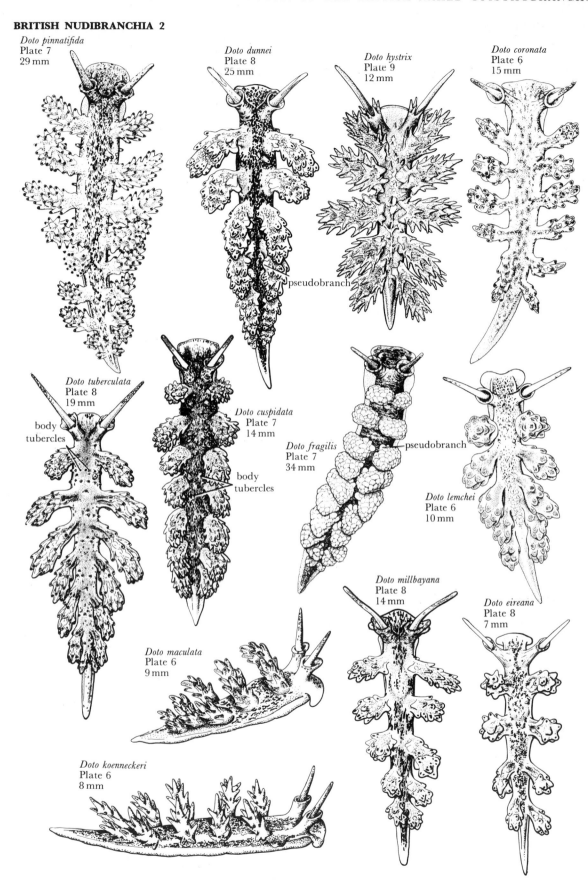

Doto pinnatifida
Plate 7
29 mm

Doto dunnei
Plate 8
25 mm

Doto hystrix
Plate 9
12 mm

Doto coronata
Plate 6
15 mm

pseudobranch

Doto tuberculata
Plate 8
19 mm

body
tubercles

Doto cuspidata
Plate 7
14 mm

body
tubercles

Doto fragilis
Plate 7
34 mm

pseudobranch

Doto lemchei
Plate 6
10 mm

Doto millbayana
Plate 8
14 mm

Doto eireana
Plate 8
7 mm

Doto maculata
Plate 6
9 mm

Doto koenneckeri
Plate 6
8 mm

215

BRITISH NUDIBRANCHIA 3

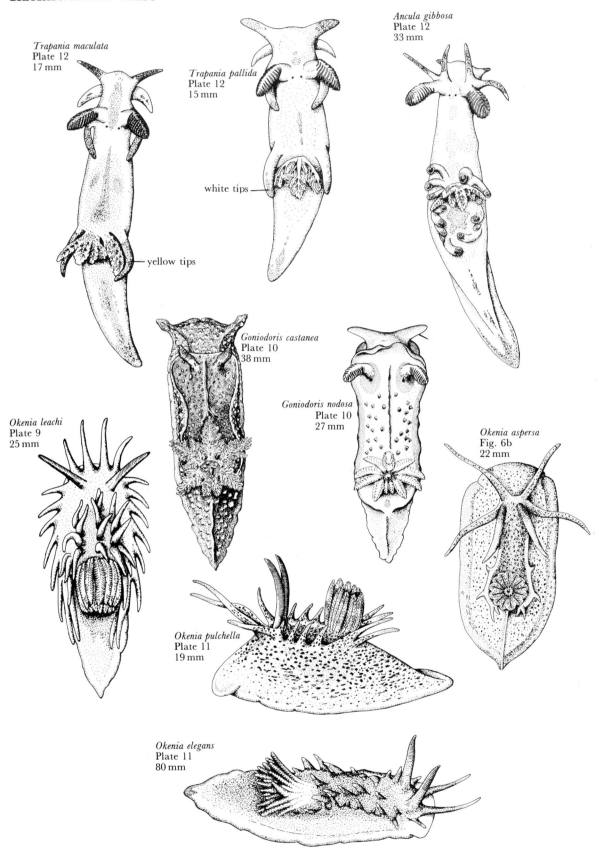

Trapania maculata
Plate 12
17 mm

Trapania pallida
Plate 12
15 mm

Ancula gibbosa
Plate 12
33 mm

white tips

yellow tips

Goniodoris castanea
Plate 10
38 mm

Goniodoris nodosa
Plate 10
27 mm

Okenia leachi
Plate 9
25 mm

Okenia aspersa
Fig. 6b
22 mm

Okenia pulchella
Plate 11
19 mm

Okenia elegans
Plate 11
80 mm

BRITISH NUDIBRANCHIA 4

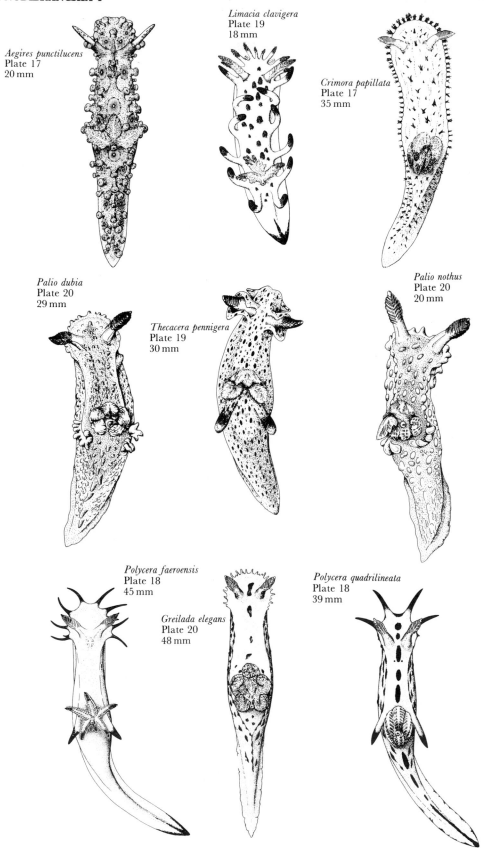

Aegires punctilucens
Plate 17
20 mm

Limacia clavigera
Plate 19
18 mm

Crimora papillata
Plate 17
35 mm

Palio dubia
Plate 20
29 mm

Thecacera pennigera
Plate 19
30 mm

Palio nothus
Plate 20
20 mm

Polycera faeroensis
Plate 18
45 mm

Greilada elegans
Plate 20
48 mm

Polycera quadrilineata
Plate 18
39 mm

BRITISH NUDIBRANCHIA 5

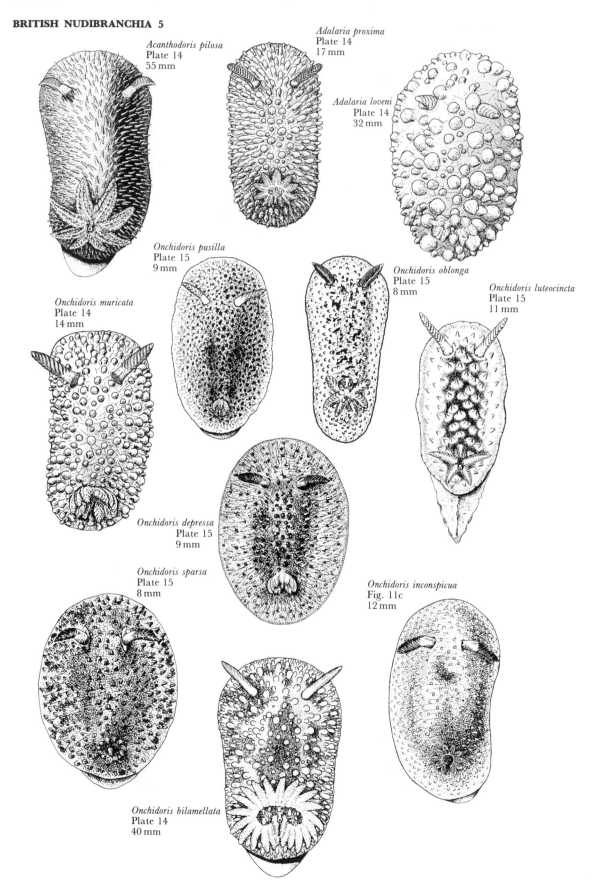

Acanthodoris pilosa
Plate 14
55 mm

Adalaria proxima
Plate 14
17 mm

Adalaria loveni
Plate 14
32 mm

Onchidoris pusilla
Plate 15
9 mm

Onchidoris oblonga
Plate 15
8 mm

Onchidoris luteocincta
Plate 15
11 mm

Onchidoris muricata
Plate 14
14 mm

Onchidoris depressa
Plate 15
9 mm

Onchidoris sparsa
Plate 15
8 mm

Onchidoris inconspicua
Fig. 11c
12 mm

Onchidoris bilamellata
Plate 14
40 mm

BRITISH NUDIBRANCHIA 6

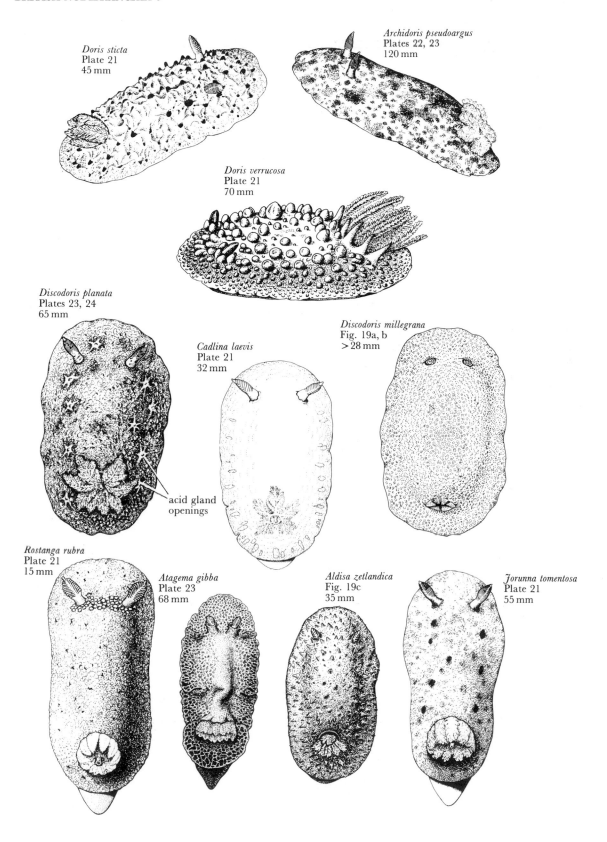

Doris sticta
Plate 21
45 mm

Archidoris pseudoargus
Plates 22, 23
120 mm

Doris verrucosa
Plate 21
70 mm

Discodoris planata
Plates 23, 24
65 mm

Cadlina laevis
Plate 21
32 mm

Discodoris millegrana
Fig. 19a, b
> 28 mm

acid gland
openings

Rostanga rubra
Plate 21
15 mm

Atagema gibba
Plate 23
68 mm

Aldisa zetlandica
Fig. 19c
35 mm

Jorunna tomentosa
Plate 21
55 mm

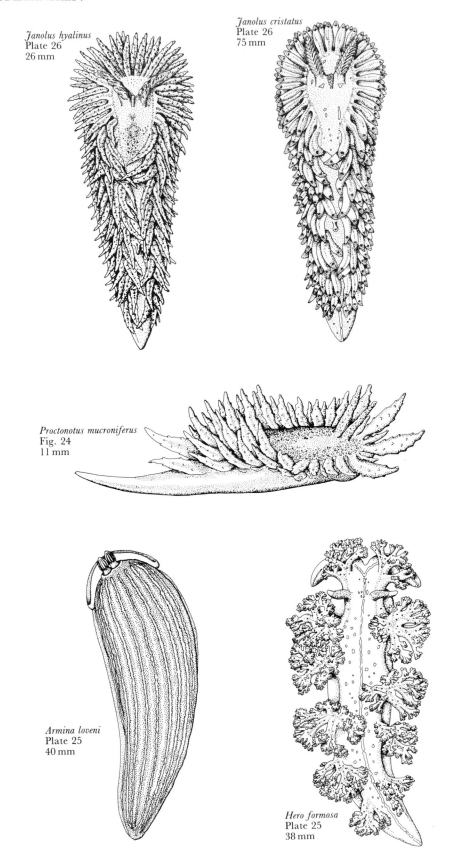

Janolus hyalinus
Plate 26
26 mm

Janolus cristatus
Plate 26
75 mm

Proctonotus mucroniferus
Fig. 24
11 mm

Armina loveni
Plate 25
40 mm

Hero formosa
Plate 25
38 mm

BRITISH NUDIBRANCHIA 8

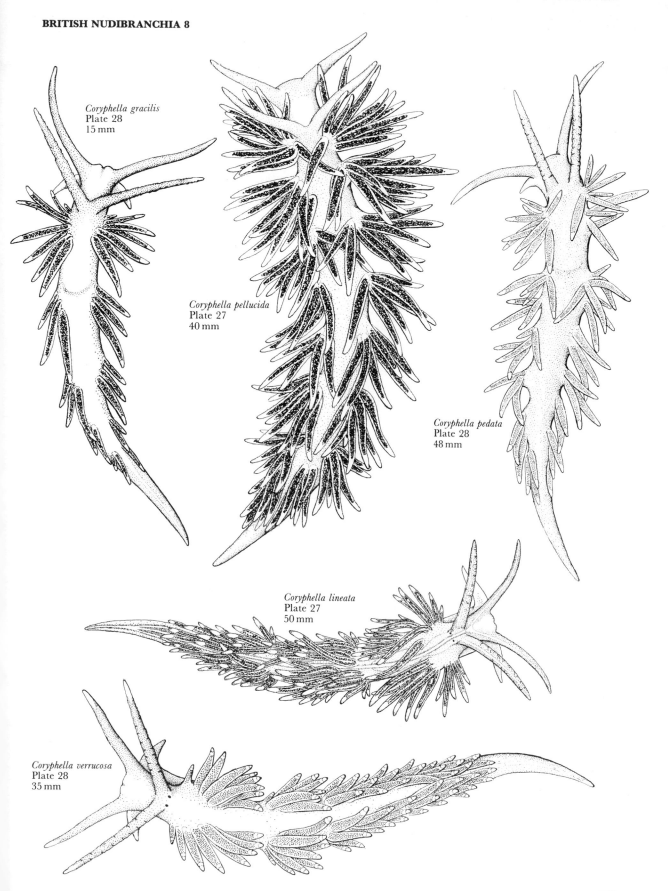

Coryphella gracilis
Plate 28
15 mm

Coryphella pellucida
Plate 27
40 mm

Coryphella pedata
Plate 28
48 mm

Coryphella lineata
Plate 27
50 mm

Coryphella verrucosa
Plate 28
35 mm

BRITISH NUDIBRANCHIA 9

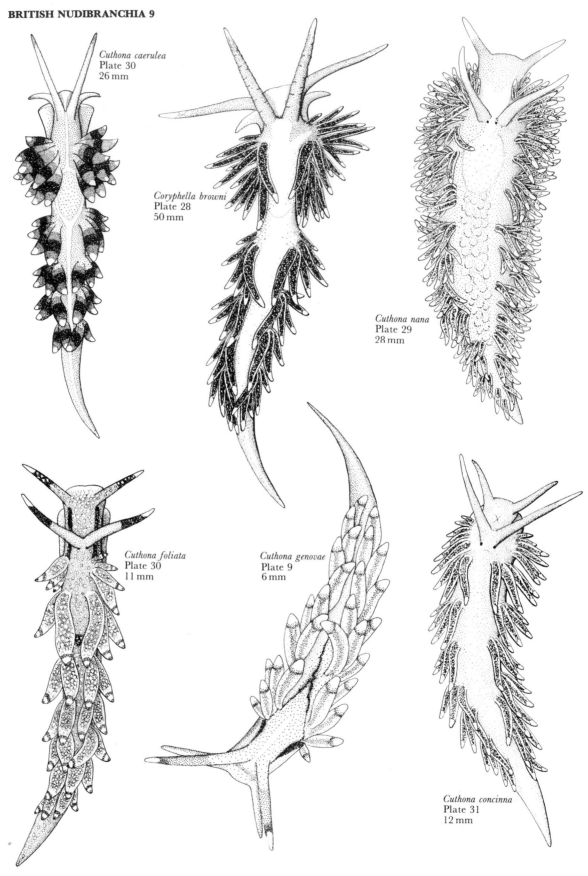

Cuthona caerulea
Plate 30
26 mm

Coryphella browni
Plate 28
50 mm

Cuthona nana
Plate 29
28 mm

Cuthona foliata
Plate 30
11 mm

Cuthona genovae
Plate 9
6 mm

Cuthona concinna
Plate 31
12 mm

BRITISH NUDIBRANCHIA 10

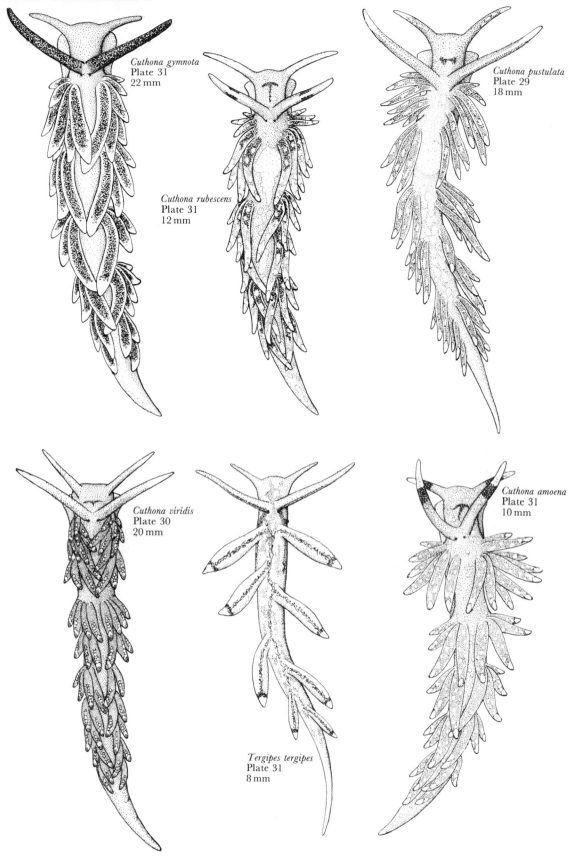

Cuthona gymnota
Plate 31
22 mm

Cuthona rubescens
Plate 31
12 mm

Cuthona pustulata
Plate 29
18 mm

Cuthona viridis
Plate 30
20 mm

Tergipes tergipes
Plate 31
8 mm

Cuthona amoena
Plate 31
10 mm

BRITISH NUDIBRANCHIA 11

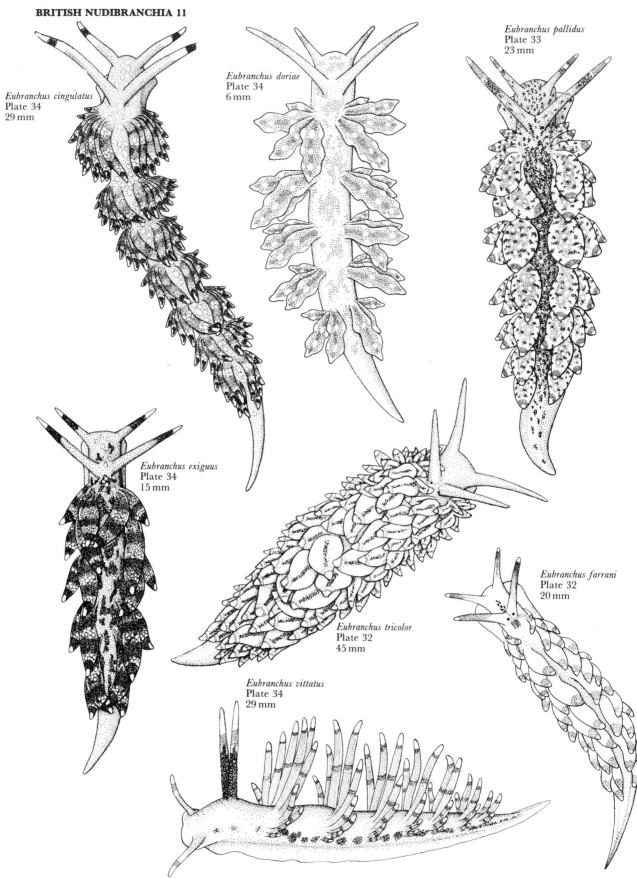

Eubranchus cingulatus
Plate 34
29 mm

Eubranchus doriae
Plate 34
6 mm

Eubranchus pallidus
Plate 33
23 mm

Eubranchus exiguus
Plate 34
15 mm

Eubranchus tricolor
Plate 32
45 mm

Eubranchus farrani
Plate 32
20 mm

Eubranchus vittatus
Plate 34
29 mm

BRITISH NUDIBRANCHIA 12

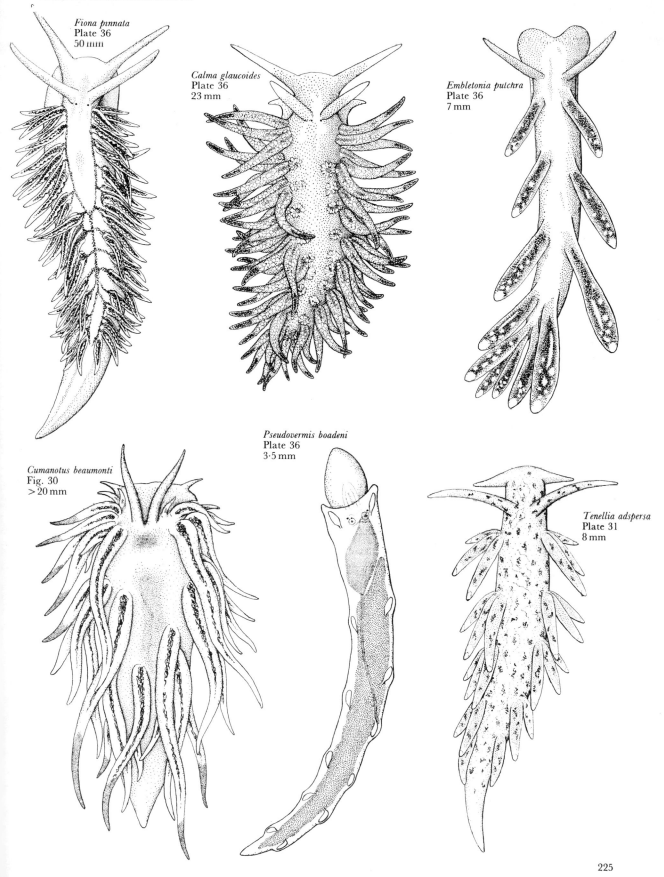

Fiona pinnata
Plate 36
50 mm

Calma glaucoides
Plate 36
23 mm

Embletonia pulchra
Plate 36
7 mm

Cumanotus beaumonti
Fig. 30
>20 mm

Pseudovermis boadeni
Plate 36
3·5 mm

Tenellia adspersa
Plate 31
8 mm

BRITISH NUDIBRANCHIA 13

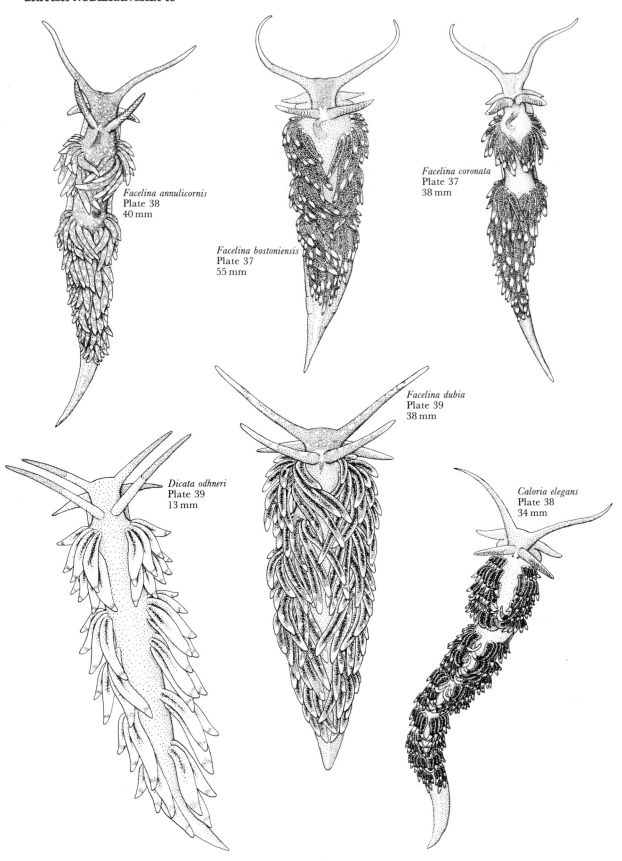

Facelina annulicornis
Plate 38
40 mm

Facelina bostoniensis
Plate 37
55 mm

Facelina coronata
Plate 37
38 mm

Facelina dubia
Plate 39
38 mm

Dicata odhneri
Plate 39
13 mm

Caloria elegans
Plate 38
34 mm

BRITISH NUDIBRANCHIA 14

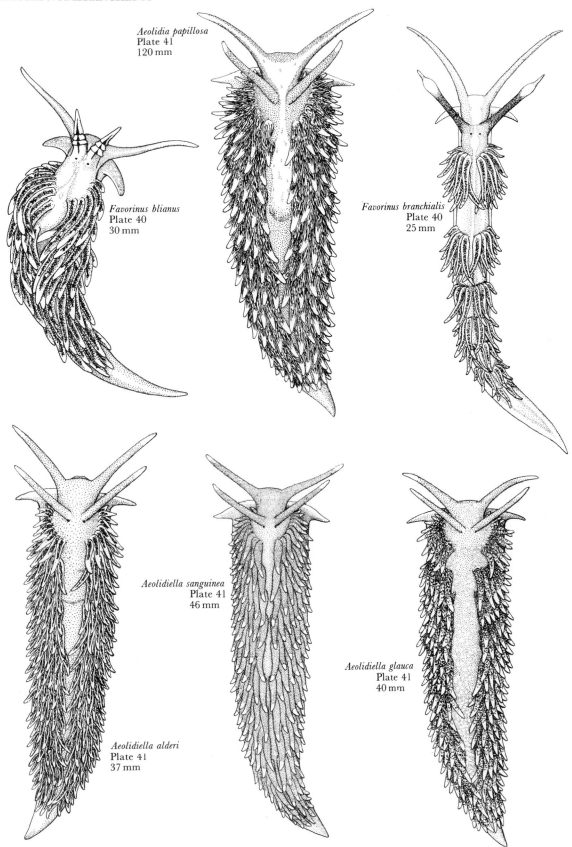

Aeolidia papillosa
Plate 41
120 mm

Favorinus blianus
Plate 40
30 mm

Favorinus branchialis
Plate 40
25 mm

Aeolidiella sanguinea
Plate 41
46 mm

Aeolidiella alderi
Plate 41
37 mm

Aeolidiella glauca
Plate 41
40 mm

BRITISH SACOGLOSSA

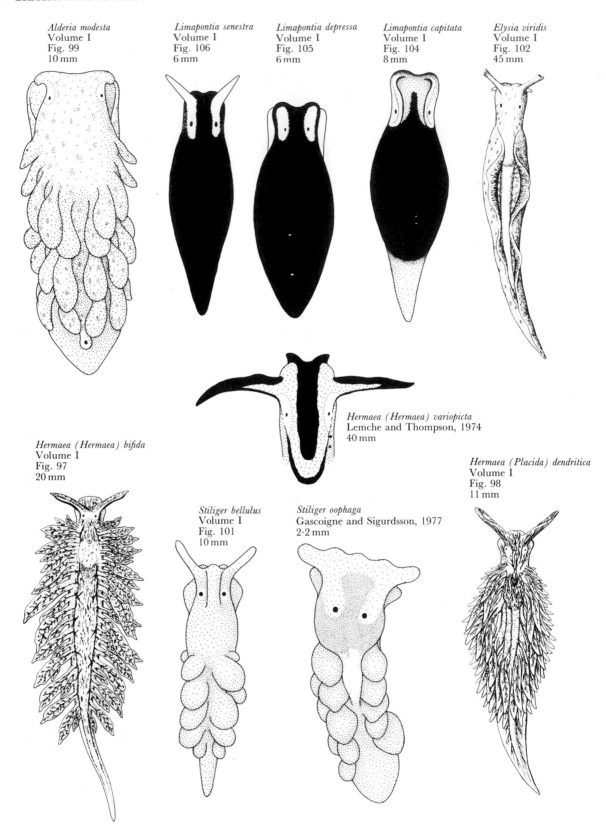

Alderia modesta
Volume I
Fig. 99
10 mm

Limapontia senestra
Volume I
Fig. 106
6 mm

Limapontia depressa
Volume I
Fig. 105
6 mm

Limapontia capitata
Volume I
Fig. 104
8 mm

Elysia viridis
Volume I
Fig. 102
45 mm

Hermaea (Hermaea) variopicta
Lemche and Thompson, 1974
40 mm

Hermaea (Hermaea) bifida
Volume I
Fig. 97
20 mm

Hermaea (Placida) dendritica
Volume I
Fig. 98
11 mm

Stiliger bellulus
Volume I
Fig. 101
10 mm

Stiliger oophaga
Gascoigne and Sigurdsson, 1977
2·2 mm

NAKED APLYSIOMORPHA, BULLOMORPHA, PLEUROBRANCHOMORPHA & ACOCHLIDIACEA

Aplysia punctata
Volume I
Fig. 87, Plate 19
200 mm

parapodial lobes

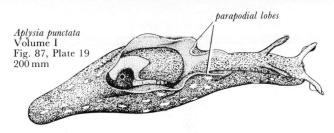

Pleurobranchus membranaceus
Volume I
Fig. 91, Plate 20
120 mm

Berthella plumula
Volume I
Fig. 94
60 mm

Philine aperta
Volume I
Fig. 68
70 mm

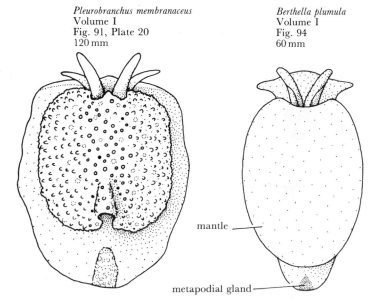

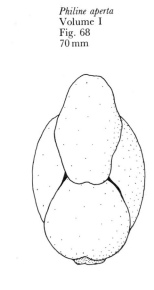

mantle

metapodial gland

Hedylopsis brambelli
Volume I
Fig. 96
2·5 mm

Runcina ferruginea
Kress, 1977
8 mm

Runcina coronata
Volume I
Fig. 77
6 mm

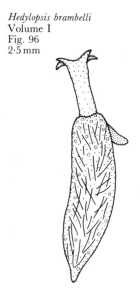

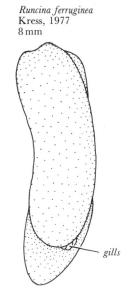

gills